Liaison Functions 2015

AA003247

Core Programming Area at the 2015 AIChE Spring Meeting and 11th Global Congress on Process Safety

Austin, Texas, USA
26-30 April 2015

ISBN: 978-1-5108-0704-4

Printed from e-media with permission by:

Curran Associates, Inc.
57 Morehouse Lane
Red Hook, NY 12571

Some format issues inherent in the e-media version may also appear in this print version.

Copyright© (2015) by AIChE
All rights reserved.

Printed by Curran Associates, Inc. (2015)

For permission requests, please contact AIChE
at the address below.

AIChE
3 Park Avenue
New York, NY 10016-5991

Phone: (203) 702-7660
Fax: (203) 775-5177

www.aiche.org

Additional copies of this publication are available from:

Curran Associates, Inc.
57 Morehouse Lane
Red Hook, NY 12571 USA
Phone: 845-758-0400
Fax: 845-758-2634
Email: curran@proceedings.com
Web: www.proceedings.com

TABLE OF CONTENTS

409505 Chemical Engineering: Is This the Golden Age?

Abstract:

Schlanger will offer his insights on the current opportunities in the chemical process industries in his talk entitled "Chemical Engineering: Is This the Golden Age?" on Monday morning, April 27. At the conclusion of his lecture, Schlanger will receive the AGILE Award in recognition of his leadership at CEVA Group, a $9 billion logistics supplier to industry worldwide; LyondellBasell, a major player in the plastics, chemicals, and refining industries; Arco Chemical; and at the other companies where he has held leadership positions.

Copyright © American Institute of Chemical Engineers. All rights reserved.

401714 Building a First of Its Kind Technology – Process Development from Idea to Commercialization

Abstract:

Building a First of Its Kind Technology – Process Development from Idea to Commercialization

Scaling a first of its kind technology is an exciting challenge that requires focus on the end goal. This presentation focuses on a case study of building the world's first commercial scale supercritical biodiesel reactor system which wone. It highlights the process development decisions required to scale from bench to pilot to commercial, including aspects of design, build, and commissioning at each scale.

The decisions made in the early stages of process development can have critical impacts on whether or not the project is able to move on to the next phase. This presentation will include key design decisions for the scaling of major unit operations throughout the process. This considers materials of construction, yield per unit operation, OPEX and CAPEX cost estimating.

Feedstock characterization, process conversion and finishing are the building blocks of biorefining production. Of the three, feedstock characterization is usually underestimated. The lack of depth in the understanding of feedstock's impact on the business plan has led to challenges that can be avoided. This presentation will take the time to indicate how feedstock characterization impacts process development.

Copyright © American Institute of Chemical Engineers. All rights reserved.

404125 Stand, Serve, Smile......Succeed

Abstract:

The Women's Initiatives Committee (WIC) luncheon will feature keynote speaker Deborah Grubbe. Grubbe was Vice President of Group Safety at BP, and held corporate director roles in safety, operations and engineering at DuPont. Now running a global consultancy, Deborah was a member of the NASA Aerospace Safety Advisory Panel for 9 years, and consulted in safety culture with the Columbia Shuttle Accident Investigation Board. She currently works around the world helping corporations address their occupational safety, operational and process safety issues. She is an Emeritus Member of the Center for Chemical Process Safety, a former AIChE Board Member, and is the past Chair of the Institute for Sustainability. Deborah is currently working with AIChE to help decide how the Institute can best support ethics in chemical engineering practice, and has co-authored an eBook and instructs on Process Safety and a Safety Orientation for undergraduates. Deborah obtained a BS in Chemical Engineering from Purdue University and attended the University of Cambridge on a Churchill Fellowship. She is a registered PE in Delaware, a chartered engineer in the UK, and is a Fellow of both the AIChE and the IChemE. She was named Engineer of the Year in Delaware in 2002, and was the first woman to receive an Honorary Doctorate in Engineering from Purdue University for being a leader in safety in the process industries and for her work in supporting faculty diversity at the University. In 2013 she was inducted into the National Academy of Construction.

This luncheon is priced at $20 and advanced registration is required.

If you haven't yet registered for the meeting, you can add this ticketed event when you register. If you have already registered for the meeting, please call Customer Service at 800-242-4363 or 203-702-7660 (outside of US) to add this event referencing item **# 102.**

Copyright © American Institute of Chemical Engineers. All rights reserved.

399632 Instrument Problem or Process Problem? That Is the Question... and I'd like the Answer Now!"

Abstract:

Instrument Problem or Process Problem? That is the question... and I'd like the answer now! This session will explore a variety of issues with process measurement and control in the process industries - from the transmitters and installation to the controllers to the control valves. From the common issues to the less well known. The key to profitability in the process industries like oil refining and petrochemicals is on-stream availability. This session is intended for both those new to process automation and the old-timers to come share their experiences.

Copyright © American Institute of Chemical Engineers. All rights reserved.

401716 Steam Load Shedding System Design Using Dynamic Simulation

Abstract:

In a refinery, steam utility is one of the most important utilities, which can be used at different temperature ranges. Proper handling of the available steam is an important factor for economic aspects of the plant. Instability in steam header pressures restricts optimum use of resources and it can also create safety issues. Constantly changing demands at various steam pressures challenge proper management and control of existing steam supply capabilities. Some emergency situations, such as boiler trip, add complexity to the above problem. Steam load shedding is required due to reduction in steam supply while demand being the same. The choice of steam load shedding location can affect economy of the process. Robust and reliable controller strategy becomes critical in such situations.

Aspen HYSYS was used as dynamic process simulation tool. A typical refinery steam system was considered as base case. A controller strategy was developed to handle various plant modes viz. normal, optimum, process upsets, and emergency (boiler trips). It also keeps the steam header pressures within specified limits. Changing steam demands for high pressure, medium pressure and low pressure steam were considered to test the controller strategy. Various situations of outage in steam supply were also considered. The methodology was to maintain header pressure, minimize steam venting and minimize steam letdowns. The target was to minimize need for external power supply while maintaining safety, and stability of the plant. This also includes reduction in wastage of steam and maximizing on-site power generation. The study has shown significant savings in the utility costs due to proper steam load shedding.

Copyright © American Institute of Chemical Engineers. All rights reserved.

Control System Migrations Gone Wild!
Best Practices & Lessons Learned

Laurie R. Ben
Director, Modernization Business Development

James Beall (acknowledge)
Principal Process Control Consultant

15 AIChE
Spring Meeting, Austin, TX

EMERSON.
Process Management

Introduction / Agenda

- What it looks like when it goes Wild?
  - Cycling Outputs
  - Controlling Super Slow or Super Fast (Unstable Control)
- Why it happens? Common Conversion Mistakes
- How to Avoid? (Tips, Tools, & Techniques)
- Summary

15 AIChE
Spring Meeting, Austin, TX

EMERSON.
Process Management

What it Looks Like when it goes Wild - Cycling Outputs

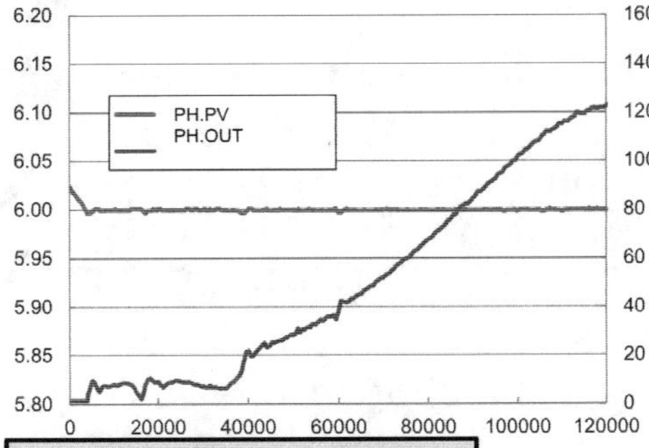

On Legacy DCS

- 14 Fermenters on Legacy System operating for years
- 2 Fermenters migrated to Modern System
- pH staying in spec, <u>but</u> Cycling - Ugly

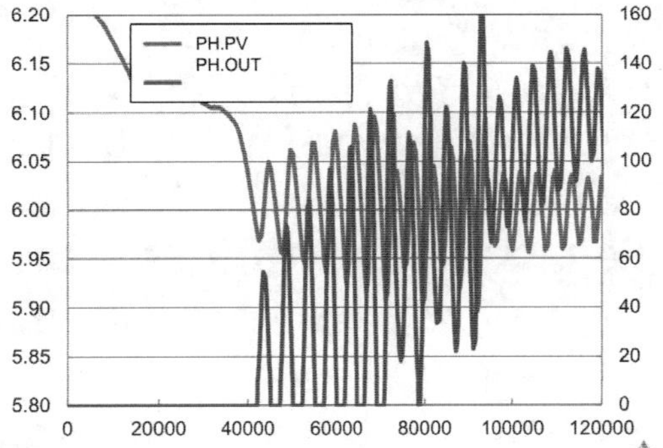

On New DCS

15 AIChE Spring Meeting, Austin, TX

EMERSON. Process Management

Why It Happens? Cycling Outputs

- Wrong PID Form Selected for Conversion
- Compounded By Derivative (Rate) Action

Control Action	Legacy Classical aka Series, as-found all fermenters	Modern Standard Form (in-correctly converted)	Modern Standard, correctly converted	estimated process dynamics
Proportional (Gain)	0.5	0.5	11.75	11.4
Integral Time (a.k.a Reset)	3 rep/min	20 sec	470 sec	2040 sec
Derivative Time (a.k.a. Rate)	7.5 min	450 sec	19 sec	0 sec

Difference suggests that even properly converted tuning may have too fast of integral action

Recommend to put in Manual for correction, OR 1st change the Reset, 2nd the Rate, 3rd the Gain!

*from analysis of as-found cycle and experience on other fermenters with aqueous solutions & weak base reagents. This should be replaced by data from step tests.

15 AIChE Spring Meeting, Austin, TX

How to Avoid?
Know the PID Forms in Your Legacy Systems

1. **Legacy DCS** = "Series" form
   - Most modern DCS's are selectable – "Series" or "Standard"
     - Verify the default selection!
   - If Derivative is not used (Rate = 0 min), then "Series" and "Standard" are the same, so choose Standard
     - "Standard" Form is more flexible from a tuning standpoint and is more common than Series (and Parallel)
   - If you have loops with Derivative action, choose = Series for ease of conversion, or consider change for plant/company practices
2. **Legacy DCS** = Standard form
   - choose Standard (typically the default) for ease of conversion
   - Then, NEW tuning will be simply a conversion of units

15 AIChE Spring Meeting, Austin, TX

EMERSON. Process Management

How to Avoid?
Know the PID Forms in Your Legacy Systems

Standard Form Platforms	Series Form Platforms
ABB Masterpiece/ADVANT	Bailey Infi90*
VALMET Damatic Classic	MOD 300*
Measurex Open, Vision 2000	Fischer-Porter, Micro DCI
Texas Instrument	MOORE-APACS
Honeywell	Honeywell
Foxboro I/A	Foxboro I/A, Spec 200
Rosemount RS3 (Emerson)	Fisher Controls PROVOX (Emerson)
Yokogawa	
Modicon 984, TI-505	L&N: 440, 446-3, Electromax V

*Note: These platforms have other PID forms available

15 AIChE Spring Meeting, Austin, TX

EMERSON. Process Management

Legacy System (Foxboro I/A) PID Multiple Function Blocks

- Some Legacy platforms have multiple function blocks as well as selectable PID forms, e.g. Foxboro I/A
- Foxboro I/A
  - PID – Basic interacting (Series) PID block
  - PIDA – Advanced PID block
  - PIDE – PID block with EXACT Tuning
  - PIDX – PID Extended block
  - PIDXE – PID Extended block with EXACT Tuning
- Must Match PID Form
  - Foxboro "Interactive"= "Series"
  - Foxboro "Non Interactive"= "Standard"

15 AIChE Spring Meeting, Austin, TX

EMERSON. Process Management

Conversion Notes for FOX I/A Example

Foxboro Feature	Modern DCS Example Features (examine closely)
PID Form • Interactive • Non-Interactive	PID Form • Series • Standard (standard)
PID Structures	PID Structures – Match
Feedforward • Additive • Multiplicative	Feedforward • Additive (standard) • Multiplicative (function blocks)
Nonlinear Gain	PID block FRSIPID_OPTS has "Nonlinear Gain"
Dead time Compensation	Can match with different options including delay block in external feedback (Dynamic Reset Limit)
Exact auto-tune	Adaptive Control
Variable Control Execution	PIDPlus feature
Measurement Filter (min) • 2nd Order Butterworth filter on signal to PID calc	PV Filter (sec) • 1st Order Filter on PV in PID block • Additional filter on signal to "D" calc

15 AIChE Spring Meeting, Austin, TX

EMERSON. Process Management

Looking in Old Documentation - Bailey

Potential Confusion: Control System Terminolo

G PID CONTROLLER

Standard form of a noninteracting PID

$$Output = \left(K + \frac{1}{T_I s} + \frac{T_D s}{T_D a s + 1}\right) error$$

where:

error SP – PV rever

error PV – SP direc

Clue: S Domain Transfer Functions

Advanced PID controller block parameters.

S11 = Gain multiplier K.
S12 = Proportional gain K_P.
S13 = Integral reset (resets per min.) K_I
S14 = Derivative rate action (min.) K_D.
S15 = Derivative lag constant (typically

NOTE: This controller works in seconds intern that K_I and K_D are in resets per minute and min The **60** term converts K_I and K_D into resets pe onds, respectively.

Clue: Parallel Blocks

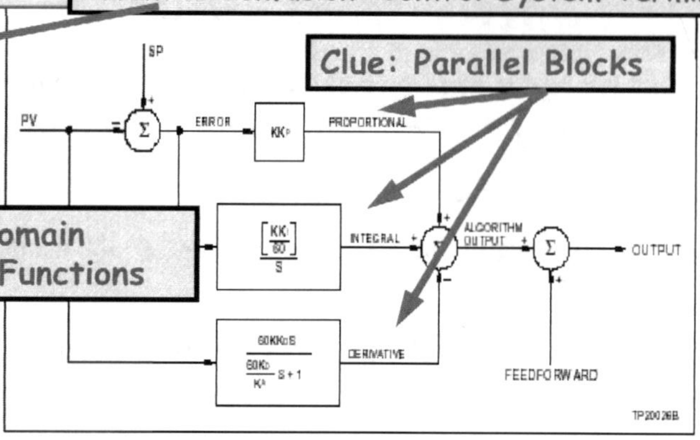

Figure 156-4. Reverse Mode Noninteracting Controller

Substituting block parameters into the original equation.

$$Output = K\left(K_P + \frac{K_I/60}{s} + \frac{60K_D s}{\frac{60K_D}{K_A}s + 1}\right) error$$

Bailey Documentation Function Code 156

EMERSON. Process Management

Spring Meeting, Austin, TX

Looking in Old Documentation Multiple Form Options may Exist ...

CLASSICAL PID CONTROLLER

Specification S18 selects the type of algorithm for the PID calculation:

0 = classical - PID output is calculated using a classical interactive controller. Tuning any of the proportional, integral or derivative terms changes the effective value of the other terms.

1 = noninteracting - PID output is calculated using a noninteracting control algorithm. Tuning the proportional, integral or derivative terms individually has no effect on the other terms. This is the same type as function code 19.

2 = classical with external reset - cascade and override configurations use this type of algorithm. The PID output is calculated using the classical interactive control algorithm. The integral contribution is calculated as a function of the external reset signal.

3 = manual reset noninteracting - PID output is calculated from the proportional and derivative terms with manual reset. For manual reset control, a manual reset time constant (S13) is used for bumpless transfer between the track and release states.

NOTE: The transfer is not bumpless if the manual reset time constant (S13) is set to zero. Any change in the manual reset is filtered by a first order lag with the manual reset time specified.

Standard form of a classical PID co

$$Output = K\left(1 + \frac{1}{T_I s}\right)\left(\frac{T_D s + 1}{T_D a s + 1}\right) error$$

where:

Bailey Documentation Function Code 156

15 AIChE Spring Meeting, Austin, TX

EMERSON. Process Management

Looking in Old Documentation - Honeywell

19.4

19.4 OPTIONS AND SPECIAL FEATURES

19.4.1 Interactive and Noninteractive PID Forms

During configuration, select one of these two forms. They differ as follows:

- Interactive (Real) Form—This form emulates traditional pneumatic-PID controllers. The P, I, and D terms are calculated as the sum of P and I, multiplied by D. D interacts in the time domain with the P and I terms. An advantage of this form is that the poles (lags) and zeros (leads) can be easily placed (See the equations under 19.5). The poles and zeros must be real.
- Noninteractive (Ideal) Form—In this form, P, I, and D are added in the time domain. D is a contro

Potential Confusion: Control System Terminology
Interactive = Real = Classical = Series
Non-interactive = Ideal = Standard

Remember Bailey Non-Interactive meant Parallel

15 AIChE
Spring Meeting, Austin, TX

EMERSON.
Process Management

What it Looks Like when it goes Wild! Control Super Slow or Super Fast

Before

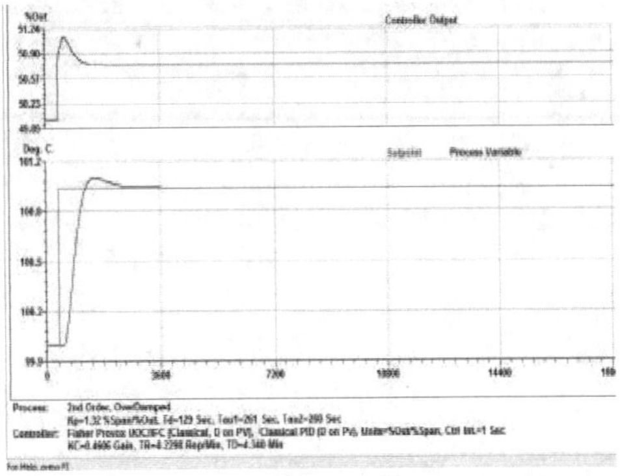

After

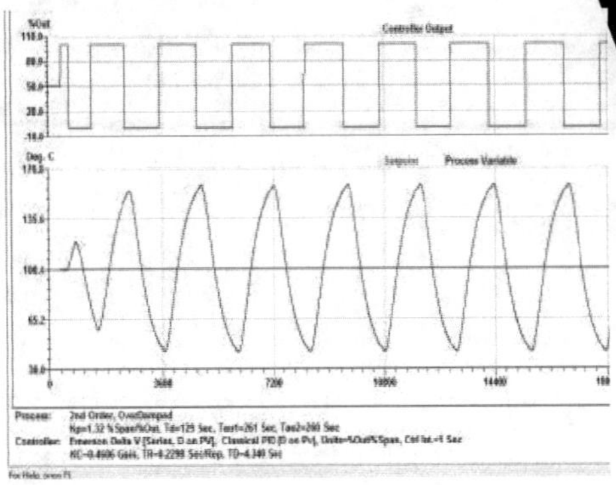

- Legacy tuning was
  - Gain=4.6, Reset=0.23 rep/min, Rate=4.3 minutes
- Modern DCS (as found) Tuning numbers, without proper units conversion
  - Gain=4.6, Reset=0.23 sec, Rate=4.3 sec

15 AIChE
Spring Meeting, Austin, TX

EMERSON.
Process Management

Why It Happens?
Control Super Slow or Super Fast

- Incorrectly Converted Parameter Units
- Some Legacy Systems allow selectable Units per Individual PID vs. System Wide

Parameter	Legacy (Standard)	Modern DCS (Standard Form)	
Gain	%PB	Gain = 100/%PB	Per System Basis
	Gain	Gain = Gain	
Reset	Ti - seconds	Reset = Ti	Per Loop Basis
	Ti - minutes	Reset = 60*Ti	
	Ti - hours	Reset = 3600*Ti	
Rate	Td - seconds	Rate = Td	
	Td - minutes	Rate = 60*Td	
	Td - hours	Rate = 3600*Td	

15 AIChE Spring Meeting, Austin, TX

EMERSON. Process Management

Legacy Conversion Units*

Platform	P Units	I Units	D Units	Other Notes
Bailey Infi90*	• Gain Multiplier (normalize EUs) • K_P (Proportional Gain)	Resets/Min	Minutes	K_I = Manual Reset Time (min) K_A = Derivative Lag Constant
MOORE-APACS	PG	Minutes	Minutes	DG (derivative gain) DeltaV α = 1/DG
Foxboro I/A*	PB %	Minutes	Minutes	KD (derivative gain) DeltaV α = 1/KD
Honeywell HPM*	Gain Value	Minutes	Minutes	Gain Value is based on chosen option
Yokogawa CS3000	PB %	Seconds	Seconds	

* Important to verify the information per platform

* Don't forget Filter Time Values

15 AIChE Spring Meeting, Austin, TX

EMERSON. Process Management

Modern DCS PID Form/Units
Right Units, Wrong Form

	Legacy DCS (Series)	Modern DCS (Series)	Modern DCS (Standard)	Difference
Gain	4	4	6	+50%
Reset	0.25 rep/min	240 sec.	360 sec.	+50%
Rate	2 minutes	120 sec.	80 sec.	-33%
PV Filter	0.1 minute	6 sec.	6 sec.	0%

15 AIChE Spring Meeting, Austin, TX

EMERSON. Process Management

PLC Conversions

- Always double check PID tuning conversions from PLC's!
  - Some do strange things such as scaling, tracking, etc.
  - Check any scaling issues and how PV filtering is handled
  - Check for any Derivative gain limiting
- Some PLCs have Standard PID Form, but
  - have a Selectable "Positional" or "Velocity" algorithm
  - Velocity selection impacts tracking and override applications
- Check your Modern DCS applications!

15 AIChE Spring Meeting, Austin, TX

EMERSON. Process Management

PLC Conversions - Example

PLC Units:

- Form is Standard
- Gain is OUT_EU/PV_EU
- Integral time (minutes)
- Derivative Time (minutes)

Thus the conversion is:

- Modern DCS PID Form = Standard (default)
- Modern DCS Gain[%OUT/%PV] = PLC_Gain*(PV_Span[1]/Out_Span[1])
- Modern DCS Reset[sec] = PLC Reset Time*60
- Modern DCS Rate [sec] = PLC Derivative Time*60

1. PV and OUT spans in Modern DCS

15 AIChE Spring Meeting, Austin, TX

EMERSON. Process Management

How to Avoid – Best Practices

- Know where you are going
  - Check the New Modern DCS System Options
- Know your Legacy PID Forms
  - Dig into old documentation if need be
  - Select to minimize effort or standardize across platform
- Know your Legacy Tuning Units
  - Convert units first
  - Convert to selected PID Form if needed
- Export Database, use appropriate tool
- Don't make system wide assumptions
  - Some DCS platform units selectable by PID
- Make sure to verify all critical loops
  - Loops with Rate/Derivative
  - Interacting loops

15 AIChE Spring Meeting, Austin, TX

EMERSON. Process Management

Summary

- What it looks like when it goes Wild?
  - Cycling Outputs
  - Controlling Super Slow or Super Fast (Unstable Control)
- Common Conversion Mistakes
- Tips, Tools, & Techniques

15 AIChE
Spring Meeting, Austin, TX

EMERSON.
Process Management

Ternary Vapor-Liquid Equilibrium at 1 Atm – The Simplest Case: A Tutorial Using Literature Data

Robert G. Kunz

RGK Environmental Consulting, L.L.C.
Hillsborough, North Carolina

Paper 28a for Presentation at

2015 AIChE Spring Meeting

April 26-30, 2016

Austin, TX

Abstract

In making the transition from two-component (binary) to three-component (ternary) distillation, it appears that an important stepping stone in the learning process has often been left out. While it may be more interesting to move directly to analyze ternary systems with unusual properties, much can be gained from a brief study of ternary distillation for cases exhibiting ideal or nearly ideal behavior. At the very least, these provide a standard for the treatment of non-ideal systems in regard to technique and presentation.

It is instructive to predict idealized vapor-liquid equilibrium (VLE) of the resulting ternary system, starting with the three constituent binaries, each assumed to follow Raoult's Law in the liquid phase with an ideal-gas vapor phase. Typical graphical presentation of numerical results for a ternary system is also explored.

Since it is unlikely that any real VLE system would ever be perfectly ideal, it is even more fruitful to evaluate any deviations between the idealized calculations and experimental data. To illustrate the principles involved, binary and related ternary VLE data at atmospheric pressure are considered for mixtures of methanol, ethanol, and the propanols and for mixtures of benzene, toluene and the isomeric xylenes, where extensive experimental VLE data are available for comparison. Relatively low-boiling components and low pressure have been selected to keep things as simple as possible in this tutorial. The minor deviations in both x-y and temperature values compare favorably with both the original investigators' raw data and with their more complicated calculations using experimentally determined activity coefficients.

The methodology can be extended to multi-component (quaternary and higher) systems and forms the basis for empirical distillation-column design methods involving families of similar molecules. These techniques are explored briefly.

Keywords. Antoine Equation; Binary and Ternary Systems; Ideal-Gas Vapor Phase; Liquid-Phase Activity Coefficients; Multi-Component Systems; Ideal Systems; Raoult's Law; Specific Binary and Ternary Systems: Benzene–Toluene, Benzene or Toluene with Ethylene Dichloride (EDC). Ethylbenzene, or Isomers of Xylene, Methanol–Water, Methanol–Ethanol–1-Propanol, and Methanol–Ethanol–2-Propanol; Quaternary Systems with the Aromatic Compounds and from Adding Butanols to the Alcohol Systems; Vapor Pressure; Vapor-Liquid Equilibrium (VLE).

Disclaimer
("Some restrictions apply; batteries not included; your mileage may vary")

The following caveat applies: The information contained herein is offered in good faith but without guarantee, warranty, or representation of any kind (expressed or implied) as to its usefulness, correctness, completeness, or fitness for any particular purpose. The user assumes all risk for its implementation and should seek independent professional verification of its accuracy. The author assumes no responsibility and shall not be liable for any loss of profit nor any special, incidental, consequential, or other damages which may result from the use of any of the information contained in this presentation, be it oral or written. Any statements concerning design, construction, operation, what constitutes regulatory compliance, and/or how to achieve such compliance should not be construed as recommendations on the part of the author and/or his organization.

Introduction

Modification of binary VLE equations for an ideal system to include a third component is straightforward. Nevertheless, no actual VLE system can be expected to be perfectly ideal over the entire range of concentrations, and one cannot be certain without experimental data the extent of any deviations from ideality. There are, however, a few systems that are likely to exhibit more or less ideal behavior. These would provide excellent test cases for validation, but one is limited to common chemicals to maximize the chances of finding appropriate data. Agreement of ideal-system calculations with experimental data depends on the degree of ideality, the precision of the data, and somewhat on the vapor pressure relationship chosen.

In the paragraphs that follow, we will address in turn the equations themselves, those binary and ternary systems likely to be ideal, and the agreement of idealized calculations with published data. The main text is supplemented by details provided in several appendices.

Summary of Vapor-Liquid Equilibrium (VLE) Relationships

Low-pressure vapor-liquid equilibrium is described by the following relationships involving temperature (t), total pressure (P_T), partial pressure (p), vapor pressure (p^{vap}), and liquid- and vapor-phase phase mol fractions (x and y) of each component, plus so-called activity coefficients (γ's) when a given system cannot be considered to be ideal. Familiar equations for binary solutions are augmented here to account for the addition of a third component.[1] Subscripts (1, 2, 3) denote ternary constituents in order of decreasing volatility.

Ideal Ternary Systems

Equations. A system is said to be ideal when it follows Raoult's Law in the liquid phase

$$p_1 = p_1^{vap} x_1 \qquad (1)$$

$$p_2 = p_2^{vap} x_2 \qquad (2)$$

$$p_3 = p_3^{vap} x_3 \qquad (3)$$

and $$P_T = p_1 + p_2 + p_3 \quad (4)$$

with $$x_1 + x_2 + x_3 = 1 \quad (5)$$

If the vapor phase is assumed to be ideal as well, the Ideal / Perfect Gas Law and Dalton's Law apply. Then vapor-phase mole fractions (y_1, y_2, and y_3) can be calculated from

$$y_1 = p_1 / P_T = p_1^{vap} x_1 / P_T \quad (6)$$

$$y_2 = p_2 / P_T = p_2^{vap} x_2 / P_T \quad (7)$$

$$y_3 = p_3 / P_T = p_3^{vap} x_3 / P_T \quad (8)$$

with $$y_1 + y_2 + y_3 = 1 \quad (9)$$

The equation below, valid for ideal binary systems[1],

$$y_1 = x_1\ \alpha_{12} / [1 + x_1 (\alpha_{12} - 1)] \quad (10)$$

is replaced by the following equations for ideal ternary systems[1,2]

$$y_1 = x_1\ \alpha_{13} / [\alpha_{13}\ x_1 + \alpha_{23}\ x_2 + x_3] \quad (11)$$

$$y_2 = x_2\ \alpha_{23} / [\alpha_{13}\ x_1 + \alpha_{23}\ x_2 + x_3] \quad (12)$$

where α_{12} is the *relative volatility* or *enrichment ratio*, defined as

$$\alpha_{12} = (y_1 / x_1) / (y_2 / x_2) = p_1^{vap} / p_2^{vap} \quad (13)$$

and α_{13} and α_{23} are similarly defined

At constant temperature, α is constant and Equations (10), (11), and (12) are exact. At constant pressure, they are approximate since α is a mild function of the varying VLE temperature.

Vapor Pressure. Pure-component vapor pressures in both ideal and non-ideal systems are conveniently calculated for any number of components by means of the Antoine Equation:

$$\log_{10} p^{vap}\ (\text{mmHg}) = A - B / [t\ (°\text{C}) + C] \quad (14)$$

Rearrangement of Equation (14) to solve for temperature yields[3]

$$t\ (°\text{C}) = B / [A - \log_{10} p^{vap}\ (\text{mmHg})] - C \quad (15)$$

from which the atmospheric boiling point can be computed by substitution of 760 mmHg for p^{vap}. Constants A, B, and C are listed in Appendix Tables A1 and B1 for materials of interest here.

Example Cases. Ideal binary systems occur with members of homologous series, isomers, and in certain other instances.[4] With similar molecules in a mixture, interactions between like and unlike species are nearly the same, and each component behaves as if independent of the other(s). Examples include ethylene dibromide–propylene dibromide, chlorobenzene–bromobenzene, propane–butane, and butane–isobutane.[4]

Some binary solutions composed of C_1 to C_3 aliphatic alcohols (methanol, ethanol, and the propanols) can be taken as idea.[4] Although one might expect that all would behave as ideal systems across the board, such is not the case. VLE x-y curves for the binary systems methanol–ethanol[3] plus methanol or ethanol with either 1-propanol or 2-propanol are drawn in Figures A1-A5, accompanied by data from numerous cited sources. Comments about how closely these real systems approach ideality are provided in Appendix A.

The classic system used to illustrate Raoult's Law for binary systems is benzene-toluene. It is highlighted as ideal in many textbooks and handbooks. The depiction of its x-y curve (Figure 1), shown with data from multiple sources too numerous to mention, has been published previously.[3] Isobaric VLE data at 1 atm pressure for several binary systems of aromatic chemicals with benzene or toluene are shown in Figures B1-B12, accompanied by data from a number of cited sources with comments given in Appendix B.

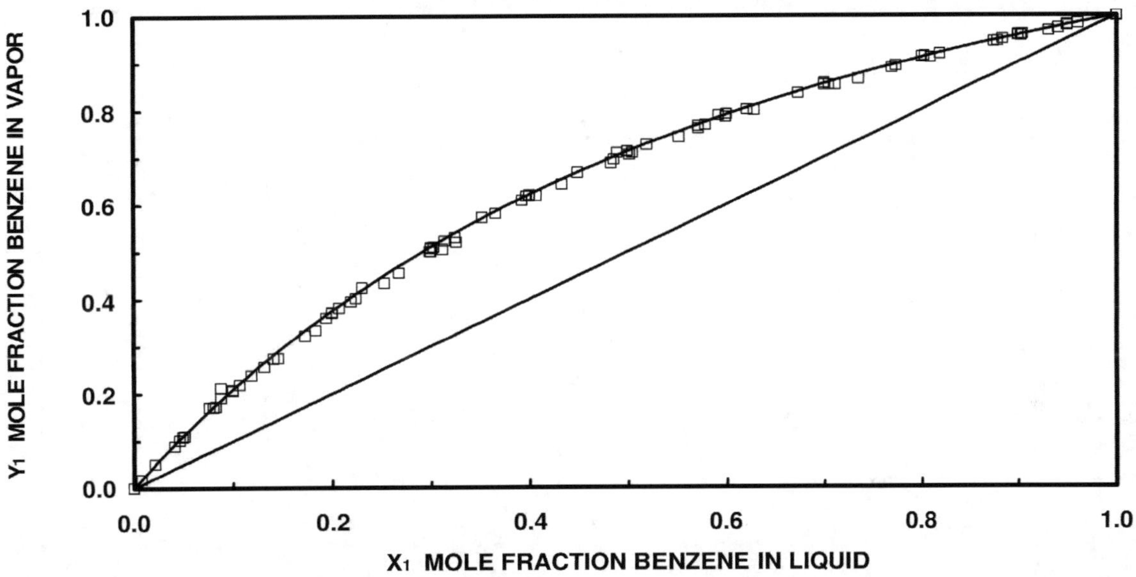

Figure 1. Vapor-Liquid Equilibrium Diagram for Benzene–Toluene at 1 Atm Pressure

Other instances of ideal behavior include carbon tetrachloride–toluene,[4] and 1.2-dichloroethane (ethylene dichloride, EDC) with either benzene[4] or toluene.[5] Figure 2 and the previously presented Figure 3[6] depict the ideal x-y curves at 1 atm for benzene–EDC and for EDC–toluene annotated with data from various investigators. The size of the "bulge" between the x-y curve and the y=x line at a 45° angle in the figures reflects the relative volatility (α) for each binary pair, around 1.1 for benzene–EDC and 2.2 - 2.3 for EDC–toluene.

The first step in assessing whether a ternary system will turn out to be ideal is to consider whether the three constituent binaries are ideal. While this condition does not necessarily guarantee an ideal ternary, it is hard to imagine a ternary system being ideal with one or more of the binaries exhibiting non-ideal behavior.

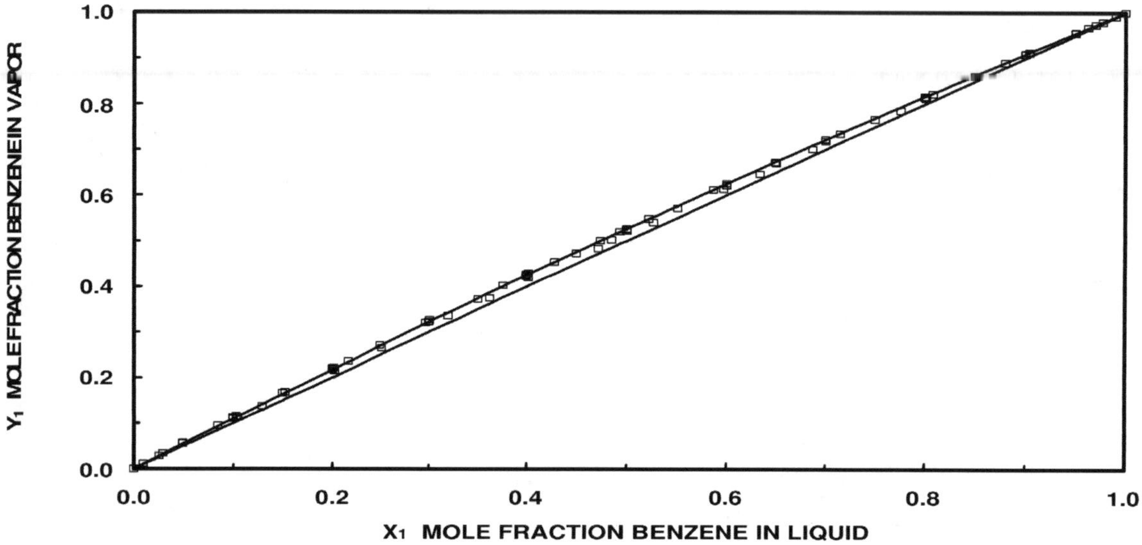

Figure 2. Vapor-Liquid Equilibrium Diagram for Benzene–EDC at 1 Atm Pressure Data from References **[9-14]**

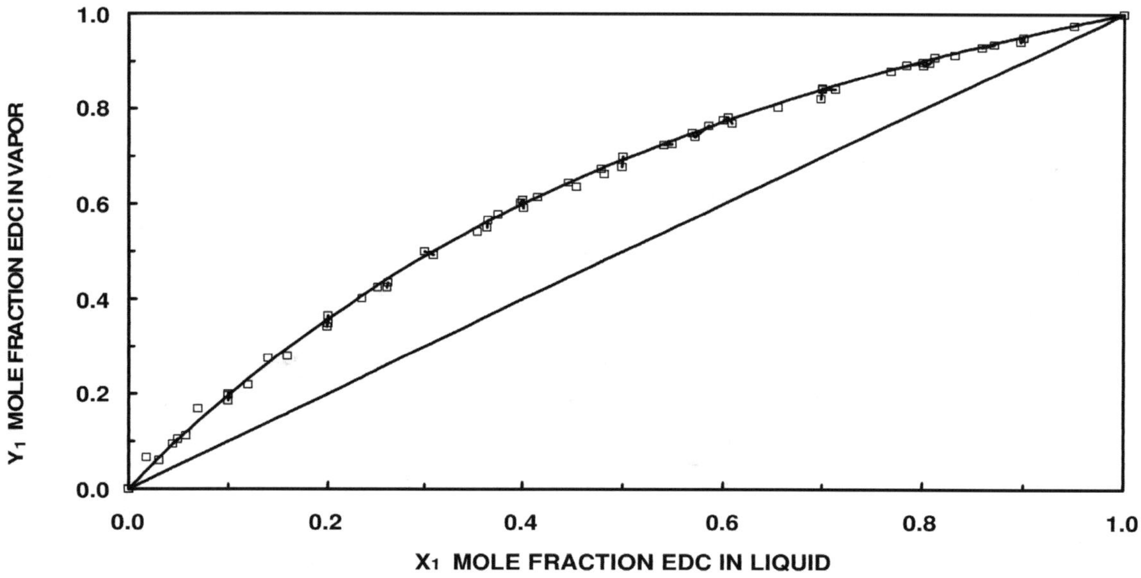

Figure 3. Vapor-Liquid Equilibrium Diagram for EDC–Toluene at 1 Atm Pressure Data from References **[5, 15-17]**

Non-Ideal Systems

For non-ideal systems, an activity coefficient (γ) is introduced into Equations (1), (2), and (3) to account for deviations from Raoult's Law in the liquid phase. In summary form,

$$p_n = p_n^{vap} \gamma_n x_n \qquad (16)$$

With an Ideal Gas / Dalton's Law vapor phase, Equations (6), (7), and (8) become

$$y_n = p_n / P_T = p_n^{vap} \gamma_n \; x_n / P_T \tag{17}$$

and upon rearrangement[18-19]

$$\gamma_n = y_n \; P_T / p_n^{vap} \; x_n{}^{*} \tag{18}$$

These latter equations allow activity coefficients to be back-calculated as point values from experimental data and to be fitted to a special type of continuous function for use in VLE calculations.

There are several empirical activity-coefficient formulations in widespread use. These include the van Laar, Margules, Wilson, NRTL (Nonrandom Two Liquid), and UNIQUAC (Universal Quasi Chemical) Equations. For any given system, some fit the deviations from ideality better than others across the range of data.

The van Laar Equations, for example, are given below in terms common logarithms (to the base 10).[3,7] Van Laar Constants are listed in Perry's Handbook for a number of VLE binary systems, including methanol-water at 1 atm pressure (temperature range ~ 64 °C to 100 °C).[4]

for methanol (component 1) $$\log_{10} \gamma_1 = A_{1\text{-}2} / [1 + (A_{1\text{-}2}\, x_1) / (A_{2\text{-}1}\, x_2)]^2 \tag{19}$$

for water (component 2) $$\log_{10} \gamma_2 = A_{2\text{-}1} / [1 + (A_{2\text{-}1}\, x_2) / (A_{1\text{-}2}\, x_1)]^2 \tag{20}$$

These equations can also be written in terms of natural logarithms (base e), using constants of $[\log_{10}(e) = 2.302585...] \times A_{1\text{-}2}$ and $A_{2\text{-}1}$.

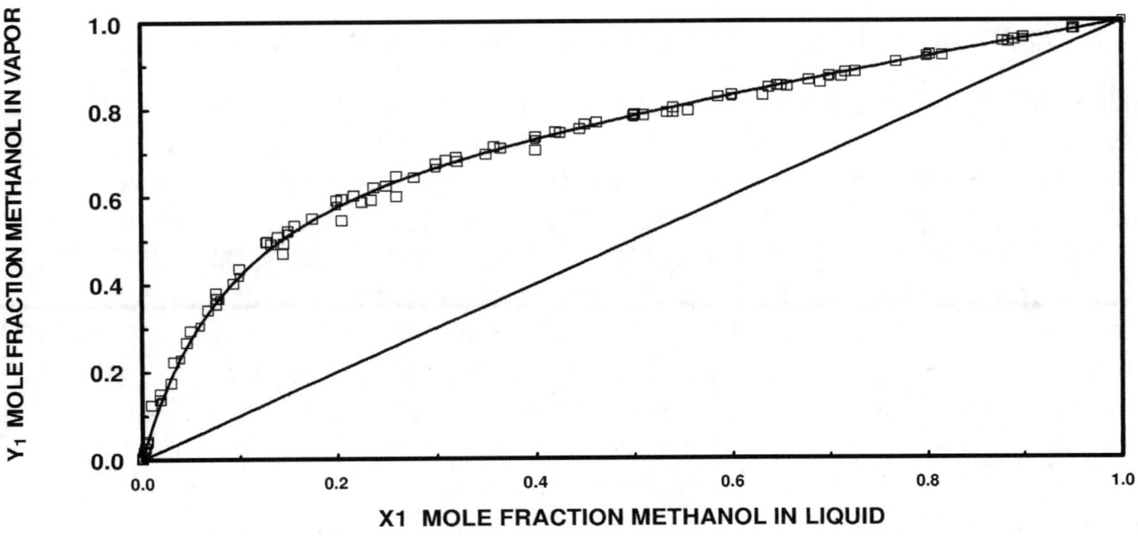

Figure 4. Vapor-Liquid Equilibrium Diagram Methanol–Water at 1 Atm Pressure

* A correction factor involving properties usually not routinely available may be included in Equation (18) to account for vapor-phase non-idealities. However. this factor, calculated from reported values for a similar system at atmospheric pressure,[20] amounts to an average difference in calculated activity coefficients of only 3%, a distinction perhaps important to a researcher but certainly within engineering accuracy.

The previously published[3] non-ideal methanol-water x-y curve with accompanying data from a number of investigators is reproduced in Figure 4. It resembles the curves for ideal systems but was constructed using van Laar Constants 0.36 and 0.22[4] to calculate activity coefficients and account for the liquid-phase non-idealities in this system. Equation (21) below

$$P_T = p_1^{vap} \gamma_1 x_1 + p_2^{vap} \gamma_2 x_2 \tag{21}$$

was then employed to produce a continuous x-y curve at 1 atm by a trial-and-error† procedure.

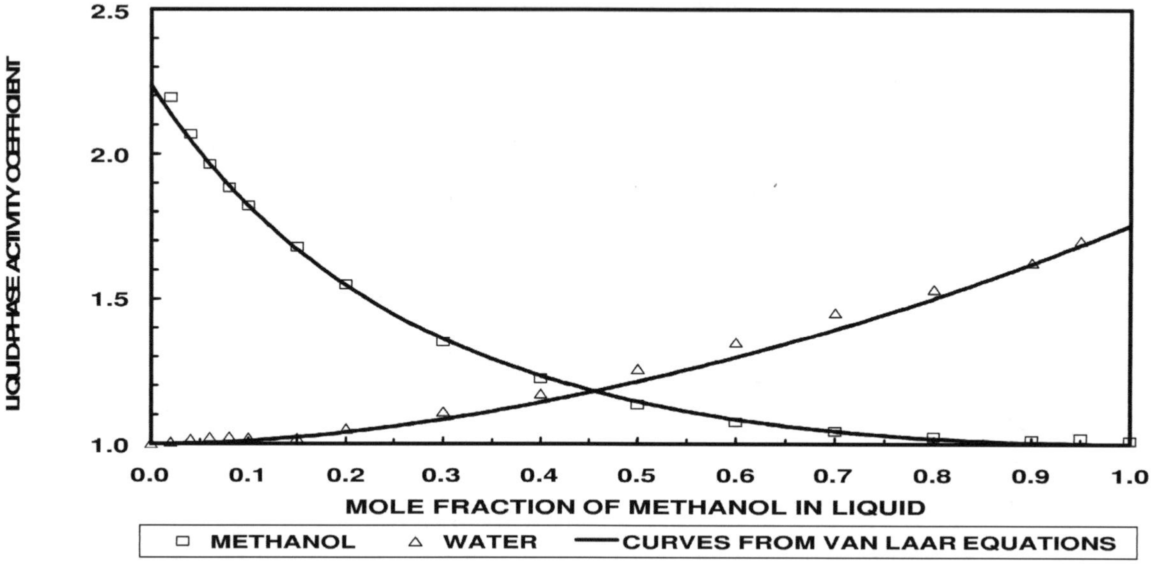

Figure 5. Activity Coefficients: Methanol–Water at 1 Atm Pressure

To illustrate the variation of activity coefficients with liquid-phase mole fraction, activity coefficients calculated for the non-ideal methanol-water system at 1 atm are plotted in Figure 5. Although some mixtures of alcohols with one another may be expected to be more or less ideal, in contrast alcohol-water solutions are never ideal, some even forming azeotropes.

One individual set of methanol-water T-x-y data is conveniently tabulated on p.13-5 of Perry's 4th ed.[4] and on p.13-3 of the 6th ed.[8] The exact same data can also be found on p. 143 of Chu et al.[9] as well. Data attributed to yet a third source are also contained on p. 147 of the DECHEMA tables,[21] along with van Laar natural-log best-fit parameters, $A_{1\text{-}2} = 0.8041$ and $A_{2\text{-}1} = 0.5619$. Those values are now listed in Perry's 6th ed.[8] On a common log basis, they correspond to 0.35 and 0.24, respectively, compared to 0.36 and 0.22 on p.528 of Perry's 3rd ed. and on p.13-7 of the 4th ed.[4] The Perry's 3rd and 4th ed. van Laar Constants are no doubt derived from a different data set or an average / combination of several.

---

† For an ideal binary system, the x-y equilibrium curve at constant total pressure (P_T) can be mapped out in a straightforward manner by means of the relationships: $x_1 = (P_T - p_2^{vap}) / (p_1^{vap} - p_2^{vap})$ and $y_1 = p_1^{vap} / P_T$. However, calculations to determine the x-y curve for a non-ideal system are always trial-and-error since the activity coefficients in Equation (21) are functions of liquid composition. An efficient calculation method is to select a value of x_1 and total pressure and then vary the temperature repeatedly at each point until the calculated total pressure obtained from Equation (21) converges as closely as desired to the total pressure selected.

Activity coefficients are not simply "fudge factors" but exhibit their own unique properties within the mathematical framework of thermodynamics. As shown in Figure 5, the binary activity coefficient (γ) for the material in greater abundance becomes 1.0 (and its logarithm zero) for the pure component. It smoothly transitions to some finite value for the dilute species, where γ extrapolates to the so-called *activity coefficient at infinite dilution* (*gamma infinity,* γ^{∞}). In essence, Raoult's Law is obeyed for each constituent when its mole fraction nears 1.0, and Henry's Law ($y\,P_T = H\,x$, with H = the Henry's Law constant) applies in the case of the other as its composition approaches zero.

Activity coefficients also display a temperature relationship with heat of mixing data. In addition, thermodynamically consistent activity coefficients produce a net zero area under the curve (+ area = – area) when plotted as $\ln(\gamma_1/\gamma_2)$ or $\log_{10}(\gamma_1/\gamma_2)$ versus x (not shown here).

Some scatter in experimentally derived values of γ is to be expected and tends to increase greatly in the immediate vicinity of zero concentration for either component. Because of mathematical problems with division by zero, gamma infinity cannot be obtained directly but must be found by extrapolation, either graphically or via an activity coefficient model.

The coefficients $A_{1\text{-}2}$ and $A_{2\text{-}1}$ for the van Laar (and Margules) Equations are the logarithms of γ^{∞} for each of the binary constituents. For the particular data set under discussion here, γ^{∞} is 2.23 for methanol and 1.75 for water. The γ^{∞}'s from other data sets at 1 atm total pressure vary from 1.91 to 2.40 for methanol and from 1.35 to 1.95 for water.[21-24]

Activity coefficients > 1 denote positive deviations from Raoult's Law as in this case, and those < 1 signify negative deviations. Larger absolute values of γ correspond to increasingly non-ideal behavior. Typical patterns of log γ are summarized graphically in the author's book.[3]

Evaluation of activity coefficients constitutes a more sensitive test of ideality than simply looking at the VLE x-y diagram alone.[19] A perfectly ideal system would show activity coefficients of exactly 1.0 for both components across the board.

Comparison of Ideal Ternary Solution Calculations with Experimental Data

Bear in mind that the validity of comparing predicted vapor-phase concentrations with values calculated for an ideal ternary system depends upon how closely any real system approximates ideality, plus the accuracy of the experimental data to which the calculated values are being compared. Selection of a relationship for vapor pressure also affects the calculations somewhat.

That said, several ternary systems come to mind to demonstrate ideal ternary VLE calculations applied to an actual case. One is benzene–EDC–toluene, in which each of the binaries is well known to be ideal. Unfortunately, no ternary data for this system could be found in the literature. Another system, benzene–toluene–ethylbenzene, also stands a good chance of being ideal, but the total number of ternary data points uncovered (eight in all) are limited: data from one investigation and ideal VLE calculations agree rather poorly; data from the other show excellent agreement, resulting in an inconclusive test case (Appendix B). In the end, comparison of calculations and data for the C_1-C_3 aliphatic alcohols (Appendix A) and for one of the xylene isomers with benzene and toluene (Appendix B) has proven successful. Results are summarized in Table 1 below; details are contained in Appendix A and Appendix B. Antoine Constants employed are listed in Tables A-1 and B-1.

Methanol–Ethanol–1-Propanol. This mixture is considered to be reasonably ideal,[25] as are all of its constituent binaries (Figures A1-A3), and literature data at 1 atm for the ternary are available for comparison, providing a suitable test case.[26] Calculated values for 45 data points at 760 mmHg pressure are compared with reported experimental data in Table A2. Absolute values of the difference between calculated vs. measured y-values for all three components average between 0.01 and 0.02 mole fraction, with some 90 % of the points (89% for y_1 methanol, 93% for y_2 ethanol, and 91% for y_3 1-propanol) showing a difference of 0.03 or less. In round numbers, maximum deviations are 0.07 (y_1), 0.05 (y_2) and 0.09 (y_3). Absolute values of temperature deviations range from 0.03 to 2.69 °C and average 0.95 °C, with 91% of the temperature deviations less than 1.8 °C. Further discussion is contained in Appendix A.

Constant temperature data at 60 °C are also available for comparison with ideal calculations.[27] Those calculations are straightforward with no trial and error necessary. Data and calculated results are listed in Table A3. As expected, values of total pressure calculated using Equations (1-4) and Equation (14) and shown in Column 3 are slightly different from the reported data reproduced in Column 2. Calculated total pressure was used to compute vapor phase composition via Equations (5-8). For the constant temperature data, maximum deviation of all y-values (absolute value of calculated vs. experimental) is less than 0.02, averaging 0.005 to 0.008, with 76-91% of deviations less than 0.01. Deviations in total pressure range from 0.4 mmHg to 9.5 mmHg, with an average deviation of 4.4 mmHg.

Methanol–Ethanol–2-Propanol. This system exhibits mild non-ideal behavior and is also discussed in Appendix A. Two sets of atmospheric pressure data[28,29] are available for comparison with ideal calculations, as well as data at 60 °C.[30] For one study, absolute deviations in all y-values average about 0.01 mole fraction, with at least 98% less than 0.03, and a maximum difference of about 0.03 for methanol, 0.02 for ethanol, and 0.04 for 2-propanol. Absolute values of temperature deviations range from 0 to 0.84 °C and average 0.27 °C, with 88% less than 0.5 °C and 96% less than 0.6 °C. For the other, the average of y-value deviations is less than 0.006, with 82-89% under 0.01 and maximums of about 0.02 for methanol, 0.01 for ethanol, and 0.015 for 2-propanol. Agreement of predicted y-values for both sets of constant pressure data is somewhat better than for the "ideal" methanol–ethanol–**1**-propanol system, suggesting a possible improvement with more precise data in that case.

For the constant temperature data, maximum absolute deviation is about 0.06 mole fraction for methanol, 0.01 for ethanol, and 0.05 for 1-propanol, averaging 0.005 – 0.008. with 96-100% of the deviations less than 0.02 and 76-91% of the deviations less than 0.01. Pressure deviations are 0.2 - 9.8 mmHg (average 3.8), similar to those exhibited by the methanol–ethanol–**1**-propanol ternary.

Benzene—Toluene—Ethylbenzene or One of the Xylene Isomers. Benzene–toluene (an aromatic hydrogen benzene ring structure with an attached methyl group) is the textbook example of an ideal binary solution. One might suspect that a ternary mixture of benzene and toluene with structurally similar aromatic compounds would also exhibit ideal behavior. These include ethylbenzene (a benzene ring with an ethyl group attached instead of a methyl group) or one of the xylene isomers (a benzene ring having two attached methyl groups at various locations).

Table 1. Summary of Ideal Calculations versus Experimental Data for Several Ternary Systems

Ternary System	Condition	Deviation Between Data and Ideal Calculations Y1	Y2	Y3	Temp/Press	Remarks
Methanol-Ethanol-1-Propanol ***45 Data Points***	760 mmHg	min 0.0002 avg 0.018 max 0.068 89% < 0.03	min 0.0004 avg 0.012 max 0.052 93% < 0.03	min 0.0007 avg 0.016 max 0.085 91% < 0.03	min 0.03 °C avg 0.95 °C max 2.69 °C 91% < 1.8 °C	"Ideal" Ternary; 3 Binaries Ideal at 760 mmHg
Methanol-Ethanol-1-Propanol ***Constant Temp Data*** ***23 Data Points***	60 °C	min 0.0000 avg 0.0076 max 0.017 76% < 0.01	min 0.0000 avg 0.0045 max 0.012 91% < 0.01	min 0.0000 avg 0.0054 max 0.018 87% < 0.01	min 0.4 avg 4.4 max 9.5 mmHg	"Ideal" Ternary; 3 Binaries Ideal at 760 mmHg
Methanol-Ethanol-2-Propanol ***1st Ternary Data Set*** ***48 Data Points***	760 mmHg	min 0.0002 avg 0.0110 max 0.0340 98% < 0.03	min 0.0000 avg 0.0065 max 0.0229 100% < 0.03	min 0.0013 avg 0.0109 max 0.0387 98% < 0.03	min 0.00 °C avg 0.27 °C max 0.84 °C 96%< 0.6 °C	Nearly Ideal; Methanol-2-Propanol Binary Not Ideal
Methanol-Ethanol-2-Propanol ***2nd Ternary Data Set*** ***28 Data Points***	757.56 mmHg	min 0.000 avg 0.006 max 0.017 82% < 0.01	min 0.000 avg 0.005 max 0.011 89% < 0.01	min 0.000 avg 0.004 max 0.015 89% < 0.03	min 0.06 °C avg 0.38 °C max 0.82 °C 93% < 0.7 °C	Nearly Ideal; Methanol-2-Propanol Binary Not Ideal
Methanol-Ethanol-2-Propanol ***Constant Temp Data*** ***23 Data Points***	60 °C	min 0.0000 avg 0.0082 max 0.0570 96% < 0.02	min 0.0000 avg 0.0047 max 0.0128 100% < 0.02	min 0.0000 avg 0.0056 max 0.0478 96% < 0.02	min 0.2 avg 3.8 max 9.8 mmHg	Nearly Ideal; Methanol-2-Propanol Binary Not Ideal
Benzene-Toluene-Ethylbenzene ***Ternary - 2 Sources*** ***(6 and 4 Data Points)***	760 mmHg	range 1st Data Set 0.04-0.16 2nd Data Set <0.01	range 1st Data Set 0.01-0.03 2nd Data Set 0.017	range 1st Data Set 0.01-0.14 2nd Data Set 0.0115	range 1st Data Set 0.24-1.39 °C 2nd Data Set 0.32-0.65 °C	Likely Ideal-Not Enough Data to Confirm; 3 Binaries Ideal
Benzene-Toluene-o-Xylene ***6 Data Points***	772 mmHg	Data Pts 1-5 $y_1 < 0.006$ Data Pt 6 $y_1 < 0.01$	Data Pts 1-5 $y_2 < 0.006$ Data Pt 6 $y_2 \sim 0.02$	Data Pts 1-5 $y_3 < 0.006$ Data Pt 6 $y_3 \sim 0.03$	min 0.4 °C avg 1.7 °C max 2.7 °C	"Ideal" at Points Investigated; Xylene Binaries Not Ideal
Benzene-Toluene-m-Xylene ***13 Data Points***	760 mmHg	min 0.0000 avg 0.015 max 0.0275 61% < 0.02 100% < 0.03	min 0.0000 avg 0.022 max 0.066 61% < 0.02 69% < 0.03	min 0.0002 avg 0.016 max 0.093 77% < 0.02 77% < 0.03	min 0.02°C avg 0.98 °C max 3.01 °C 61% < 1.0°C 92% < 2.0°C	Not Ideal for All Compositions; Xylene Binaries Not Ideal
Benzene-Toluene-p-Xylene ***1st Ternary Data Set*** ***26 Data Points***	760 mmHg	min 0.000 avg 0.005 max 0.013 81%< 0.01	min 0.000 avg 0.003 max 0.008 100%< 0.01	min 0.000 avg 0.002 max 0.011 96%<0.01	min 0.00 °C avg 0.27 °C max 0.83 °C 87% < 0.7°C	"Ideal" Ternary Comparable to Alcohols; Ideal Binaries
Benzene-Toluene-p-Xylene ***2nd Ternary Data Set*** ***18 Data Points***	760 mmHg	min 0.0001 avg 0.015 max 0.04 72% < 0.02 89%< 0.03	min 0.0012 avg <0.01 max 0.02 94% < 0.02 100% < 0.03	min 0.0018 avg 0.01 max 0.04 89% < 0.02 95% < 0.03	min 0.01 °C avg 0.93 °C max 4.54 °C 56% < 0.7°C 72% < 1.0	"Ideal" Ternary Comparable to Alcohols; Ideal Binaries

The three constituent binaries of the **benzene–toluene–ethylbenzene** ternary are regarded as ideal. Based on data found in the literature, the ternary also appears to be ideal, but the limited number of ternary data points uncovered and lack of agreement between the data of two independent investigations [31,32] results in an inconclusive test case (Table B2). The y-value data from one investigation and ideal VLE calculations agree rather poorly, differing by up to 0.16 mole fraction. In the other, all measured y-values deviate from ideal calculations by about 0.01 or less. See Appendix B for details.

Minimal ternary data from a single source for the **benzene-toluene-*o*-xylene** ternary were found for this system at 772 mmHg,[33] along with binary data from the same laboratory for benzene–*o*-xylene at 761 mmHg[34] and toluene–*o*-xylene at 770 mmHg.[35] Ternary data are listed in Table B3. Deviations from ideal temperatures average 1.7 °C and range from 0.4 °C to 2.7 °C. Except for one point at the extremes of benzene and *o*-xylene investigated, maximum deviation in the absolute value of $y_{ideal}–y_{data}$ for all species is less than 0.006 mole fraction, and at that point maximum deviation in y_1 is less than 0.01 and about 0.02 for y_2 and 0.03 for y_3. The data closer to the center of a triangular ternary VLE diagram have the greatest likelihood of being ideal. In an example problem dealing with a mid-range liquid composition of 50% benzene and 25% each of toluene and *o*-xylene, Treybal[2] assumes the benzene–toluene–*o*-xylene system at atmospheric pressure to be ideal. Further details are provided in Appendix B.

Nevertheless, binary x-y data plotted in Figures B3 and B5 and the corresponding activity coefficients (Figures B4 and B6) are clearly non-ideal and therefore inconsistent with an ideal ternary system. The γ of *o*-xylene above a benzene mole fraction of 0.7 seriously deviates from ideality. Activity coefficients in toluene–*o*-xylene solutions also deviate significantly when the other constituent is in the majority. Activity coefficients from more recent data (not shown) have been found to exhibit large negative deviations from ideality.[36]

The system **benzene–toluene–*m*-xylene** and its three constituent binaries have been studied in a single investigation at 760 mmHg,[37] with more data entries than the two previous systems but still relatively few. Table B4 summarizes the ternary data and ideal calculations. Benzene–toluene x-y data plot as ideal (Figure B1), and activity coefficients computed from them using Equation (18) are virtually flat at 1.0 between x=0 and x=1.0 (Figure B2). However, based on binary x-y data (Figures B7 and B8) and activity coefficients (Figures B9, B10) for the two binaries containing *m*-xylene, both binaries are non-ideal, more so at lower benzene or toluene concentrations. Conclusion is that this ternary is not completely ideal for all compositions. Details are given in Appendix B.

The ternary of **benzene and toluene with *p*-xylene** and its three constituent binaries have been studied at atmospheric pressure.[38] Forty-four ternary data points from this plus a second study[39] are listed in Table B5 of Appendix B, along with calculated values assuming ideal conditions. Binary data are plotted in Figures B11-B12. As shown in the figures, the x-y data for benzene–toluene and the other binaries follow the curves for Raoult's Law. To repeat, the benzene–toluene x-y data of Figure B1 would fit in nicely with the other data in Figure 1. Activity coefficients for benzene–toluene computed from Equation (18) are virtually flat at 1.0 between x = 0 and x = 1.0, with the greatest deviation occurring for toluene in the benzene–toluene system at the lowest toluene concentration. Activity coefficients for the other binaries scatter around 1.0 over the range of the data.

Average absolute value of the difference in y-values for the first study[38] ranges from 0.002 to 0.005 mole fraction (m.f.) depending on the species, with a maximum difference of approximately 0.01. Maximum deviation for all y-values is 0.0135, with 81-96% of the deviations less than 0.01. Average temperature difference is 0.27 °C; maximum is 0.83 °C, with 85% less than 0.5 °C. These minor deviations from the data for both x-y and temperature values compare favorably with the original investigators' more complicated calculations using experimentally determined activity coefficients.

Deviations for the second study are greater but tolerable. Average deviation is greater, up to 0.015 m.f. Maximum deviation is 0.04, with 89% less than 0.03 for y_1 (benzene) and about 90% less than 0.02 for y_2 (toluene) and y_3 (p-xylene). Temperature difference ranges from a minimum of 0.14 °C to a maximum of 4.5 °C, with 56% less than 0.7 °C and an average of 1.0 °C. Conclusion is that the **benzene–toluene–*p*-xylene** system is comparable to the alcohol systems described above.

In summary, agreement shown here between VLE calculations and published ternary data for the alcohol systems and for benzene–toluene–*p*-xylene is considered satisfactory overall.

Graphical Presentation

An equilateral triangular diagram (Figures 6, 7, and 8) is often employed as a graphical two-dimensional representation of the concentrations in a ternary VLE system. An equilateral triangle contains three vertices and three sides, with adjacent sides separated from each other by a 60° angle. In an equilateral triangle, the sum of the perpendicular distances to the sides from any interior point is equal to the altitude of the triangle (the perpendicular distance from one side to the opposite vertex),[2,40] and the coordinates of points within the triangle can each denote a unique percentage composition in the ternary mixture. Each vertex stands for 100% of a given constituent, falling to zero percent at the opposite side. One hundred percent of the second constituent is plotted at the second vertex, and 100% of the third occurs at the remaining vertex. The side opposite any vertex corresponds to the binary system containing zero percent of the constituent being depicted on that vertex. Further comments on the triangular diagram are contained in the cited references.[41,42]

There are several ways of graphing a VLE diagram for a ternary system.[1] One additional method[43] is to plot curves of constant y-values on the x-coordinates of the triangular diagram, as illustrated in the cited figures. The calculations are simpler when the system is assumed to be ideal. To construct such a diagram where the three constituent binaries follow Raoult's Law, one can proceed as follows without trial and error, using Component 1 (methanol: x_1, y_1) as an example:

- Choose a total pressure (P_T), and a desired value of y_1;
- Select a value for x_1. (Limiting values of x_1 at y_1 and P_T occur for the binary along the methanol–ethanol side of the triangle where 1-propanol (x_3) is zero and along the methanol–1-propanol side of the triangle where ethanol (x_2) is zero.)
- Calculate $p_1^{vap} = y_1 \, P_T / x_1$ from Equation (6) rearranged.
- Substitute A, B, and C for methanol into Equation (15) repeated below

$$t\ (^{\circ}C) = B/[A - \log_{10} p_1^{vap}\ (mmHg)] - C \qquad (15)$$

- At t (°C), calculate $\log_{10} p_2^{vap}$ and $\log_{10} p_3^{vap}$ from Equation (14) and take the anti-logs.
- Calculate $y_1 P_T = p_1^{vap} x_1$ from Equation (6).
- Combine Equations (1-4) and substitute for $p_1^{vap} x_1 = y_1 P_T$ to obtain

$$P_T = p_1^{vap} x_1 + p_2^{vap} x_2 + p_3^{vap} x_3 = y_1 P_T + p_2^{vap} x_2 + p_3^{vap} x_3 \qquad (22)$$

- Substitute for x_3 from Equation (5) and rearrange Equation (22) Part 2 to solve for x_2

$$x_2 = [(P_T (1 - y_1) - p_3^{vap} (1 - x_1)] / (p_2^{vap} - p_3^{vap}) \qquad (23)$$

- With x_1 and x_2, solve Equation (5) for x_3 and check using Part 1 of Equation (22).
- Repeat for different values of x_1 within its range to map out locus of points for y_1.
- Select another value for y_1 and repeat as necessary.
- When finished with y_1, modify equations and repeat for y_2 and y_3.

This technique can also be used to plot lines of constant temperature, pressure, liquid density, etc.

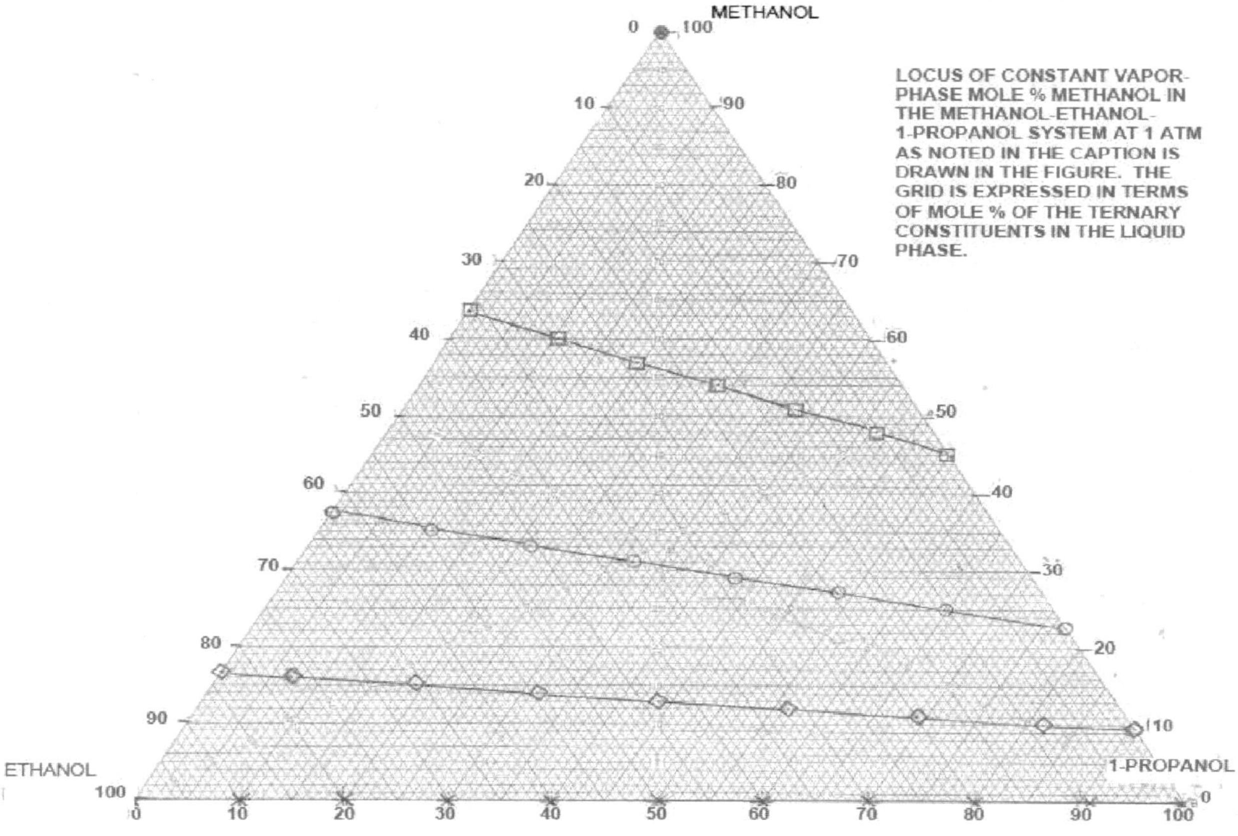

Figure 6. Ideal Vapor Phase Methanol in Methanol–Ethanol–1-Propanol System -at 1 Atm (Mole % Methanol as noted: ● 100, □ 75, o 50, ◇ 25, **X** 0)

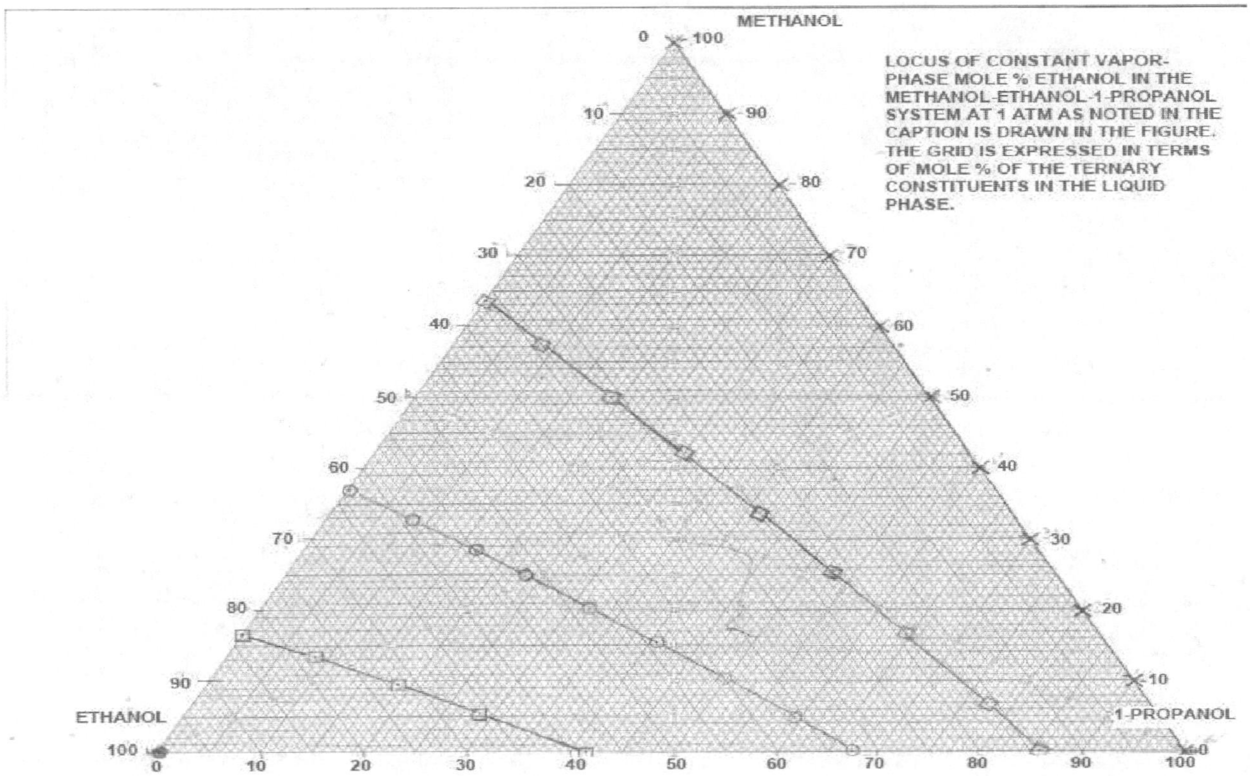

Figure 7. Ideal Vapor Phase Ethanol in Methanol–Ethanol–1-Propanol System at 1 Atm (Mole % Ethanol as noted: ● 100, □ 75, o 50, ◇ 25, **X** 0)

Multi-component Systems

In an academic sense, ideal VLE calculations can be extended to any number of components merely by writing additional equations. In practice, however, ideal multi-component systems are not common. Chances of finding a system approaching ideality are enhanced when dealing with similar molecules,‡ like the systems discussed in this paper and families of hydrocarbons in petroleum distillation.

In accordance with custom in the petroleum industry,[42] Equation (18) can be recast into

$$\gamma\ p^{vap} / P_T = y / x = K \tag{24}$$

in which the ratio K is known as the *distribution coefficient*,[44] *K-value*[45] or *K-factor*.[42]

‡ But even then, one cannot be sure *a priori* that behavior would be perfectly ideal. For example, there are four isomers of butanol: the straight-chain 1-butanol and three others with various degrees of branching – *sec*- or 2-butanol, isobutanol or 2-methyl-1-propanol, and *tert*-butanol or 2-methyl-2-propanol. The only ones that appear likely to form an ideal or nearly ideal quaternary VLE solution with methanol–ethanol–1-propanol are isobutanol and 1-butanol because they are the only ones whose binary solutions with all three of the existing components can be shown to be ideal or nearly ideal (Appendix C).

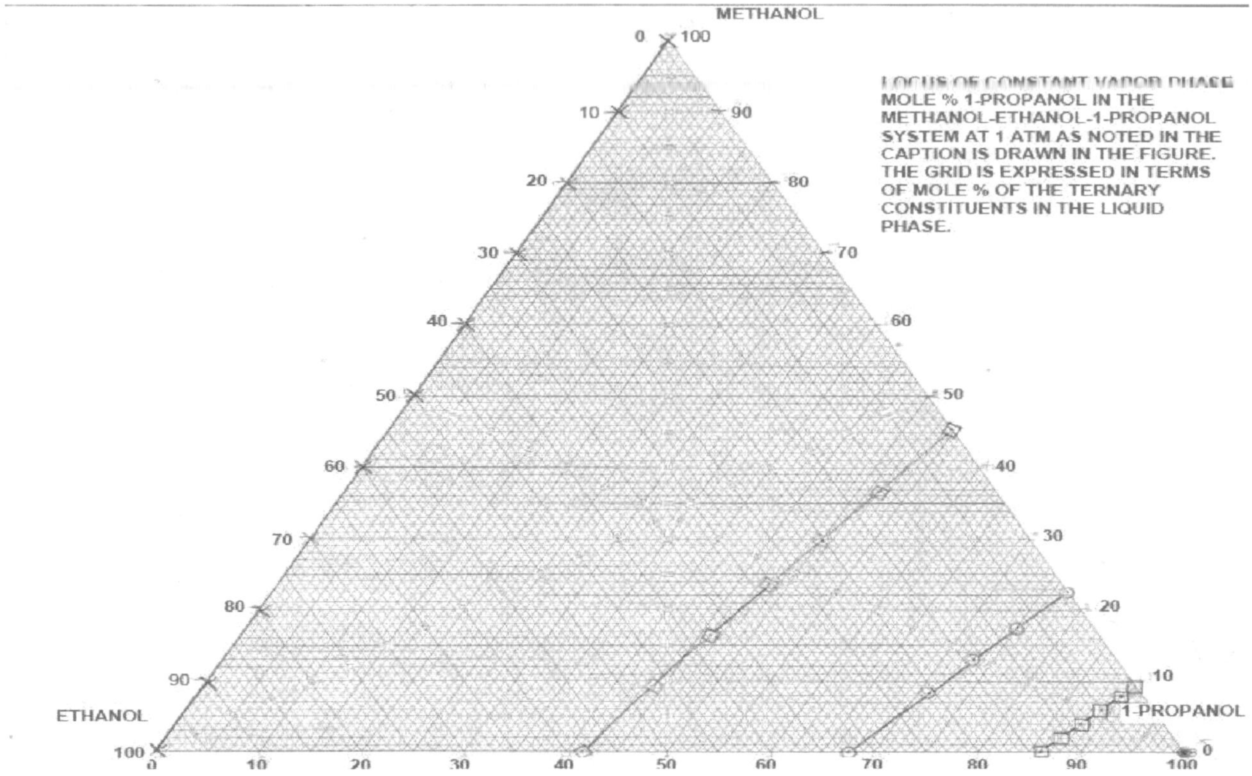

Figure 8. Ideal Vapor Phase 1-Propanol in Methanol–Ethanol–1-Propanol System at 1 Atm (Mole % 1-Propanol as noted: ● 100, □ 75, o 50, ◇ 25, **X** 0)

The concept of a K-factor applies to any system, ideal or non-ideal.[42] In non-ideal systems, K depends on temperature, pressure, and composition through γ. When both liquid and vapor phases form ideal solutions, rigorous thermodynamic treatment reveals that K should be a function only of temperature and pressure, independent of composition, and dependent on the properties of the pure species alone.[45] In the simplest case corresponding to Equation (25) below,

$$p^{vap} / P_T = y / x = K \tag{25}$$

The numerator is the vapor pressure-temperature relationship, and the pressure dependence in the denominator is the total pressure itself.

To simplify the more complicated iterative calculations when Equation (24) would ordinarily apply, K-factors have been correlated for light hydrocarbons. One such correlation involves a series of nomographs to evaluate K at different temperatures and pressures for normal alkanes methane through n-decane, branched alkanes isobutane and isopentane, and the alkenes ethylene and propylene.[44-46] It is based on experimental data for such mixtures and VLE relationships derived from the extremely accurate Benedict-Webb-Rubin (BWR) Equation of State,[44] in widespread use in the petroleum and natural gas industries.[45] The nomographs are a function of temperature and pressure only and account for an "average" effect of composition. They are said to be suitable for use in preliminary[44] or approximate[45] calculations.

These K-values can be employed in the calculation of bubble points and dew points and in equilibrium flash§ calculations.[42,45,46] Example problems are solved in the standard textbooks cited.

Summary and Conclusions

- Theoretically, ideal binary VLE calculations can be extended to ternary systems and beyond.
- The method is straightforward and denotes *the simplest case*.
- Results can be presented graphically on a two-dimensional diagram.
- In practice, however, truly ideal ternary and multi-component systems are not easy to find.
- Ideal systems occur in homologous series, with isomers, and in certain other instances.
- But even then, one cannot be sure *a priori* that behavior would be perfectly ideal.
- Idea/nearly ideal ternary systems identified here include:
  + C_1 to C_3 and perhaps to C_4 aliphatic alcohols,
  + Benzene–toluene–*p*-xylene,
  + plus possibly benzene–EDC–toluene and benzene–toluene–ethylbenzene.

About The Author

Robert G. Kunz was an environmental engineering manager at Air Products and Chemicals, Inc., Allentown, PA before retiring after 26+ years of service. He then joined Cormetech, Inc., Durham, NC as Technical Project Manager in support of sales and marketing efforts in the petroleum refining and petrochemical industries. He held engineering positions previously at Esso Research and Engineering Company, Florham Park, NJ and The M.W. Kellogg Company, New York, NY and is currently an independent environmental consultant. "Dr. Bob" has earned a BChE degree in Chemical Engineering from Manhattan College, a PhD in Chemical Engineering from Rensselaer Polytechnic Institute, an MS in Environmental Engineering from Newark College of Engineering, and an MBA from Temple University. He has contributed numerous publications to the technical literature, including topics in phase equilibria, and is a recipient of the Water Pollution Control Federation's Harrison Prescott Eddy Medal for noteworthy research in wastewater treatment. He is author or co-author of two books and holder of one U.S. patent. He is a member of the American Institute of Chemical Engineers (AIChE), the American Chemical Society (ACS), and the Air & Waste Management Association (A&WMA) and is a licensed professional engineer in several states.

Acknowledgements

The author is grateful to Roberta Kunz Fox, AIA, of fox² design for assistance in preparing Figures 6-8 for presentation. The author would also like to thank the library staff at North Carolina State University, Raleigh, NC; Lehigh University, Bethlehem, PA; and the Johns Hopkins University, Baltimore, MD for their help in locating reference materials.

§ When pressure is sufficiently reduced on a liquid whose pressure is at or above its bubble point, the liquid is said to "flash" or partially vaporize into a two-phase mixture of vapor and liquid in equilibrium.

References

1. Hala, E., J. Pick, V. Fried, and O. Vilim, "Vapour-Liquid Equilibrium, (translated by G. Standart)" pp. 21-23, 61-63, Pergamon Press, New York (1958).
2. Treybal, R.E., "Mass-Transfer Operations," pp. 171-173, 271, 385, McGraw-Hill, New York (1955).
3. Kunz, R.G., "Environmental Calculations A Multimedia Approach," pp.29, 41-52, Wiley, Hoboken, NJ (2009).
4. Perry, J.H., editor, "Chemical Engineers' Handbook," 3 ed., pp. 526-528, 578, McGraw-Hill, New York (1950) and Perry, R.H., C.H. Chilton, and S.D. Kirkpatrick, editors, "Chemical Engineers' Handbook," 4 ed., pp. 13-2 to 13-7, McGraw-Hill, New York (1963, renewed 1969).
5. Jones, C.A., E.M. Schoenborn, and A.P. Colburn, "Equilibrium Still for Miscible Liquids Data on Ethylene Dichloride–Toluene and Ethanol–Water," *Ind. Eng. Chem.*, **35(6)**, 666-672 (June 1943), cited by Gmehling, J., U. Onken, and W. Arlt, "Vapor-Liquid Equilibrium Data Collection Aromatic Hydrocarbons," Chemistry Data Series, Vol. I, Part 7, p.380, DECHEMA, Frankfurt/Main, Federal Republic of Germany (1980).
6. Kunz, R.G., "The Rayleigh Equation Revisited: What to Do When Alpha (α) Isn't Constant," Paper 118b presented at the 2012 A.I.Ch.E. Spring Meeting, Houston, TX (Apr. 1-5, 2012).
7. Carlson, H.C. and A.P. Colburn, "Vapor-Liquid Equilibrium of Nonideal Solutions – Utilization of Theoretical Methods to Extend Data," *Ind. Eng. Chem.*, **34**(**5**), 581-589 (May 1942).
8. Green, D.W. and J.O. Maloney, editors, "Chemical Engineers' Handbook," 6 ed., pp. 13-10 to 13-22, McGraw-Hill, New York (1984).
9. Ju Chin Chu, "Distillation Equilibrium Data," p.33, Reinhold, New York (1950) and Gmehling, J., U. Onken, and W. Arlt, *op. cit.*, Vol. I, Part 7, p.152 (1980), citing Smith, E.R. and H. B. Matheson, "Boiling Points of Benzene, Ethylene Chloride, *n*-Heptane, and 2,2,4-Trimethyl Pentane over the Range 660- to 860-mm Pressure," *J. Res. Nat. Bur. Stand. Sect. A*, **20**, 641-650 (1938).
10. Gmehling, J., U. Onken, and W. Arlt, *op. cit.*, Vol. I, Part 7, p. 138 (1980), citing Canjar, L.N., E.C. Horni, Jr., and R.R. Rothfus, "Vapor-Liquid Equilibrium in Acetone–Benzene–Ethylene Dichloride System," *Ind. Eng. Chem.*, **48(3)**, 427-430 (Mar. 1956).
11. Gmehling, J., U. Onken, and W. Arlt, *op. cit.*, Vol. I, Part 7, p. 142 (1980), citing Kirschbaum, E., H. Fischer, and D. Wolf, *Chem,-Eng.-Tech.*, **6**, 423 (1962).
12. Gmehling, J., U. Onken, and W. Arlt, *op. cit.*, Vol. I, Part 7, p. 146 (1980), citing Rollet, A.P., P. Toledano, G. Elkaim, and M. Senez, *Alger Sci. Phys.*, **2**, 403 (1956).
13. Gmehling, J., U. Onken, and W. Arlt, *op. cit.*, Vol. I, Part 7, p. 147 (1980), citing Rosanoff, M.A. and C.W. Easley, "On the Partial Vapor Pressures of Binary Mixtures," *J. Am. Chem.* Soc., **31(9)**, 953-987 (Sept. 1909). (corrected).
14. Bragg, L.B. and A.R. Richards, "Binary Mixtures for Testing Fractionating Columns at Atmospheric and Reduced Pressures, *Ind. Eng. Chem.*, **34(9)**, 1088-1091 (Sept. 1942).
15. Alpert, N. and P.J. Elving, "Vapor-Liquid Equilibria in Binary Systems," *Ind. Eng. Chem.*, **41(12)**, 2864-2867 (Dec. 1949), cited by Gmehling, J., U. Onken, and W. Arlt, *op. cit.*, Vol. I, Part 7, p.377 (1980).
16. Gmehling, J., U. Onken, and W. Arlt, *op. cit.* Vol. I, Part 7, p.377 (1980), citing Rollet, A.P., P. Toledano, G. Elkain, and M. Selez, *Alger Sci. Phys.*, **2**, 403 (1956).
17. Gmehling, J. and U. Onken, "Vapor-Liquid Equilibrium Data Collection Aromatic Hydrocarbons Supplement 1," Chemistry Data Series, Vol. I, Part 7a, p.241, DECHEMA, Frankfurt/Main, Federal Republic of Germany (2000), citing Rivenq, F., *Bull. Soc. Chim. Fr.*, **0**, 2429 (1974).
18. Brunjes, A.S. and M.J.P. Bogart, "Vapor-Liquid Equilibria for Commercially Important Systems of Organic Solvents," *Ind. Eng. Chem.*, **35(2)**, 255-260 (Feb. 1943), cited by Gmehling, J. and U.

Onken, "Vapor-Liquid Equilibrium Data Collection Alcohols: Ethanol and 1,2-Ethanediol (Supplement 6)," Chemistry Data Series Vol. I, Part 2h, p.229, DECHEMA, Frankfurt/Main, Federal Republic of Germany (2006).
19. Amer, H.H., R.R. Paxton, and M. Van Winkle, "Methanol–Ethanol–Acetone Vapor-Liquid Equilibria, *Ind. Eng. Chem.*, **48(1)**, 142-146 (Jan. 1956).
20. Dribika, M.M., I.G. Rashed, and M.W. Biddulph, "Vapor-Liquid Equilibria in the Ternary System Methyl Alcohol–Ethyl Alcohol–Isopropyl Alcohol at Atmospheric Pressure," *J. Chem. Eng. Data,* **30(2)**, 146-149 (April 1985).
21. Gmehling, J. and U. Onken, "Vapor-Liquid Equilibrium Data Collection Aqueous-Organic Systems," Chemistry Data Series, Vol. I, Part 1, DECHEMA, Frankfurt/Main, Federal Republic of Germany (1977).
22. Gmehling, J., U. Onken, and W. Arlt, "Vapor-Liquid Equilibrium Data Collection Aqueous-Organic Systems (Supplement 1)," Chemistry Data Series, Vol. I, Part 1a, DECHEMA, Frankfurt/Main, Federal Republic of Germany (1981).
23. Gmehling, J., U. Onken, and J.R. Rarey-Nies, "Vapor-Liquid Equilibrium Data Collection Aqueous Systems (Supplement 2)," Chemistry Data Series, Vol. I, Part 1b, DECHEMA, Frankfurt/Main, Federal Republic of Germany (1988).
24. Gmehling, J. and U. Onken, "Vapor-Liquid Equilibrium Data Collection Aqueous Systems (Supplement 3 Br_2-$C_4H_{10}O$)," Chemistry Data Series, Vol. I, Part 1c, DECHEMA, Frankfurt/Main, Federal Republic of Germany (2003).
25. Rooks, R.E., "Draw Insights on Distillation," Chemical Processing.com (May 15, 2006).
26. Gmehling, J. and U. Onken, "Vapor-Liquid Equilibrium Data Collection Organic Hydroxy Compounds Alcohols," Chemistry Data Series, Vol. I, Part 2a, pp. 620-621, DECHEMA, Frankfurt/Main, Federal Republic of Germany (1977), citing Hirata, M. and T. Matsubayashi, *Kagaku Kogaku*, **30**, 29 (1966).
27. Gmehling, J. and U. Onken, "Vapor-Liquid Equilibrium Data Collection Alcohols Methanol (Supplement 5)," Chemistry Data Series, Vol. I, Part 2g, p. 598, DECHEMA, Frankfurt/Main, Federal Republic of Germany (2005), citing Nagata, I., K. Tamara, and K. Miyai, *Fluid Phase Equlib.*, **170**, 37 (2000).
28. Gmehling, J, U. Onkun, and J.R. Rarey-Nies, "Vapor-Liquid Equilibrium Data Collection Organic Hydroxy Compounds Alcohols, (Supplement 3)" Chemistry Data Series, Vol. I, Part 2e, pp.533-534, DECHEMA, Frankfurt/Main, Federal Republic of Germany (1988), citing Dribika, M.M., I.G. Rashed. and M.W. Biddulph, *J. Chem. Eng. Data*, **30**, 146-149 (1985).
29. Gmehling, J. and U. Onken, *op. cit.*, Vol. I, Part 2g, p. 599 (2005), citing Jan, D.-S., H.-Y. Shiau, and F.-N. Tsai, J. Chin. Inst. Chem. Eng., 25(2), 137 (1994).
30. Gmehling, J. and U. Onken, *op. cit.*, Vol, I, Part 2g, p. 600 (2005), citing Nagata, I., K. Tamara, and K. Miyai, *Fluid Phase Equlib.*, **170**, 37 (2000).
31. Gmehling, J., U. Onken, and W. Arlt, *op.cit.*, Vol. I, Part 7, p.539 (1980), citing Makh, G.M. and Z.N. Azarova, *Zh. Prikl. Khim.*, **19**, 585 (1946).
32. Gmehling, J. and U. Onken, "Vapor-Liquid Equilibrium Data Collection Aromatic Hydrocarbons (Supplement 1 C_8-C_{14})," Chemistry Data Series, Vol. I, Part 7b, p. 724, DECHEMA, Frankfurt/Main, Federal Republic of Germany (2000), citing Beregovykh, V.V., M.M. Korabelnikov, and N.V. Ermak, *Prom. Sin. Kauch. Nauch. –Tekh. Sb.*, **5**, 4 (1977).
33. Gmehling, J. and U. Onken, *op. cit.*, Vol. I, Part 7b, p. 728 (2000), citing Kutsarov, R.K. and V.K. Sharlopov, *Zh. Prikl. Khim. (Leningrad)*, **66(5)**, 1048 (1993).
34. Gmehling, J. and U. Onken, *op. cit.*, Vol. I, Part 7a, p.191 (2000), citing Kutsarov, R.K., N.D. Ralev, and V.K. Sharlopov, *Zh. Prikl. Khim. (Leningrad),* **66**, 567 (1993).
35. Gmehling, J. and U. Onken, *op. cit.*, Vol. I, Part 7b, p. 364, (2000), citing Kutsarov, R.K., N.D. Ralev, and V.K. Sharlopov, *Zh. Prikl. Khim. (Leningrad),* **66**, 567 (1993).

36. Gupta, B.S. and M.-J. Lee, "Isobaric Vapor-Liquid Equilibrium for Binary Systems of Toluene + *o*-Xylene, Benzene + *o*-Xylene, Nonane + Benzene and Nonane + Heptane at 101.3 kPa, *Fluid Phase Equlib.*, **352(25)**, 00-92 (Aug. 2013).
37. Gültekin, N., "Vapor-Liquid Equilibria at 1 Atm for Binary and Ternary Systems Composed of Benzene, Toluene, and *m*-Xylene," *J. Chem. Eng. Data*, **35**, 130-132 (1990), ternary system cited by Gmehling, J. and U. Onken, *op. cit.*, Vol. I, Part 7b, p. 725 (2000).
38. Jin Zhangli, Hu Aibao, and Liu Kunyuan, "Vapor-Liquid Equilibrium for Ternary Mixtures of Benzene, Toluene, and p-Xylene," *Chinese J. of Chem. Eng.*, 1(1), 47-51 (1993), cited by Gmehling, J. and U. Onken, *op. cit.*: Vol. I, Part 7b, p.726 (2000) for ternary; Vol. I, Part 7a, p.192 (2000) for benzene–*p*-xylene; Vol. I, Part 7b, p.365 (2000) for toluene–*p*-xylene; and Vol. I, Part 7a, p.177 (2000) for benzene–toluene.
39. Gmehling, J. and U. Onken, *op. cit.*, Vol. I, Part 7b, p. 727 (2000), citing Schmelzer, J. and C. Wolf, *Chem. Techn. (Leipzig)*, **30**, 305 (1978).
40. Treybal, R.E., "Liquid Extraction," 1 ed., pp. 12-13, McGraw-Hill, New York (1951).
41. Prutton, C.F. and S.H. Maron, "Fundamental Principles of Physical Chemistry," revised (2) ed., 6 printing, pp. 417-418, The Macmillan Company, New York (1957).
42. Sandler, S.I., "Chemical and Engineering Thermodynamics," pp.393-397, 430, 470, 474, Wiley, New York (1989).
43. Canjar, L.N., E.C. Horni, Jr., and R.R. Rothfus, "Vapor-Liquid Equilibrium in Acetone–Benzene–Ethylene Dichloride System," *Ind. Eng. Chem.*, **48(3)**, 427-430 (Mar. 1956).
44. DePriester, C.L., "Light-Hydrocarbon Vapor-Liquid Distribution Coefficients Pressure-Temperature-Composition Charts and Pressure-Temperature Nomographs," in "Applied Thermodynamics," *CEP Symp. Ser. No. 7*, **49**, 1-43, A.I.Ch.E. New York (1953).
45. Smith, J.M., H.C. Van Ness, and M.M. Abbott, "Introduction to Chemical Engineering Thermodynamics," 5 ed., pp. 84, 459-461, 493-497, McGraw-Hill, New York (1996).
46. Smith, J.M. and H.C. Van Ness, "Introduction to Chemical Engineering Thermodynamics," 2 ed., pp. 368-378, McGraw-Hill, New York (1959).
47. Delzenne, A.O., "Vapor-Liquid Equilibrium Data for Ternary System Methanol–Ethanol–Water," *Ind. Eng. Chem. Chem. & Eng. Data Series*, **3(2)**, 224-230 (Oct. 1, 1958), cited by Gmehling, J. and U. Onken, "Vapor-Liquid Equilibrium Data Collection Organic Hydroxy Compounds: Alcohols," Chemistry Data Series, Vol. I, Part 2a, p. 50, DECHEMA, Frankfurt/Main, Federal Republic of Germany (1977).
48. Kurihara, K., M. Nakamichi, and K. Kojima, "Isobaric Vapor-Liquid Equilibria for Methanol + Ethanol + Water and the Three Constituent Binary Systems, *J. Chem. Eng. Data*, **38(3)** (1993).
49. Gmehling, J. and U. Onken, *op. cit.*, Vol. I, Part 2a, p. 60 (1977), citing Slobodyanik, I.P. and E.M. Babuskhina, *Zh. Prikl. Khim.*, **39**, 1555 (1966).
50. Gmehling, J., U. Onken, and W. Arlt, "Vapor-Liquid Equilibrium Data Collection Organic Hydroxy Compounds: Alcohols (Supplement 1)," Chemistry Data Series, Vol. I, Part 2c, p.56, DECHEMA, Frankfurt/Main, Federal Republic of Germany (1982), citing Butcher, K.L., M.S. Medani, and W.I. Robinson, "Recommended Test Mixtures for Dist. Columns," Inst. of Chem. Eng., London (1969).

51. Ju Chin Chu, *op.cit.*, p.140, citing Hausbrand, E., “Die Wirkungswise der Rektifizer und Destillierapparate,” pp. 206, 209 (1921).
52. Hughes, H. E. and J. O. Maloney, "The App;ication of Radioactive Tracers to Diffusional Operations Binary and Ternary Equilibrium Data," *Chem. Eng. Progr.*, **48(4)**, 192-200 (plotted VLE x-y data of Fig. 16) (April 1952).
53. Gültekin, N., "Vapor-Liquid Equilibria at 1 Atm for Ternary and Quaternary Systems Composed of Acetone, Methanol, 2-Propanol, and 1-Propanol," *J. Chem. Eng. Data*, **35(2)**, 132-136 (April 1990), cited by Gmehling, J., and U. Onken, *op. cit.*, Vol. **I**, Part 2g, p.171 (2005).
54. Gmeling, J. and U. Orken, *op. cit.*, Vol. **I**, Part 2c, p.94 (1982), citing Kojima, K., K. Toghigi, H. Seki, and K. Watasi, *Kagaku Kogaku*, **32**, 149 (1968).
55. Gmehling, J., U. Onken, and J.R. Rarey-Nies, *op. cit.*, Vol. **I**, Part 2e, p.91 (1988), citing Patlasov, V.P., A.V. Grishunin, and M.I. Balashov, *Osnovnoi Organ.Sintez I Neftikhim*, **8**, 103 (1977).
56. Gmehling, J. and U. Onken, *op. cit.*, Vol. **I**, Part 2a, p.336 (1977), citing Ochi, K, and K. Kojima, *Kagaku Kogaku*, **33**, 352 (1969).
57. Gmehling, J. and U. Onken, *op. cit.*, Vol. **I**, Part 2h, p. 108 (2006), citing Belousov, V.P., KH.F. Gkhoneimi, and I.L. Shulgin, *Viniti* 6296-86 (1986).
58. Gmehling, J. and U. Onken, *op. cit.*, Vol. **I**, Part 2h, pp. 109-110, (2006), citing Bereznikova, D.S., V.G. Artyukhov, and A.S. Egorov,, *Tr.Ukr.Nii.Spirt.Liker.Vodoch.Prom.*,**15**,130 (1973).
59. Ballard,L.H. and M. Van Winkle, Vapor-Liquid Equilibria at 760 Mm Pressure 2-Propanol-Methanol, 2-Propanol-Ethyl Alcohol, 2-Propanol-Propanol,and 2-Propanol-2-Butyl Alcohol Systems," *Ind. Eng. Chem.*, **44(10)**, 2450-2453 (Oct. 1952).
60. Gültekin, N., "Vapor-Liquid Equilibria at 1 Atm for Ternary and Quaternary Systems Composed of Acetone, Methanol, 2-Propanol, and 1-Propanol," *J. Chem. Eng. Data*, **35(2)**, 132-136 (April 1990).
61. Gmehling, J. and U. Onken, Vapor-Liquid Equilibrium Data Collection Organic Hydroxy Compounds: Alcohols." op. cit., Vol. I, Part 2a, p.124 (1977), citing Dunlop, J.G., Thesis Brooklyn (1948).
62. Kohoutova, J., J. Suska, J.P. Novak, and J. Pick, "Liquid-Vapour Equilibrium.XLV. System Methanol–2-Propanol–Water," *Collect. Czech. Chem, Commun.,* **.35**, 3210 (1970).
63. Gmehling, J. and U. Onken, *op. cit.*, Vol. **I**, Part 2a, p.128 (1977), citing Verhoeye, L. and H. DeSchepper, *J. Appl. Chem. Biotechn.*, **23** ,607(1973).
64. Gmehling, J. and U. Onken, *op. cit.*, Vol. **I**, Part 2a, p.126 (1977), citing Ochi, K. and K. Kojima, *Kagaku Kogaku*, **33**,352 (1969).
65. Gmehling, J. and U. Onken, *op. cit.*, Vol. **I**, Part 2g, p.180 (2005), citing Lenka, M.O. Persson, and P. Rasmussen, *J. Chem. Thermodynamics*, **23**, 851(1991).
66. Gmehling, J. and U. Onken, *op. cit.*, Vol. **I**, Part 2g, p.179 (2005), citing Jan, D,.-S., H.-Y. Shiau, and F.-N. Tsai, *J. Chin. Inst. Chem. Eng.*, **25(2)**, 137(1994).
67. Freshwater, D.C. and K.A. Pike, “Vapor-Liquid Equilibrium Data for Systems of Acetone-Methanol-Isopropanol,” *J. Chem. Eng. Data*, **12(2)**, 179-182(April 1967).
68. Allen, P. and K.A. Pike, “Experimental Determination of the Vapour-Liquid Equilibrium Data for the System Acetone-Methanol-*Iso*propanol, *Industrial Chemist*, **40**, 260-266 (May 1964).
69. Kojima, K,. K. Ochi, and Y. Nakazawa, “Relationship between Liquid Activity Coefficient and Composition for Ternary Systems,” *Int. Chem. Eng.*, **9(2)**, 342-347 (April 1969).
70. Gültekin, N., "Vapor-Liquid Equilibria at 1 Atm for Binary and Ternary Systems Composed of Acetone,Methanol,2-Propanol, and 1-Propanol,"*J. Chem. Eng. Data*, **34(2)**, 168-171 (April 1989).
71. Dean, J.A., editor, “Lange’s Handbook of Chemistry,” 11 ed., McGraw-Hill, New York (1973).
72. Dean, J.A., editor, “Lange’s Handbook of Chemistry,” 12 ed., McGraw-Hill, New York (1978).

73. Weast, R. C. and M. J. Astle, "CRC Handbook of Chemistry and Physics," 62 ed., CRC Press, Inc., Boca Raton, FL (1981).
74. Dean, J.A., editor, "Lange's Handbook of Chemistry," 13 ed., McGraw-Hill, New York (1985).
75. Kesselman, W.D., G.E. Hollenbach, A.L. Myers, and A.E. Humphrey, "Vapor-Liquid Equilibrium Data for Benzene-Alkylbenzene Systems," *J. Chem. Eng. Data*, **13(1)**, 34-36 (Jan.1968), cited by Gmehling, J., U. Onken, and W. Arlt, *op.cit.*, Vol. I, Part 7, p.306 (1980).
76. Stephan, K. and H. Hildwein, "Recommended Data of Selected Compounds and Binary Mixtures," Chemistry Data Series, Vol. IV, Parts 1 + 2, DECHEMA, Frankfurt am Main, Federal Republic of Germany (1987).
77. Gmehling, J., U. Onken, and W. Arlt, *op.cit.*, Vol. I, Part 7, p.310 (1980), citing Michishita, T., Y. Arai, and S. Saito, *Kagaku Kogaku*, **35**, 111 (1971).
78. Gmehling, J. and U. Onken, *op. cit.*, Vol. I, Part 2a, p.171 (1977), citing Jaenecke, E., *Z. Phys. Chem.*, **164**, 401 (1933).
79. Suska, J., R. Holub, P. Vonka, and J. Pick, "Liquid-Vapour Equilibrium. XLII. Systems: Ethyl Alcohol–Water–Tert-Butyl Alcohol and Ethyl Alcohol–Water–Isobutyl Alcohol," *Collect. Czech. Chem, Commun.,* **35**, 385-395 (1970), cited by Gmehling, J. and U. Onken, *op. cit.*, Vol. I, Part 2a, p.376 (1977).
80. Gmehling, J., U. Onken, and J.R. Rarey-Nies, *op. cit.*, Vol. I, Part 2e, p.91 (1988), citing Villa Rivera, J.E., Thesis Paris (1983).
81. Gmehling, J. and U. Onken, *op. cit.*, Vol. I, Part 2g, p. 262 (2005), citing Arce, A., J. Martinez-Agitos, E. Rodil, and A. Sotoi, *Fluid Phase Equlib.*, **171**, 207 (2000).
82. Gmehling, J. and U. Onken, *op. cit.*, Vol. I, Part 2a, p. 365 (1977), citing Gay, L., *Chim. Ind.*, **18**, 187 (1927).
83. Gmehling, J., U. Onken, and J.R. Rarey-Nies, *op. cit.*, Vol. I, Part 2e, p.305 (1988), citing Kharin, S.E., V.M. Perelygin, and G.P. Remizov, *Izv. Vyssh. Ucheb. Zaved., Khim. Khim. Tekhnol.*, 12, 424 (1969).
84. Gmehling, J. and U. Onken, *op. cit.*, Vol. I, Part 2a, p. 539 (1977), citing Gay, L., *Chim. Ind.*, **18**, 187 (1927).
85. Arce, H.S., E. Rodil, and A. Soto, "Extractive Distillation of 2-Methoxy-2-methyl propane + Ethanol Using 1-Butanol as Entrainer: Equilibria and Simulation," *Can. J. Chem. Eng.*, **77**, 1135-1140 (Dec. 1999), cited by Gmehling, J. and U. Onken, *op. cit.*, Vol. I, Part 2h, p. 227 (2006).
86. Gmehling, J. and U. Onken, *op. cit.*, Vol. I, Part 2h, p. 228 (2006), citing citing Belousov, V.P., KH.F. Gkhoneimi, and I.L. Shulgin, *Viniti* 6296-86 (1986).
87. Gmehling, J. and U. Onken, *op. cit.*, Vol. I, Part 2g, p. 265 (2005), citing Wohland, R. and T. Roscher, *Fiz. Report*, 12151 (1975).
88. Gmehling, J. and U. Onken, *op. cit.*, Vol. I, Part 2a, p. 176 (1977), citing Pascal P., DuPey, ERO, and Garnier, *Bull. Soc. Chim. Fr.*, **29**, 1555 (1921).
89. Gmeling, J. and U. Orken, *op. cit.*, Vol. I, Part 2c, p.129 (1982), citing Lesteva, T. M., S. K. Ogorodnikov and S.V. Kazakova, *Zh. Prikl. Khim (Leningrad)*, **43**, 1574 (1970).
90. Gmeling, J. and U. Orken, *op. cit.*, Vol. I, Part 2c, p.131 (1982), citing Ziak, J., M. Zavodska and J. Jurik, *Petrochemia*, **14**, 41 (1970).
91. Gmehling, J. and U. Onken, *op. cit.*, Vol. I, Part 2g, p. 266 (2005), citing Komatsu, H., M. Nakamura, Y. Yamashita, and C. Hirai, *Kagaku Kogaku Ronbonshu*, **23(6)**, 983 (1997).

APPENDIX A

The Low Molecular Weight Alcohol Systems at 760 mmHg Total Pressure

Idealities / Non-Idealities in the Alcohol Binary Systems

Isobaric VLE x-y data for alcohol binary systems are shown in Figures A1-A5. The solid y-value curves are calculated from Raoult's Law for the liquid phase with an Ideal-Gas) for vapor phase; the points reproduce experimental data from the literature. The line y=x is shown as the straight diagonal line in each of the figures. Antoine Constants used are listed in Table A-1.

- Data for the methanol–ethanol system from a number of investigators [19,47-52] are depicted in Figure A1.[3] It is among the systems listed as ideal mixtures in Perry's 3rd and 4th eds.[4] Values of the activity coefficients from one investigation range from 0.958 to 1.018 for both species, close enough to 1.0 for the methanol-ethanol system at atmospheric pressure to be treated as an ideal solution in that study.[48]
- Figure A2 displays x-y data from several investigations[53-55] for methanol–1-propanol. This system is also listed as an ideal mixture in Perry's.[4] Tabulated activity coefficients from one investigation range from 0.977 to 1.001 for methanol and 0.987 to 0.999 for 1-propanol,[53] again the mark of a system extremely close to ideal.
- The only three sets of x-y data found for ethanol–1-propanol at 1 atm [56-58] are plotted together in Figure A3. One[58] contains 44 closely spaced experimental points. Amid scatter, activity coefficients derived from the data of that study range from 0.991 to 1.134 for ethanol and from 0.943 to 1.109 for 1-propanol, as calculated here using the temperatures, compositions, and Antoine Constants reported in the DECHEMA tabulation.[58] A fit shown there to the van Laar equation predicts γ^{∞}'s of 1.120 and 1.036 for ethanol and 1-propanol, respectively. Calculated γ's from the other two, less comprehensive studies are similar. This system is characterized in Perry's[4] as ideal.
- Numerical data for the methanol–2-propanol system of Figure A4 come from numerous investigations,[59-68] agreeing with one another and the ideal x-y curve at about 0.4 mole fraction methanol and above. At lower methanol concentrations, the majority of the data in the figure lies below the ideal curve, indicating negative deviations from Raoult's Law. One study reports γ^{∞}'s of 0.82 for methanol and 0.93 for 2-propanol, as read from a small graph contained in the journal article.[59] Those investigators point out that the methanol activity-coefficient deviations from 1.0 are greater than the 2-propanol deviations. This system is <u>not</u> listed among the ideal mixtures in Perry's 3rd and 4th eds.[4]
- Data from two different investigations are combined in Figure A5 for the ethanol–2-propanol system.[59,69] Tabulated activity coefficients for both species vary from 0.97-1.02 and plot no worse than 0.95 for 2-propanol and 0.98 for ethanol at the ends of the concentration range.[59] This system is therefore concluded to approach ideality and is listed as an ideal mixture in Perry's Third and Fourth Editions.[4]
- The 1-propanol–2-propanol binary is not needed for this analysis, and its x-y diagram is not shown. One study concludes that it deviates from ideal solution behavior;[59] another[70] shows a plot of activity coefficients clearly inconsistent with an ideal VLE system. This system did not make the ideal-mixture list in Perry's[4] as well.

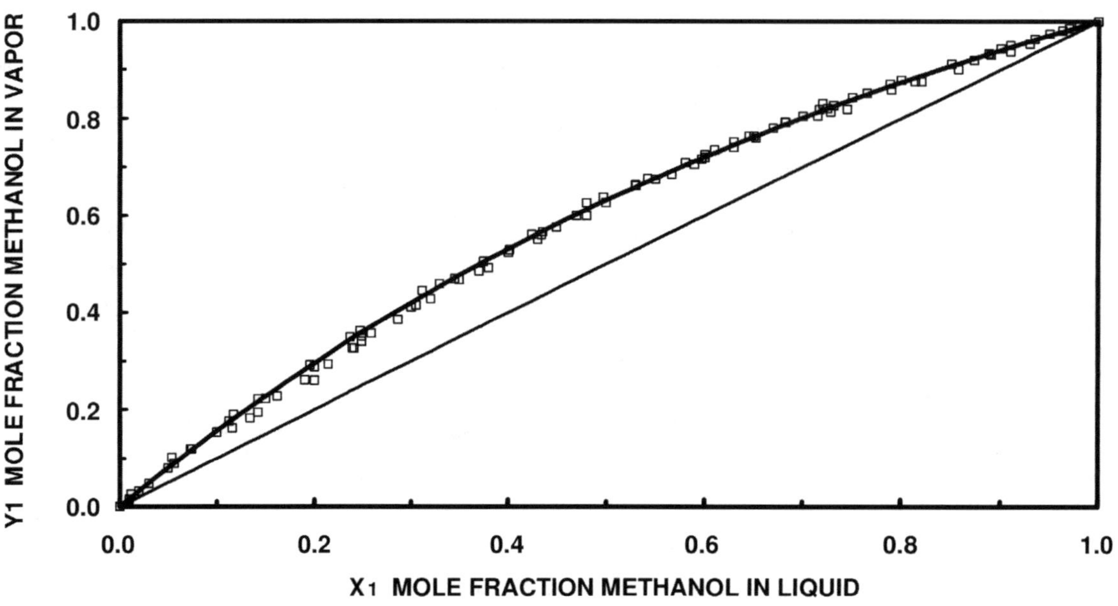

Figure A!. Vapor-Liquid Equilibrium Diagram Methanol–Ethanol at 1 Atm Pressure Data from References **[19, 47-52]**

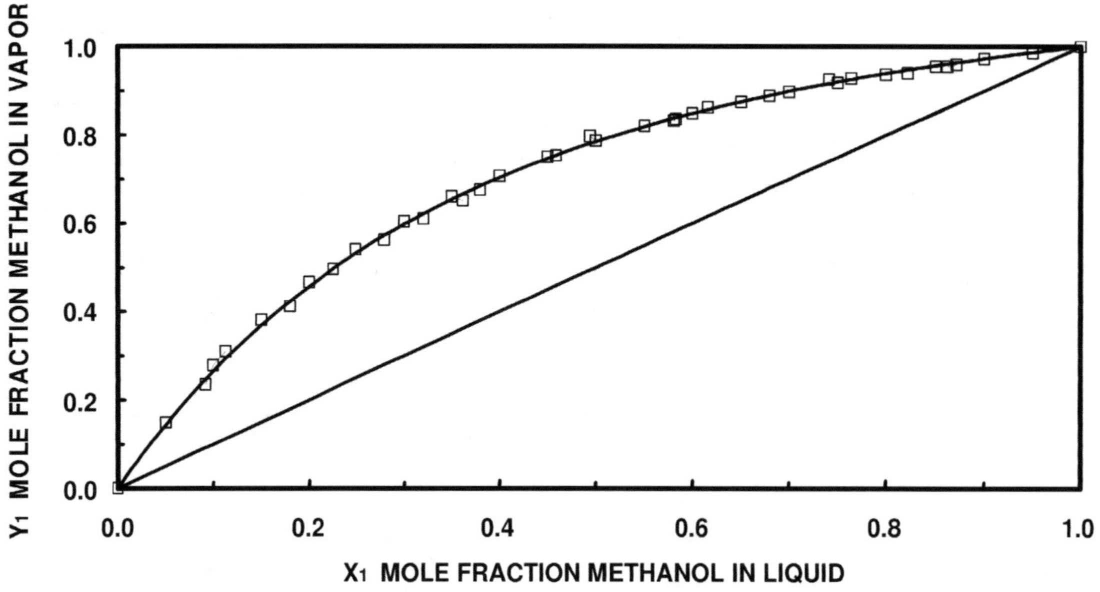

Figure A2. Vapor-Liquid Equilibrium Diagram for Methanol–1-Propanol at 1 Atm Pressure Data from References **[53-55]**

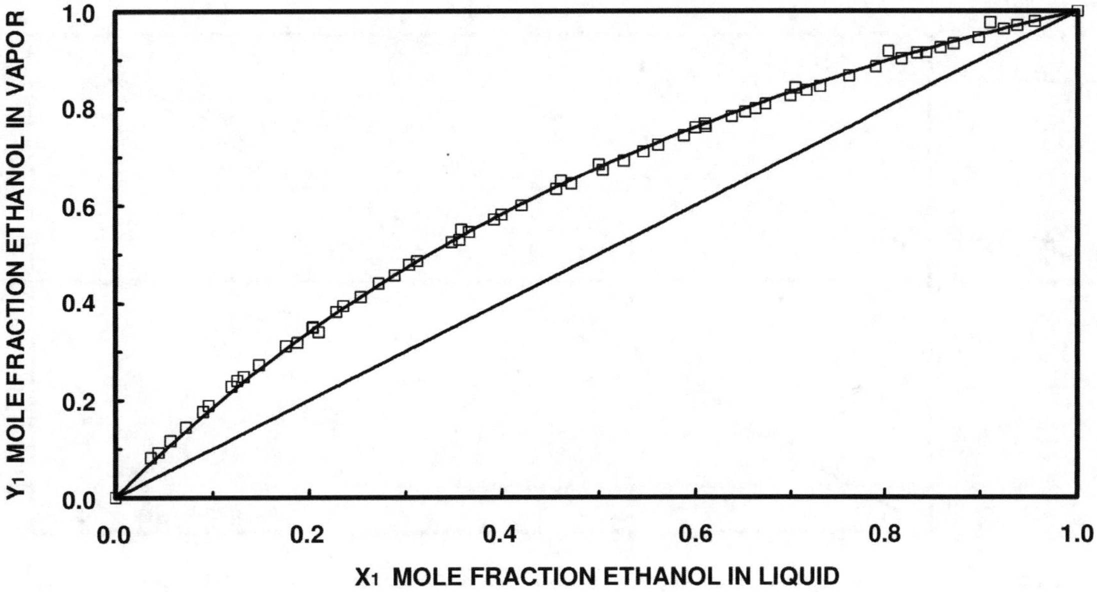

Figure A3. Vapor-Liquid Equilibrium Diagram Ethanol–1-Propanol at 1 Atm Pressure Data from References **[56-58]**

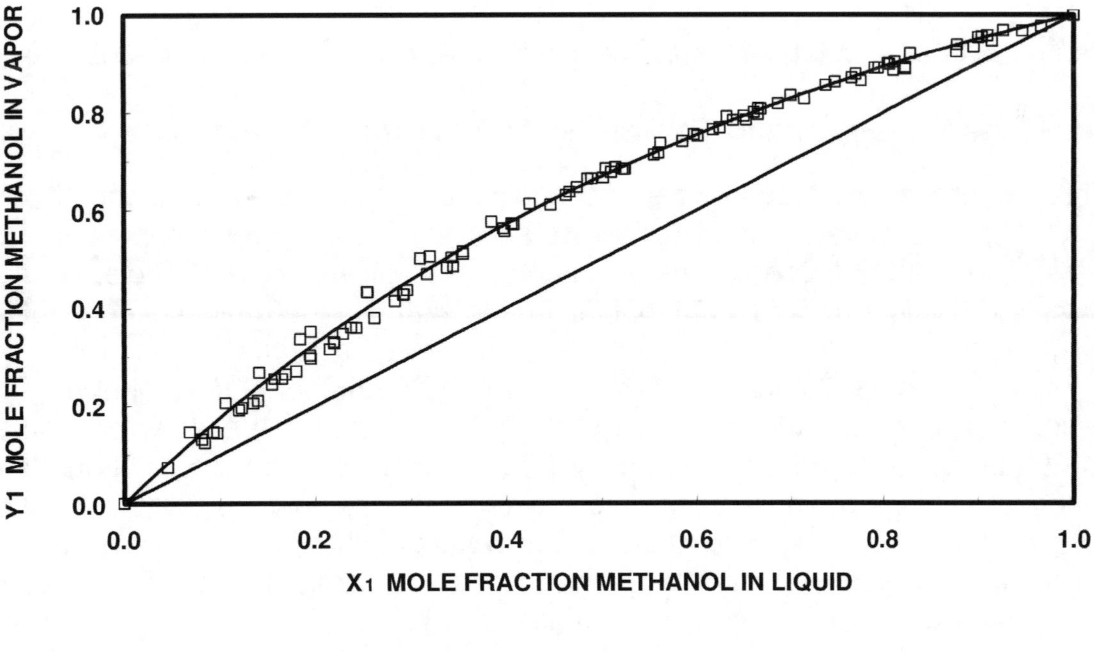

Figure A4. Vapor-Liquid Equilibrium Diagram for Methanol–2-Propanol at 1 Atm Data from References **[59-68]**

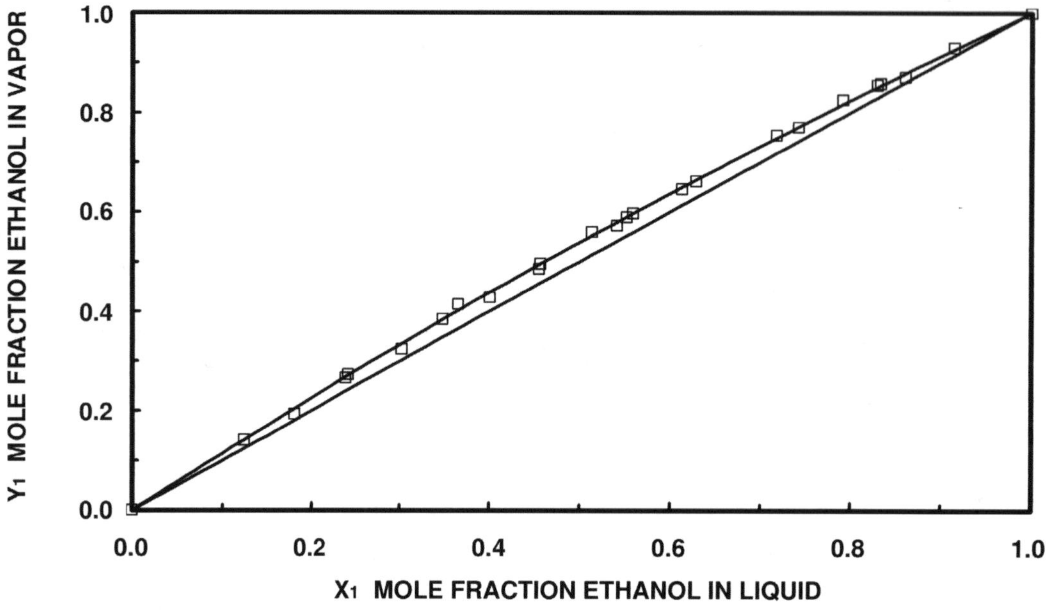

Figure A5. Vapor-Liquid Equilibrium Diagram Ethanol–2-Propanol at 1 Atm Pressure Data from References **[59, 69]**

Ternary Data and Calculations

Comparison of calculations and data for the C_1-C_3 aliphatic alcohols is discussed below.

Methanol–Ethanol–1-Propanol System at 760 mmHg Total Pressure

This mixture is considered to be reasonably ideal,[25] as are all of its constituent binaries (Figures A1-A3), and literature data for the ternary are available for comparison, providing a suitable test case. Antoine Coefficients are among those given below in Table A1.

Calculated values for 45 data points in the methanol–ethanol–1-propanol system at 760 mmHg Pressure are compared with reported experimental data in Table A2. Data reported for this data set[26] consist of temperature and methanol and ethanol mol fractions in both the liquid (x_1 and x_2) and the vapor (y_1 and y_2). Those data are shown in Table A2 along with mole fractions for 1-propanol (x_3 and y_3) obtained by difference from Equations (5) and (9) rearranged. Values of vapor-phase mol fractions (y_1, y_2, and y_3) were then calculated at the stated temperatures using calculated total pressure [Equations (1) to (4) and (6) to (8)]. Those calculated pressures (P_T) are listed in Column 3. Vapor pressures were computed by means of Equation (14), using the constants listed in Table A1. A minor change to a slightly different temperature (not shown) is required to adjust calculated total pressure to 760 mmHg (by trial and error). The resulting y-values calculated before and after the temperature adjustment are virtually identical, and calculated y-values listed in Table A2 correspond to the original temperatures.

Table A1. Antoine Equation Constants for Water and Alcohols of Interest[(a)]

Chemical Name and Synonyms	Formula	MW (b)	BP (°C) (c)	Applicable Range(°C)	A	B	C
Water Hydrogen Oxide	HOH (H_2O)	18.02	100.00	0 to 60 60 to 150 11 ed. p.10-45	8.10765 7.96681	1750.286 1668.21	235.0 228.0
Methanol Methyl Alcohol Wood Alcohol Carbinol	CH_3OH	32.04	64.7	–20 to +140 11 ed. p.10-39	7.87863	1473.11	230.0
Ethanol Ethyl Alcohol Grain Alcohol Methyl Carbinol	C_2H_5OH	46.07	78.3	— 11 ed. p.10-36	8.04494	1554.3	222.65
1-Propanol Propanol Propanol-1 *n*-Propyl Alcohol Ethyl Carbinol	C_3H_7OH	60.11	97.2	— 11 ed. p.10-43	7.99733	**1569.70**	209.5
2-Propanol Propanol-2 Isopropanol *Iso*-Propyl Alcohol Dimethyl Carbinol	$(CH_3)_2CHOH$	60.11	82.4	0 to 113 11 ed. p.10-43	6.66040	813.055	132.93
1-Butanol *n*-Butyl Alcohol *n*-Propyl Carbinol	C_4H_9OH	74.12	118.0	15 to 131 12 ed. p.10-38	7.47680	1362.39	178.77
2-Butanol *sec*-Butyl Alcohol Methyl Ethyl Carbinol	$CH_3CH_2CHOHCH_3$	74.12	99.5	25 to 120 12 ed. p.10-38	7.47431	1314.19	186.55
***iso*-Butanol** Isobutyl Alcohol 2-methyl-1-propanol Isopropyl Carbinol	$(CH_3)_2CHCH_2OH$	74.12	107.9	20 to 115 12 ed. p.10-48	7.32705	1248.48	172.92
***tert*-Butanol** *tert*-Butyl Alcohol 2-methyl-2-propanol Trimethyl Carbinol	$(CH_3)_3COH$	74.12	82.3	— 11 ed. p.10-33	8.13596	1582.4	218.9

Notes:
(a) Antoine Constants from "Lange's Handbook of Chemistry", 11th Ed.[71] [Table 10-10] or 12th Ed.[72] [Table 10-8]. Sources for Applicable Temperature Ranges of Antoine Constants as shown above.
(b) Molecular Weight (MW) from "Handbook of Chemistry and Physics," 62nd Ed.[73] [pp. C-65 to C-576].
(c) Boiling point (BP) at 760 mmHg from "Lange's Handbook," 11th Ed.[71] [Table 7-4], 12th Ed.[72] [Table 7-4], or 13th Ed.[74] [Table 7-15], or "Handbook of Chemistry and Physics," 62nd Edition[73] [pp. C-65 to C-576], chosen to be the most consistent with boiling point calculated using the Antoine Constants listed.

Table A2. Comparison of Ideal Ternary VLE Calculations with Experimental Data – Methanol(1)–Ethanol(2)–1-Propanol(3) at 760 mmHg Pressure (Data from Ref. **[26]**)

#	T °C	P_T at T mmHg	X_1 meas.	X_2 meas.	X_3 meas.	Y_1 meas.	Y_1 calc.	Y_2 meas.	Y_2 calc.	Y_3 meas.	Y_3 calc.
1	67.30	750.759	0.8110	0.1120	0.0770	0.9030	0.9061	0.0730	0.0721	0.0240	0.0217
2	68.60	753.465	0.7600	0.1030	0.1370	0.8890	0.8892	0.0680	0.0698	0.0430	0.0410
3	70.00	816.502	0.7590	0.1760	0.0650	0.8520	0.8641	0.1280	0.1168	0.0200	0.0192
4	70.30	736.951	0.6570	0.1170	0.2260	0.8130	0.8381	0.0940	0.0871	0.0930	0.0748
5	70.10	769.210	0.6830	0.1830	0.1340	0.8030	0.8285	0.1470	0.1294	0.0500	0.0421
6	69.10	737.330	0.6380	0.2770	0.0850	0.7620	0.7774	0.2020	0.1960	0.0360	0.0266
7	72.60	724.486	0.5490	0.1200	0.3310	0.7520	0.7762	0.1030	0.0999	0.1450	0.1238
8	71.60	718.432	0.5410	0.2050	0.2540	0.7340	0.7432	0.1700	0.1652	0.0960	0.0915
9	71.30	757.358	0.5780	0.2700	0.1520	0.7370	0.7449	0.2090	0.2039	0.0540	0.0513
10	70.50	734.376	0.5300	0.3800	0.0900	0.6760	0.6836	0.2910	0.2862	0.0330	0.0302
11	75.20	753.726	0.4940	0.1170	0.3890	0.7200	0.7387	0.1020	0.1041	0.1780	0.1572
12	74.00	723.694	0.4620	0.2000	0.3380	0.6630	0.6886	0.1770	0.1766	0.1600	0.1348
13	73.10	735.251	0.4740	0.2800	0.2460	0.6580	0.6727	0.2380	0.2345	0.1040	0.0928
14	72.40	753.824	0.4960	0.3460	0.1580	0.6570	0.6690	0.2840	0.2747	0.0590	0.0563
15	71.50	743.605	0.4780	0.4360	0.0860	0.6080	0.6321	0.3560	0.3381	0.0360	0.0298
16	78.80	712.558	0.3340	0.1110	0.5550	0.5820	0.6013	0.1200	0.1206	0.2980	0.2781
17	77.40	765.529	0.3930	0.2070	0.4000	0.6010	0.6265	0.2080	0.1981	0.1910	0.1755
18	77.10	740.499	0.3300	0.3060	0.3640	0.4910	0.5380	0.2740	0.2991	0.2350	0.1629
19	74.70	734.701	0.3650	0.3940	0.2410	0.5840	0.5498	0.3190	0.3525	0.0970	0.0977
20	74.20	728.560	0.3280	0.5000	0.1720	0.4800	0.4891	0.4500	0.4421	0.0700	0.0688
21	73.90	738.565	0.3240	0.5620	0.1140	0.4550	0.4714	0.4970	0.4842	0.0480	0.0444
22	82.20	738.526	0.2640	0.1120	0.6240	0.4900	0.5169	0.1360	0.1340	0.3740	0.3491
23	81.60	717.557	0.2120	0.2170	0.5710	0.4060	0.3184	0.2460	0.2612	0.3480	0.3205
24	79.70	744.021	0.2460	0.3210	0.4330	0.4450	0.4379	0.3310	0.3460	0.2240	0.2160
25	78.50	729.819	0.2280	0.4080	0.3640	0.3890	0.3965	0.4170	0.4277	0.1940	0.1757
26	77.30	725.912	0.2200	0.5000	0.2800	0.3620	0.3685	0.4950	0.5025	0.1430	0.1290
27	76.50	686.671	0.1560	0.5990	0.2450	0.3360	0.2684	0.5640	0.6164	0.1000	0.1152
28	75.80	759.230	0.2220	0.6950	0.0830	0.3140	0.3368	0.6470	0.6290	0.0390	0.0342
29	87.50	723.094	0.1260	0.0960	0.0778	0.3460	0.3021	0.1260	0.1434	0.5280	0.5545
30	86.30	721.582	0.0970	0.2160	0.6870	0.2280	0.2238	0.2850	0.3092	0.4870	0.4671
31	83.70	727.659	0.1180	0.3220	0.5600	0.2330	0.2470	0.4070	0.4142	0.3600	0.3388
32	82.60	732.905	0.1050	0.4250	0.4700	0.2240	0.2101	0.4920	0.5205	0.2840	0.2695
33	81.20	748.537	0.1240	0.5030	0.3730	0.2150	0.2313	0.5610	0.5714	0.2240	0.1973
34	79.80	772.222	0.1450	0.6030	0.2520	0.2330	0.2496	0.6330	0.6288	0.1340	0.1217
35	78.80	764.918	0.1310	0.6870	0.1820	0.2090	0.2197	0.7020	0.6953	0.0890	0.0849
36	80.40	843.650	0.1100	0.8010	0.0890	0.1660	0.1770	0.7900	0.7826	0.0440	0.0404
37	90.00	738.351	0.0740	0.1150	0.8110	0.1770	0.1889	0.1810	0.1845	0.6420	0.6266
38	89.40	786.828	0.0700	0.2320	0.6980	0.1230	0.1643	0.2980	0.3417	0.5790	0.4940
39	88.00	800.755	0.0670	0.3340	0.5990	0.1380	0.1475	0.4340	0.4590	0.4280	0.3935
40	85.00	766.967	0.0720	0.4260	0.5020	0.1370	0.1495	0.5430	0.5463	0.3200	0.3042
41	84.90	813.554	0.0710	0.5240	0.4050	0.1210	0.1385	0.6190	0.6311	0.2600	0.2304
42	83.50	814.261	0.0650	0.6250	0.3100	0.1110	0.1207	0.6870	0.7131	0.2020	0.1662
43	80.50	767.340	0.0690	0.7120	0.2190	0.1120	0.1225	0.7660	0.7678	0.1220	0.1097
44	81.30	817.992	0.0590	0.7920	0.1490	0.0910	0.1011	0.8290	0.8265	0.0800	0.0724
45	79.40	806.680	0.0690	0.8800	0.0510	0.1000	0.1121	0.8750	0.8647	0.0250	0.0232

Note: A minor change in temperature is required to adjust calculated total pressure to 760 mmHg; the resulting y-values before-and-after are virtually identical. Calculated y-values in table correspond to original temperatures.

Absolute differences between calculated vs. measured y-values for all three components average 0.01 to 0.02 mol fraction. Some 90 % of the points show a difference of 0.03 or less. (Three differences of 0.034 barely missed the cutoff and are included among the remaining five experimental points.) Differences of up to 1½ to 3 times greater occur at those remaining points – for all constituents at one point, for two of three constituents at three points, and for one of them at one point.** In round numbers, maximum deviations are 0.07 for methanol, 0.05 for ethanol, and 0.09 for 1-propanol. Absolute values of temperature deviations range from 0.03 to 69 °C and average 0.95 °C, with 91% of them less than 1.8 °C.

Methanol–Ethanol–1-Propanol System at 60 °C

Constant temperature data at 60 °C are also available for comparison with ideal calculations.[27] Those calculations are straightforward with no trial and error necessary. Data and calculated results are listed in Table A3. As expected, values of calculated total pressure shown in Column 3 are slightly different from the reported data reproduced in Column 2. Calculated total pressure was used to compute vapor phase composition via Equations (6-9).

Table A3. Comparison of Ideal Ternary VLE Calculations with Experimental Data – Methanol(1)–Ethanol(2)–1-Propanol(3) at 60 °C (Data from Ref. **[27]**)

#	PTmeas mmHg	PT calc mmHg	X1 meas.	X2 meas.	X3 meas.	Y1 meas.	Y1 calc.	Y2 meas.	Y2 calc.	Y3 meas.	Y3 calc.
1	151.95	148.88	0.0000	0.0000	1.0000	0.0000	0.0000	0.0000	0.0000	1.0000	1.0000
2	351.88	351.49	0.0000	1.0000	0.0000	0.0000	0.0000	1.0000	1.0000	0.0000	0.0000
3	633.70	629.42	1.0000	0.0000	0.0000	1.0000	1.0000	0.0000	0.0000	0.0000	0.0000
4	425.20	423.43	0.3310	0.5700	0.0990	0.4850	0.4920	0.4760	0.4732	0.0390	0.0348
5	602.60	600.15	0.9180	0.0500	0.0320	0.9520	0.9628	0.0340	0.0293	0.0140	0.0079
6	324.10	316.26	0.0810	0.6340	0.2850	0.1700	0.1612	0.6930	0.7046	0.1370	0.1342
7	577.50	574.20	0.8640	0.0500	0.0860	0.9360	0.9491	0.0330	0.0306	0.0310	0.0223
8	566.30	570.42	0.8380	0.0930	0.0690	0.9090	0.9247	0.0640	0.0573	0.0270	0.0180
9	547.20	543.14	0.7180	0.2430	0.0390	0.8260	0.8321	0.1600	0.1573	0.0140	0.0107
10	428.70	424.69	0.4500	0.2940	0.2560	0.6620	0.6629	0.2410	0.2433	0.0970	0.0897
11	537.80	535.26	0.6940	0.2610	0.0450	0.8110	0.8161	0.1750	0.1714	0.0140	0.0125
12	535.30	529.40	0.7050	0.2060	0.0890	0.8300	0.8382	0.1410	0.1368	0.0290	0.0250
13	521.40	518.30	0.6330	0.3220	0.0450	0.7630	0.7687	0.2220	0.2184	0.0150	0.0129
14	527.60	532.42	0.6860	0.2660	0.0480	0.8010	0.8110	0.1840	0.1756	0.0150	0.0134
15	525.20	517.01	0.7450	0.0500	0.0250	0.8990	0.9070	0.0350	0.0340	0.0660	0.0590
16	462.10	459.44	0.4460	0.4750	0.0790	0.6070	0.6110	0.3590	0.3634	0.0340	0.0256
17	361.40	354.06	0.2440	0.4340	0.3220	0.4180	0.4338	0.4290	0.4308	0.1530	0.1354
18	376.50	374.26	0.1490	0.7590	0.0920	0.2400	0.2506	0.7200	0.7128	0.0400	0.0366
19	592.30	596.13	0.9050	0.0610	0.0340	0.9420	0.9555	0.0450	0.0360	0.0130	0.0085
20	518.10	512.22	0.6410	0.2730	0.0860	0.7830	0.7877	0.1800	0.1873	0.0370	0.0250
21	530.60	522.90	0.7210	0.1360	0.1430	0.8510	0.8679	0.0960	0.0914	0.0530	0.0407
22	575.10	565.63	0.8310	0.0860	0.0830	0.9210	0.9247	0.0650	0.0534	0.0140	0.0218
23	538.09	535.26	0.6940	0.2610	0.0450	0.8110	0.8161	0.1750	0.1714	0.0140	0.0125

Note: Calculated y-values in the table correspond to the measured constant temperature and the total pressure calculated from Raoult's Law and the Antoine vapor-pressure relationships for mixture components.

** Those five points plot as outliers when cast in the form of activity coefficients (Figures A6-A8).

For the methanol–ethanol–1-propanol 60 °C constant temperature data, maximum deviation of all y-values (absolute value of calculated vs. experimental) is less than 0.02, averaging 0.005 to 0.008, with 76-91% of deviations less than 0.01. Deviations in total pressure range from 0.4 mmHg to 9.5 mmHg, with an average deviation of 4.4 mmHg.

Methanol–Ethanol–2-Propanol System at Atmospheric Pressure

Data for this system at atmospheric pressure[28,29] and at 60 °C[30] are also available for comparison with ideal calculations with ideal calculations (Tables A4-A6). Binary VLE x-y diagrams for methanol–2-propanol and ethanol–2-propanol are plotted in Figures A4 and A5. The methanol–2-propanol binary is classified as non-ideal, as explained above.

The ternary itself also exhibits non-ideal behavior over the range investigated, as shown by the activity coefficient plots of Figures A6-A8. In these plots, the original authors' tabulated activity coefficients[28] for each component are plotted to the same scale versus the mole fraction of that component. The few points those authors have declared to be <u>not</u> thermodynamically consistent are omitted from the figures.

In general, the data points follow expected patterns for activity coefficients, with a finite value above or below 1.0 for a dilute solution and smoothly approaching unity for the pure component. The methanol activity coefficient shows positive deviations from Raoult's Law; those of ethanol and 2-propanol indicate negative deviations. In addition to the usual scatter, the data spread in the graphs reflects the varying composition of the balance of the liquid mixture (1 minus the mole fraction of the given species).

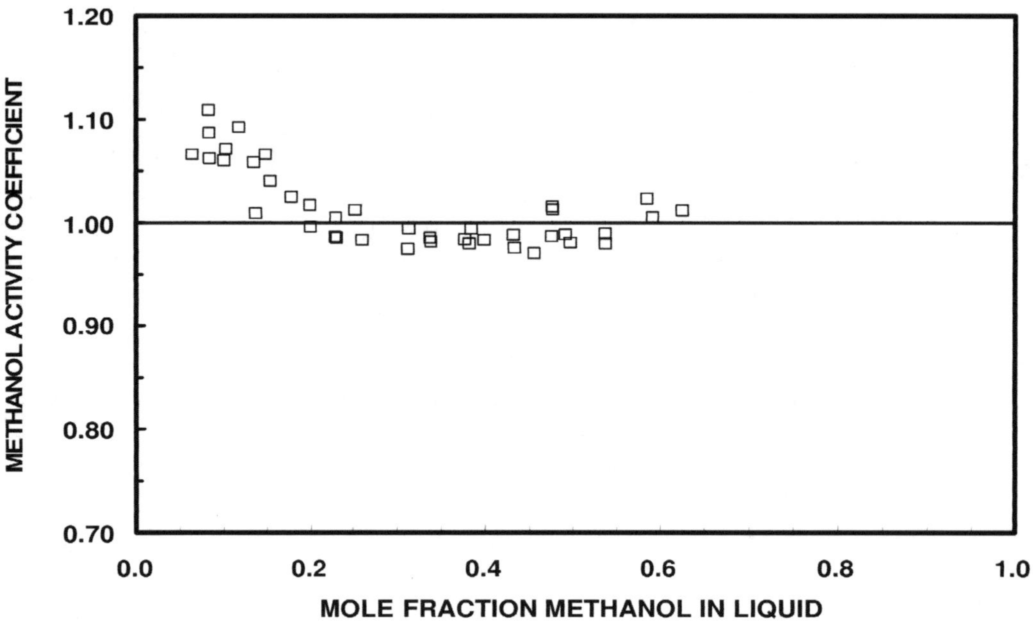

Figure A6. Methanol Activity Coefficient in Methanol-Ethanol-2-Propanol System at 1 Atm

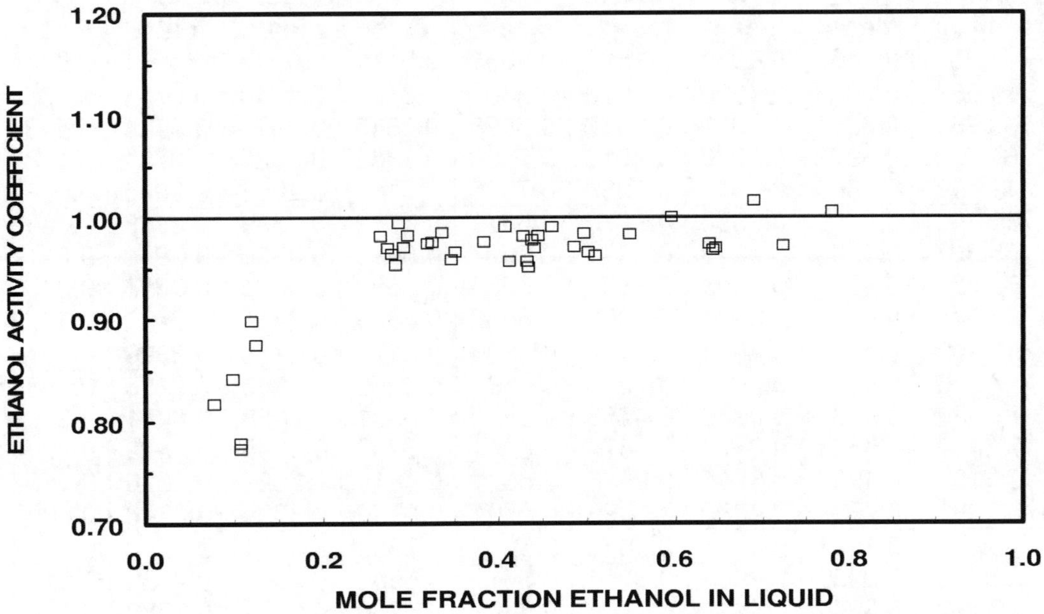

Figure A7. Ethanol Activity Coefficient in Methanol-Ethanol-2-Propanol System at 1 Atm

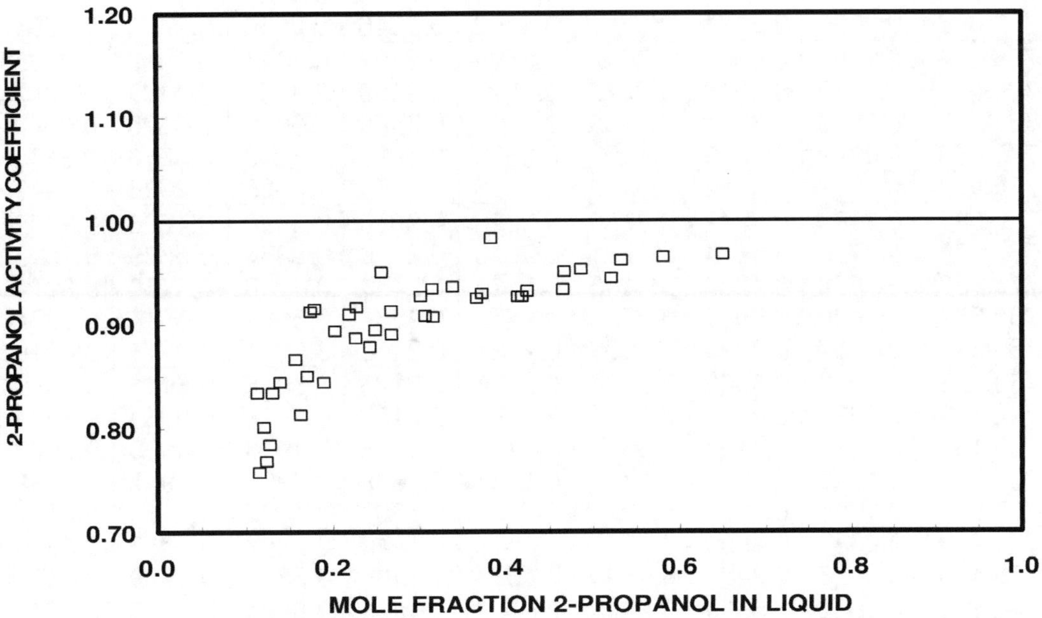

Figure A8. **2-**Propanol Activity Coefficient in Methanol–Ethanol–2-Propanol System at 1 Atm

Table A4. Comparison of Ideal Ternary VLE Calculations with Experimental Data – Methanol(1)–Ethanol(2)–2-Propanol(3) at 760 mmHg Pressure (Data from Ref. [28])

#	T °C	P_T at T mmHg	X_1 meas.	X_2 meas.	X_3 meas.	Y_1 meas.	Y_1 calc.	Y_2 meas.	Y_2 calc.	Y_3 meas.	Y_3 calc.
1	68.20	761.750	0.7467	0.0967	0.1576	0.8851	0.8511	0.0444	0.0631	0.0705	0.0858
2	69.00	760.062	0.6897	0.1030	0.2073	0.8401	0.8122	0.0540	0.0704	0.1059	0.1174
3	70.00	759.761	0.6251	0.1072	0.2677	0.7898	0.7648	0.0616	0.0764	0.1488	0.1588
4	70.60	761.270	0.5912	0.1070	0.3018	0.7586	0.7383	0.0625	0.0781	0.1789	0.1838
5	69.55	743.959	0.5846	0.2918	0.1236	0.7362	0.7181	0.2056	0.2085	0.0582	0.0733
6	70.95	766.366	0.5369	0.3356	0.1275	0.6811	0.6749	0.2539	0.2468	0.0650	0.0783
7	71.60	765.837	0.5366	0.1243	0.3391	0.7027	0.6916	0.0856	0.0940	0.2117	0.2145
8	72.50	772.520	0.4973	0.1193	0.3834	0.6666	0.6580	0.0874	0.0928	0.2460	0.2502
9	71.80	767.545	0.4914	0.3191	0.1895	0.6466	0.6366	0.2461	0.2427	0.1073	0.1207
10	71.25	752.343	0.4770	0.4078	0.1152	0.6300	0.6176	0.3128	0.3093	0.0572	0.0730
11	71.45	747.536	0.4765	0.2968	0.2267	0.6380	0.6256	0.2285	0.2285	0.1335	0.1459
12	72.90	771.630	0.4761	0.0986	0.4253	0.6521	0.6391	0.0688	0.0781	0.2791	0.2829
13	73.52	777.489	0.4554	0.0772	0.4674	0.6261	0.6207	0.0535	0.0622	0.3204	0.3171
14	73.00	774.182	0.4330	0.3244	0.2426	0.5871	0.5815	0.2627	0.2570	0.1502	0.1615
15	72.30	762.821	0.4322	0.4285	0.1393	0.5794	0.5740	0.3399	0.3348	0.0807	0.0913
16	73.97	766.372	0.4161	0.0396	0.5443	0.5916	0.5850	0.0211	0.0330	0.3873	0.3820
17	73.20	758.555	0.3990	0.2860	0.3150	0.5511	0.5509	0.2388	0.2332	0.2101	0.2159
18	73.09	763,297	0.3846	0.4449	0.1705	0.5331	0.5256	0.3642	0.3588	0.1027	0.1156
19	73.20	770.443	0.3822	0.4969	0.1209	0.5230	0.5196	0.4083	0.3988	0.0687	0.0816
20	73.25	765.575	0.3769	0.4606	0.1625	0.5216	0.5166	0.3830	0.3728	0.0954	0.1106
21	73.60	764.271	0.3383	0.5486	0.1131	0.4724	0.4705	0.4593	0.4512	0.0683	0.0783
22	74.67	775.572	0.3371	0.3464	0.3165	0.4895	0.4805	0.2940	0.2932	0.2165	0.2263
23	74.48	765.963	0.3131	0.4381	0.2488	0.4560	0.4487	0.3772	0.3726	0.1668	0.1787
24	74.60	773.756	0.3119	0.4856	0.2025	0.4471	0.4445	0.4167	0.4108	0.1362	0.1447
25	76.22	772.055	0.2605	0.2734	0.4661	0.3996	0.3946	0.2500	0.2475	0.3504	0.3579
26	75,58	765.614	0.2525	0.4404	0.3071	0.3894	0.3769	0.3921	0.3918	0.2185	0.2314
27	76,05	774.045	0.2310	0.5018	0.2672	0.3523	0.3469	0.4530	0.4499	0.1947	0.2032
28	75.30	765.146	0.2300	0.6391	0.1309	0.3491	0.3400	0.5662	0.5625	0.0847	0.0975
29	76.65	775.688	0.2299	0.3509	0.4192	0.3583	0.3520	0.3242	0.3216	0.3175	0.3264
30	77.30	770.027	0.1999	0.2782	0.5219	0.3223	0.3157	0.2635	0.2636	0.4142	0.4208
31	76.85	772.582	0.1989	0.4342	0.3669	0.3220	0.3080	0.3983	0.4028	0.2797	0.2892
32	76.15	761.278	0.1775	0.6434	0.1791	0.2830	0.2720	0.5854	0.5889	0.1316	0.1391
33	77.70	771.132	0.1532	0.4323	0.4145	0.2611	0.2451	0.4118	0.4156	0.3271	0.3394
34	76.12	738.369	0.1480	0.5956	0.2564	0.2452	0.2336	0.5594	0.5614	0.1954	0.2050
35	78.40	765.847	0.1367	0.2825	0.5808	0.2321	0.2258	0.2761	0.2811	0.4918	0.4931
36	76.90	759.163	0.1346	0.6466	0.2188	0.2277	0.2125	0.6075	0.6116	0.1648	0.1759
37	79.87	772.549	0.1239	0.0157	0.8604	0.2255	0.2137	0.0171	0.0164	0.7574	0.7699
38	77.87	763.138	0.1172	0.5101	0.3727	0.2110	0.1906	0.4918	0.4988	0.2972	0.3106
39	77.25	758.751	0.1020	0.7239	0.1741	0.1767	0.1632	0.6901	0.6947	0.1332	0.1421
40	78.71	767.270	0.1000	0.4127	0.4873	0.1798	0.1667	0.4085	0.4150	0.4117	0.4184
41	79.23	756.205	0.0837	0.2656	0.6507	0.1538	0.1442	0.2857	0.2766	0.5605	0.5793
42	78.78	755.446	0.0828	0.3837	0.5335	0.1534	0.1405	0.3894	0.3929	0.4572	0.4666
43	76.97	735.111	0.0823	0.6899	0.2278	0.1462	0.1345	0.6804	0.6758	0.1734	0.1897
44	77.45	747.871	0.0635	0.7799	0.1566	0.1102	0.1038	0.7751	0.7654	0.1147	0.1304
45	79.68	749.718	0.0558	0.2446	0.6996	0.1143	0.0985	0.2844	0.2615	0.6013	0.6400

Note: A minor change in temperature is required to adjust calculated total pressure to 760 mmHg; the resulting y-values before-and-after are virtually identical. Calculated y-values in table correspond to original temperatures.

Table A5. Comparison of Ideal Ternary VLE Calculations with Experimental Data – Methanol(1)–Ethanol(2)–2-Propanol(3) at 757.56 mmHg Pressure (Data Ref. **[29]**)

#	T °C meas	P_T at T mmHg	X_1 meas	X_2 meas	X_3 meas	Y_1 meas	Y_1 calc	Y_2 meas	Y_2 calc	Y_3 meas	Y_3 calc
1	76.36	772.116	0.2295	0.4026	0.3679	0.3391	0.3494	0.3771	0.3664	0.2838	0.2842
2	76.22	769.414	0.2105	0.5207	0.2688	0.3085	0.3200	0.4837	0.4729	0.2078	0.2071
3	75.90	771.233	0.2463	0.4352	0.3185	0.3571	0.3692	0.3999	0.3893	0.2430	0.2415
4	75.85	771.465	0.2545	0.4073	0.3382	0.3745	0.3807	0.3726	0.3635	0.2529	0.2558
5	75.39	775.973	0.3055	0.3079	0.3866	0.4414	0.4468	0.2759	0.2682	0.2827	0.2850
6	75.04	774.720	0.3121	0.3633	0.3246	0.4439	0.4514	0.3217	0.3125	0.2344	0.2361
7	75.03	771.583	0.3074	0.3612	0.3314	0.4438	0.4462	0.3197	0.3118	0.2365	0.2419
8	74.76	763.919	0.2713	0.5612	0.1675	0.3868	0.3939	0.4913	0.4841	0.1219	0.1221
9	73.80	777.905	0.4103	0.2378	0.3519	0.5641	0.5648	0.2000	0.1937	0.2359	0.2415
10	73.60	764.609	0.3332	0.5794	0.0874	0.4573	0.4632	0.4805	0.4763	0.0622	0.0605
11	73.39	768.113	0.3761	0.4480	0.1759	0.5169	0.5164	0.3669	0.3635	0.1162	0.1201
12	73.29	768.180	0.3930	0.3885	0.2185	0.5334	0.5376	0.3217	0.3139	0.1449	0.1485
13	73.02	771.523	0.4278	0.3189	0.2533	0.5739	0.5769	0.2605	0.2537	0.1656	0.1694
14	72.55	773.837	0.4780	0.2203	0.3017	0.6410	0.6316	0.1718	0.1714	0.1872	0.1970
15	72.40	762.410	0.4127	0.4973	0.0900	0.5450	0.5504	0.3942	0.3903	0.0608	0.0593
16	72.02	764.638	0.4508	0.4355	0.1137	0.5895	0.5911	0.3366	0.3356	0.0739	0.0734
17	71.70	768.286	0.5010	0.3075	0.1915	0.6448	0.6460	0.2374	0.2327	0.1178	0.1213
18	71.16	767.418	0.5357	0.2862	0.1781	0.6791	0.6777	0.2150	0.2120	0.1059	0.1102
19	70.98	782.426	0.6106	0.1087	0.2807	0.7697	0.7526	0.0761	0.0784	0.1542	0.1690
20	70.58	774.088	0.6112	0.1467	0.2421	0.7576	0.7501	0.1081	0.1052	0.1343	0.1447
21	70.35	759.313	0.5507	0.3800	0.0693	0.6839	0.6831	0.2726	0.2751	0.0435	0.0418
22	70.10	765.611	0.5962	0.2905	0.1133	0.7262	0.7266	0.2051	0.2064	0.0687	0.0670
23	69.95	766.410	0.6223	0.2113	0.1664	0.7578	0.7533	0.1501	0.1491	0.0921	0.0976
24	69.71	764.795	0.6356	0.2028	0.1616	0.7699	0.7641	0.1414	0.1419	0.0887	0.0940
25	68.45	779.470	0.7758	0.0610	0.1632	0.8852	0.8724	0.0387	0.0397	0.0761	0.0879
26	67.18	759.910	0.8037	0.1310	0.0653	0.8831	0.8831	0.0816	0.0829	0.0353	0.0340
27	66.90	760.187	0.8336	0.0782	0.0882	0.9138	0.9058	0.0464	0.0489	0.0398	0.0453
28	66.11	759.543	0.8873	0.0746	0.0381	0.9297	0.9360	0.0484	0.0451	0.0219	0.0189

Note: A minor change in temperature is required to adjust calculated total pressure to 760 mmHg; the resulting y-values before-and-after are virtually identical. Calculated y-values in table correspond to original temperatures.

For the first set of constant pressure data (Table A4), absolute values of temperature deviations range from 0 to 0.84 deg C and average 0.27 deg C, with 88% less than 0.5 °C and 96% less than 0.6 °C. Average absolute value of the deviation in all y-values is about 0.01 mole fraction, with at least 98% of the deviations under 0.03. Maximum deviation is about 0.03 for methanol, 0.02 for ethanol, and 0.04 for 2-propanol.

For the second set (Table A5), absolute values of temperature deviations range from 0.06 to 0.82 °C and average 0.38 °C, with 86% less than 0.6 °C, 93% less than 0.7 °C, and 96% less than 0.8 °C. Average absolute value of the deviation in all y-values is less than 0.006 mole fraction, with 82-89% under 0.01. Maximum deviation is about 0.02 for methanol, 0.01 for ethanol, and 0.015 for 2-propanol. Despite the non-ideal methanol–2-propanol binary, predicted ideal vapor composition for the ternary with added ethanol compares even more favorably with data within the experimental range than that observed for the essentially ideal methanol–ethanol–1-propanol system. Such good agreement despite non-ideal behavior suggests that both sets of constant pressure data for methanol–ethanol–2-propanol may be of somewhat higher quality than the data for the essentially ideal methanol–ethanol–1-propanol system.

For the constant temperature data at 60 °C (Table A6), maximum absolute deviation in y-values is about 0.06 mole fraction for methanol, 0.01 for ethanol, and 0.05 for 1-propanol, averaging 0.005 – 0.008. At least 96% of the deviations are less than 0.02, with 76-91% of deviations less than 0.01. Average deviation in calculated vs. measured total pressure is 3.8 mmHg, with a range of 0.2 (min) to 9.8 (max), similar to those exhibited by the methanol–ethanol–**1**-propanol ternary.

Once again, how well predicted vapor-phase concentrations compare with values calculated assuming an ideal system depends upon how closely any real system approximates ideality, plus the accuracy of the experimental data to which the calculated values are being compared. The vapor pressure relationship chosen also affects the calculations somewhat.

Table A6. Comparison of Ideal Ternary VLE Calculations with Experimental Data – Methanol(1)–Ethanol(2)–2-Propanol(3) at 60 °C (Data from Ref. **[30]**)

#	$P_T$meas mmHg	P_T calc mmHg	X_1 meas.	X_2 meas.	X_3 meas.	Y_1 meas.	Y_1 calc.	Y_2 meas.	Y_2 calc.	Y_3 meas.	Y_3 calc.
1	289.11	279.35	0.0000	0.0000	1.0000	0.0000	0.0000	0.0000	0.0000	1.0000	1.0000
2	351.88	351.49	0.0000	1.0000	0.0000	0.0000	0.0000	1.0000	1.0000	0.0000	0.0000
3	633.70	629.42	1.0000	0.0000	0.0000	1.0000	1.0000	0.0000	0.0000	0.0000	0.0000
4	582.96	577.45	0.8330	0.0900	0.0770	0.9000	0.9080	0.0570	0.0548	0.0430	0.0373
5	506.78	508.54	0.6110	0.2120	0.1770	0.7460	0.7562	0.1530	0.1465	0.1010	0.0972
6	604.15	601.59	0.9100	0.0510	0.0390	0.9460	0.9521	0.0340	0.0298	0.0200	0.0181
7	582.38	577.90	0.8250	0.1350	0.0400	0.8980	0.8986	0.0830	0.0821	0.0190	0.0193
8	579.19	577.09	0.8400	0.0510	0.1090	0.9020	0.9162	0.0340	0.0311	0.0640	0.0528
9	548.93	543.92	0.7030	0.2560	0.0410	0.8050	0.8135	0.1720	0.1654	0.0230	0.0211
10	539.39	539.59	0.7300	0.0650	0.2050	0.8490	0.8515	0.0450	0.0423	0.1060	0.1061
11	549.26	545.14	0.7170	0.2050	0.0780	0.8160	0.8279	0.1430	0.1322	0.0410	0.0400
12	548.57	546.23	0.7430	0.0940	0.1630	0.8520	0.8562	0.0630	0.0605	0.0850	0.0834
13	593.90	590.48	0.8700	0.0910	0.0390	0.9240	0.9274	0.0570	0.0542	0.0190	0.0185
14	592.83	590.11	0.8770	0.0520	0.0710	0.9310	0.9354	0.0340	0.0310	0.0350	0.0336
15	376.64	374.64	0.0960	0.8550	0.0490	0.1480	0.1613	0.8150	0.8022	0.0370	0.0365
16	361.98	362.70	0.0920	0.7090	0.1990	0.1450	0.1597	0.6940	0.6871	0.1610	0.1533
17	324.90	316.04	0.0840	0.1010	0.8150	0.1670	0.1673	0.1020	0.1123	0.7310	0.7204
18	419.13	413.71	0.3290	0.2660	0.4050	0.4900	0.5005	0.2230	0.2260	0.2870	0.2735
19	511.34	509.00	0.5880	0.3300	0.0820	0.7350	0.7271	0.2200	0.2279	0.0450	0.0450
20	530.40	524.31	0.6540	0.2220	0.1240	0.7800	0.7851	0.1520	0.1488	0.0680	0.0661
21	506.83	503.19	0.5780	0.2980	0.1240	0.7800	0.7230	0.1990	0.2082	0.0210	0.0688
22	338.92	335.32	0.0890	0.3440	0.5670	0.1620	0.1671	0.3530	0.3606	0.4850	0.4724
23	376.93	371.62	0.1420	0.5900	0.2680	0.2390	0.2405	0.5560	0.5580	0.2050	0.2015

Note: Calculated y-values in the table correspond to the measured constant temperature and the total pressure calculated from Raoult's Law and the Antoine vapor-pressure relationships for mixture components.

APPENDIX B

Systems of Aromatic Compounds with Benzene and Toluene at 1 Atm

Idealities / Non-Idealities in Some Binary Systems Containing Aromatic Chemicals

Isobaric VLE data at 1 atm pressure for several binary systems of aromatic chemicals with benzene or toluene are shown in Figures B1-B11. These figures include x-y data and activity coefficients. The solid y-value curves are calculated from Raoult's Law for the liquid phase and Dalton's Law (ideal gas) for the vapor phase; the points reproduce experimental data from the literature. The line y=x is shown as the straight diagonal line in each of the figures.

- Data for the benzene–toluene system from a great number of investigators are depicted in Figure 1.[3] It is among the systems listed as ideal mixtures in Perry's 3rd and 4th eds.[4] The benzene–toluene x-y data[37,38] of Figure B1 from the m-xylene and p-xylene studies discussed here would fit in nicely with other investigators' data depicted in Figure 1 of the main text. Activity coefficients (Figure B2) are flat across the board at 1.0.
- Investigators of the benzene–ethylbenzene binary have concluded that this system obeys Raoult's Law within experimental error.[75] (VLE x-y data not shown here)
- Recommended data for toluene–ethylbenzene assumes an ideal mixture in the liquid phase and an ideal gas vapor phase.[76] (x-y data also not shown)
- The benzene–*o*-xylene system at 761 mmHg[34] is clearly non-ideal, as shown by its x-y diagram (Figure B3) and its activity coefficient curves (Figure B4). In Figure B4, both benzene and *o*-xylene appear to behave in a fairly ideal manner up to a benzene mole fraction of 0.6-0.7. However, the system shows a highly non-ideal benzene γ^{∞} of 1.87. Toluene–*o*-xylene at 770 mmHg[35] deviates from ideality as well (Figure B5, x-y curve). Activity coefficients of both components confirm non-ideal behavior (Figure B6), with γ^{∞}'s of 1.24 and 1.39 for benzene and o-xylene, respectively. Both sets of x-y data were obtained by the same investigators, with activity coefficients fitted by DECHEMA.
- The benzene–toluene–*m*-xylene ternary and its three constituent binaries have been studied in a single investigation at 760 mmHg.[37] Both benzene–*m*-xylene and toluene–*m*-xylene at 760 mmHg are non-ideal, especially for lower benzene or toluene concentrations (Figures B7 and B8, x-y curves). Reported activity coefficients derived from the data reported indicate that both binaries containing *m*-xylene are non-ideal (Figures B9 and B10). The γ^{∞} for benzene is 1.35, and that for toluene is 2.20, while γ^{∞} for *m*-xylene in benzene solution is 1.25 and 1.70 in toluene solution, extrapolated by "eyeball" for both solutions. Data for the ternary system are discussed below.
- The x-y diagram for the nearly ideal benzene–*p*-xylene system at 760 mmHg is drawn in Figure B11 along with data from two sources.[38,77] Activity coefficients from one of these[77] scatter close to 1.0, with an estimated γ^{∞} of 0.931 for benzene and 0.964 for *p*-xylene. The ideal x-y diagram for toluene–p-xylene is displayed with experimental data[38] in Figure B12. It, too, closely approximates ideality, with nearly flat activity coefficient curves and a calculated γ^{∞} of 1.01 for toluene and 1.003 for *p*-xylene (activity coefficient curves not shown for the p-xylene binaries).

Table B1. Antoine Equation Constants for Aromatics and a Related Compound of Interest [(a)]

Chemical Name and Synonyms	Formula	MW [(b)]	BP (°C) [(c)]	Applicable Range(°C)	A	B	C
Benzene Benzol	C_6H_6	78.12	80.1	8 to 103	6.90565	1211.033	220.790
Toluene Methyl Benzene Phenyl Methane	$C_6H_5CH_3$	92.14	110.6	6 to 137	6.95464	1344.800	219.482
Ethylbenzene Phenyl Ethane	$C_6H_5C_2H_5$	106.17	136.2	26 to 164	6.95719	1424.255	213.206
o-Xylene o-Xylol 1,2-Xylene 1.2-Dimethylbenzene 2-Methyl Toluene	$C_6H_4(CH_3)_2$	106.17	144.4	32 to 172	6.99891	1474.679	213.686
m-Xylene m-Xylol 1,3-Xylene 1.3-Dimethylbenzene 3-Methyl Toluene	$C_6H_4(CH_3)_2$	106.17	139.1	28 to 166	7.00908	1462.266	215.105
p-Xylene p-Xylol 1,4-Xylene 1.4-Dimethylbenzene 4-Methyl Toluene	$C_6H_4(CH_3)_2$	106.17	138.3, 138.4 (138.35 calc)	27 to 166	6.99052	1453.430	215.307
Ethylene Dichloride (EDC) Ethylene Chloride 1,2-Dichloroethane	$ClCH_2CH_2Cl$	98.96	83.5	—	7.18431	1358.46	232.2

Notes:

(a) Antoine Constants from "Lange's Handbook of Chemistry", 11th Edition[71] [Table 10-10]. Temperature Ranges for Antoine Constants from 12th Edition[72] [Table 10-8].

(b) Molecular Weight (MW) from "Handbook of Chemistry and Physics," 62nd Ed.[73] [pp. C-65 to C-576].

(c) Boiling point (BP) at 760 mmHg from "Lange's Handbook," 11th Ed.[71] [Table 7-4], 12th Ed.[72] [Table 7-4], or 13th Ed.[74] [Table 7-15], or "Handbook of Chemistry and Physics," 62nd Edition[73] [pp. C-65 to C-576], chosen to be the most consistent with boiling point calculated using the Antoine Constants listed.

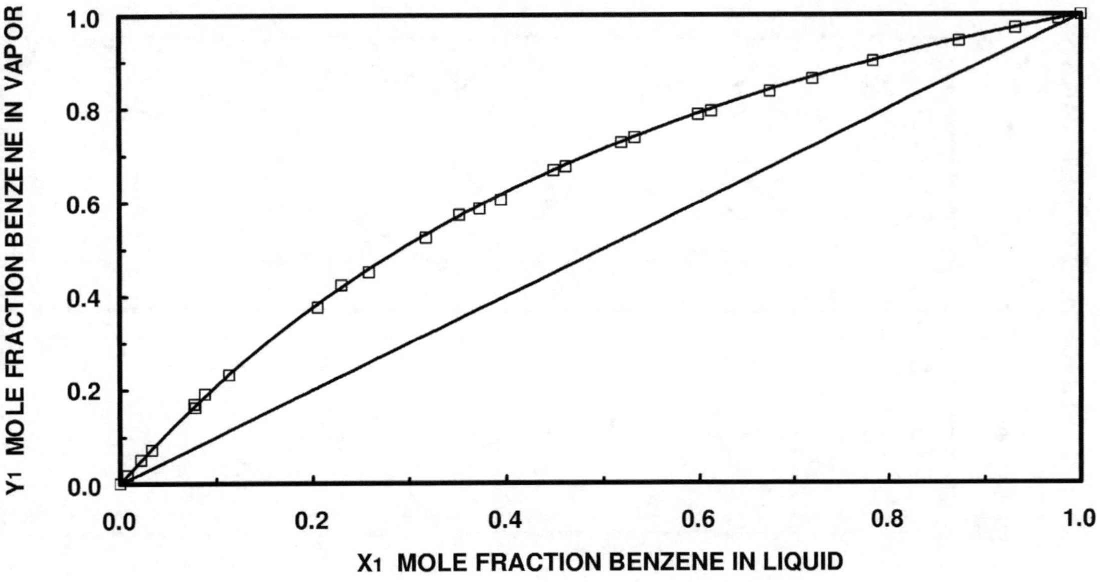

Figure B!. Vapor-Liquid Equilibrium Diagram Benzene–Toluene at 1 Atm Pressure Supplemental Data from References **[37, 38]**

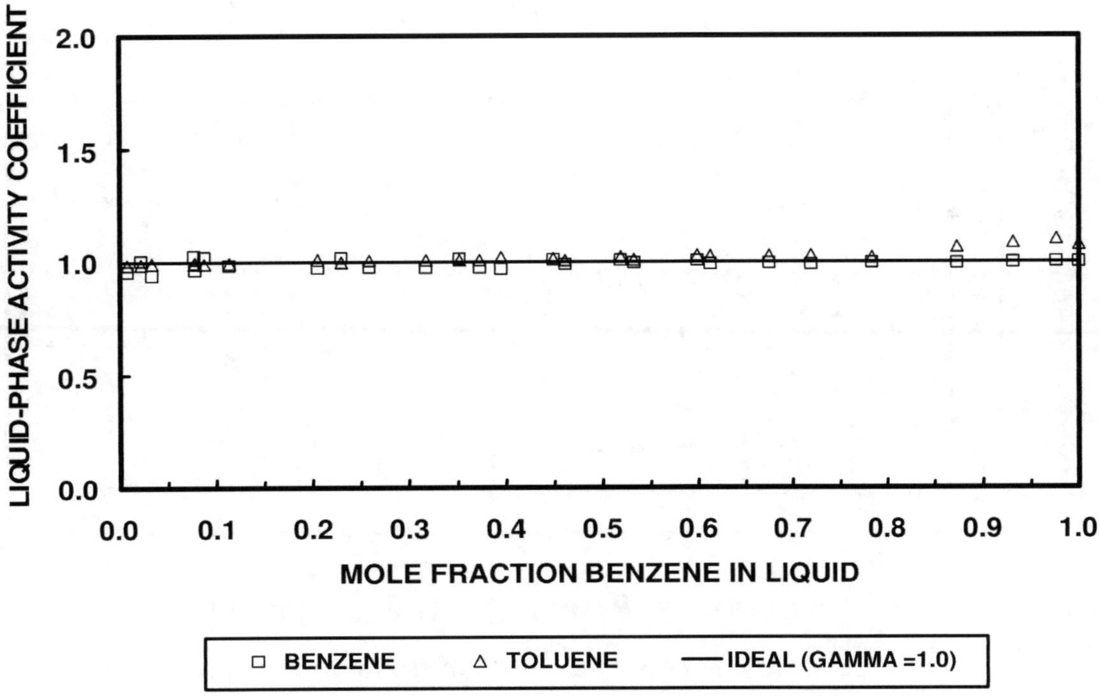

Figure B2. Activity Coefficients: Benzene–Toluene at 1 Atm Pressure Calculated from Data in References **[37, 38]**

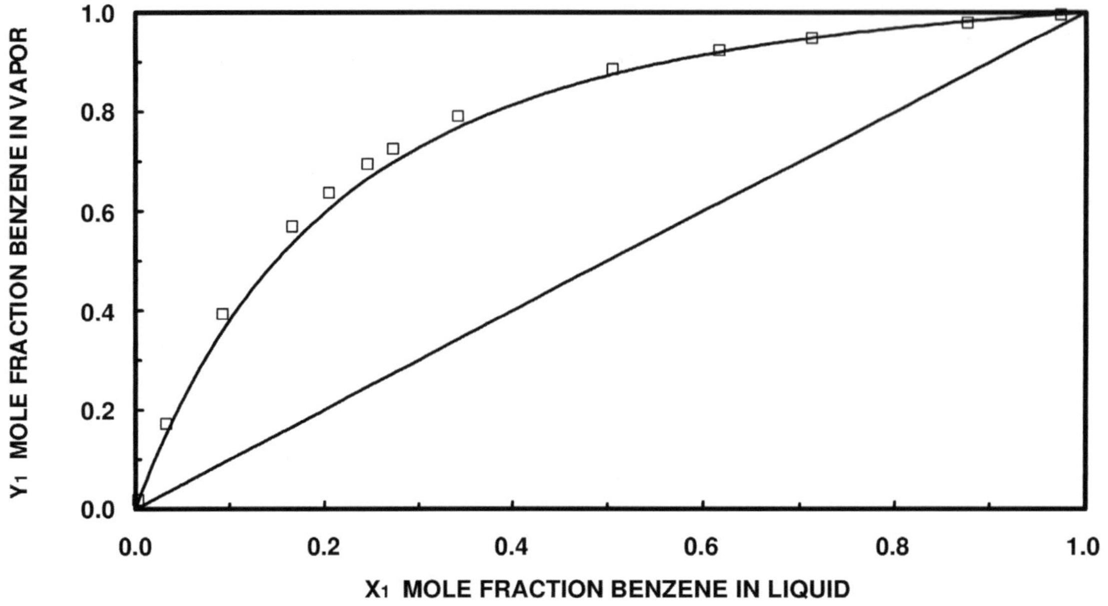

Figure B3. Vapor-Liquid Equilibrium Diagram Benzene–*o*-Xylene at 761 mmHg Data from Reference **[34]**

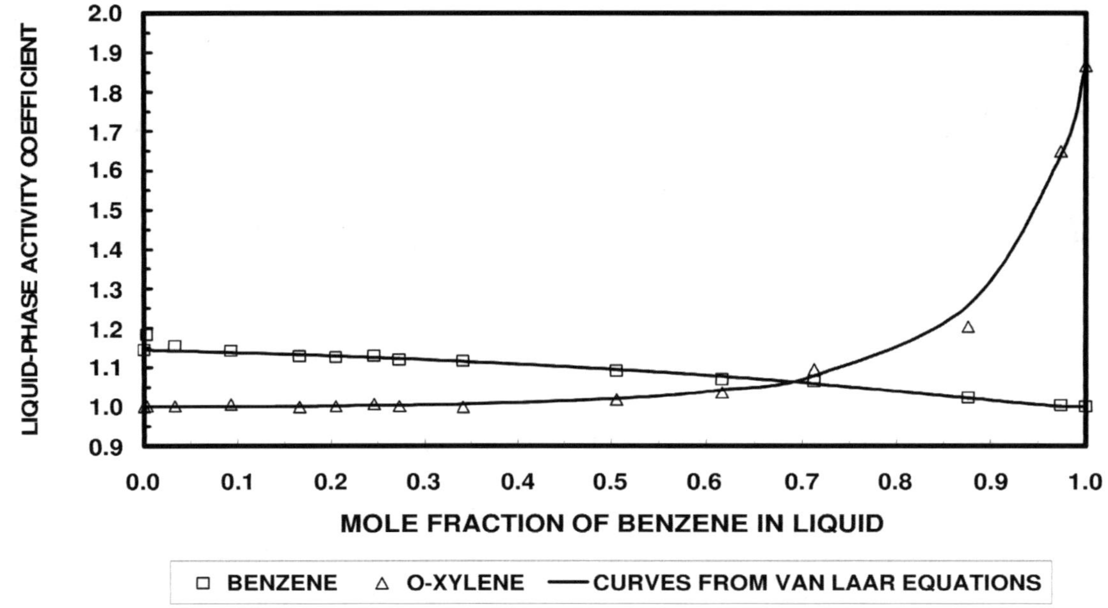

Figure B4. Activity Coefficients: Benzene–*o*-Xylene at 761 mmHg Calculated from Data in Reference **[34]**

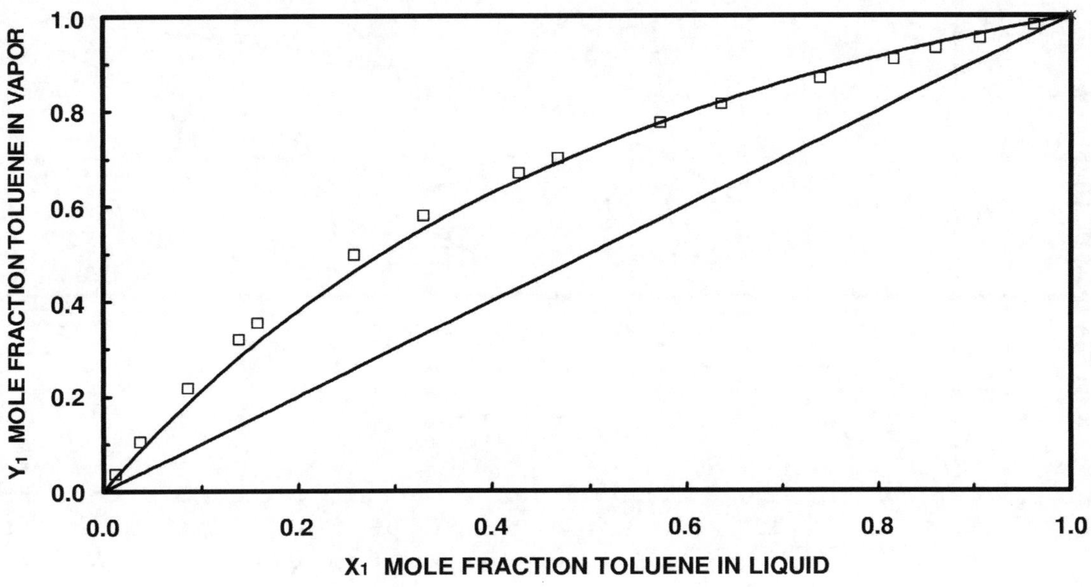

Figure B5. Vapor-Liquid Equilibrium Diagram Toluene–*o*-Xylene at 770 mmHg
Data from Reference **[35]**

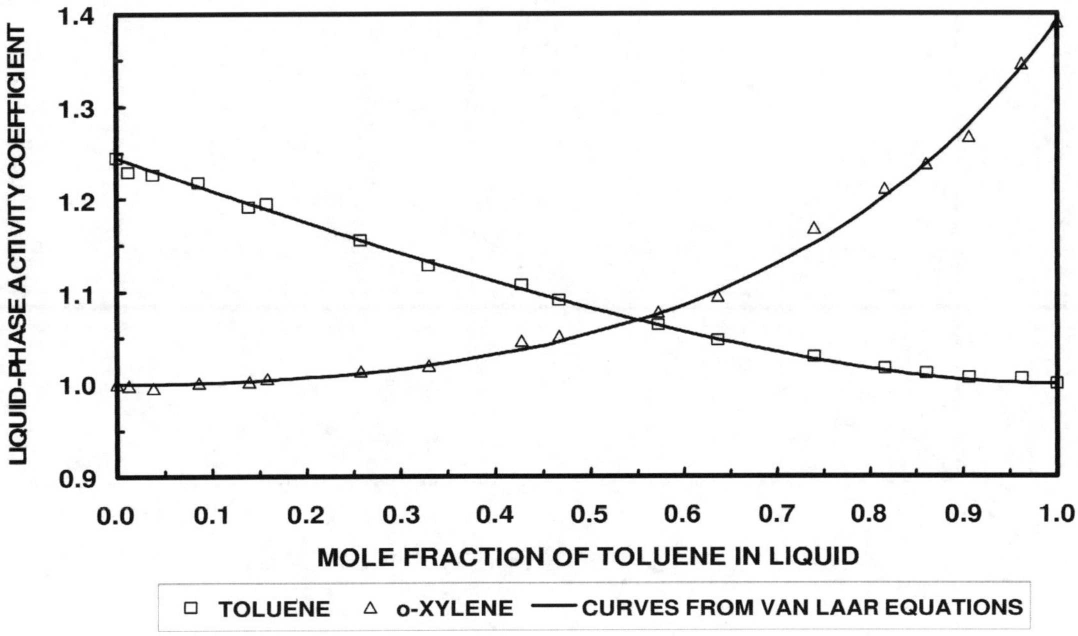

Figure B6. Activity Coefficients: Toluene–*o*-Xylene at 770 mmHg
Calculated from Data in Reference **[35]**

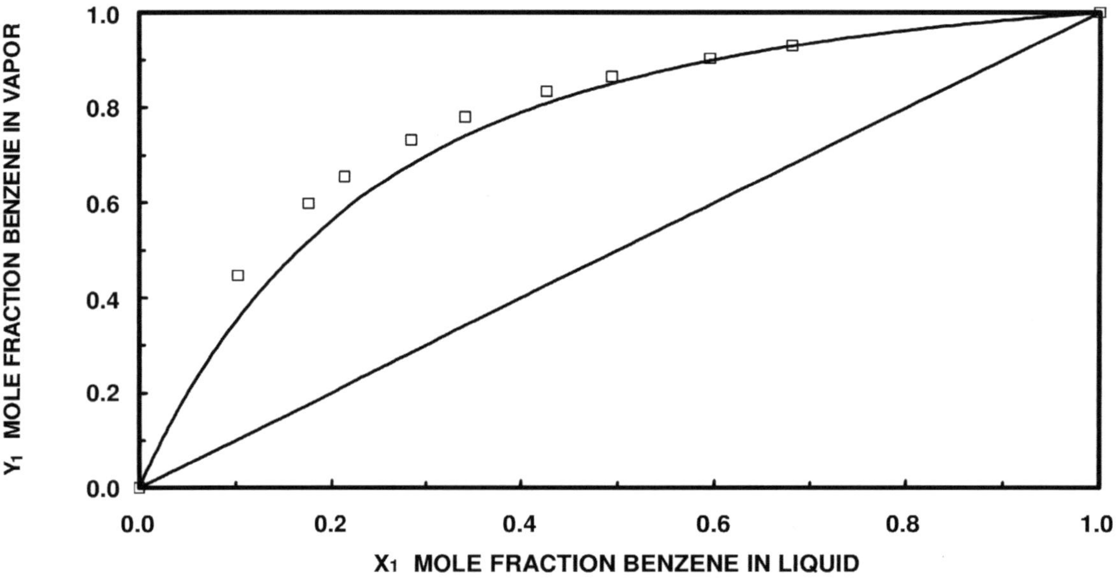

Figure B7. Vapor-Liquid Equilibrium Diagram Benzene–*m*-Xylene at 760 mmHg
Data from Reference **[37]**

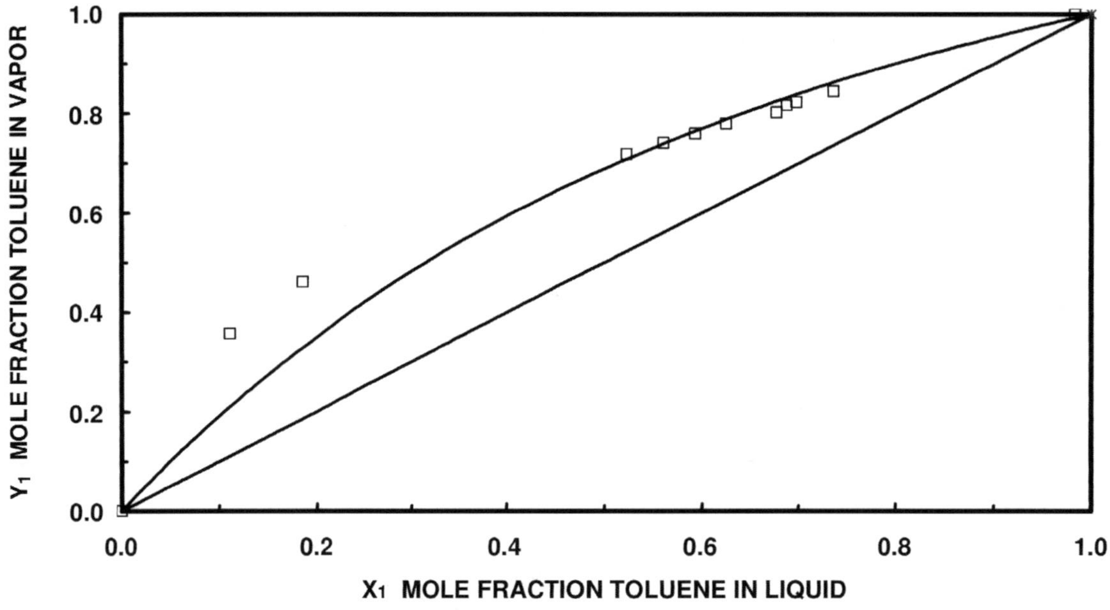

Figure B8. Vapor-Liquid Equilibrium Diagram Toluene–*m*-Xylene at 760 mmHg
Data from Reference **[37]**

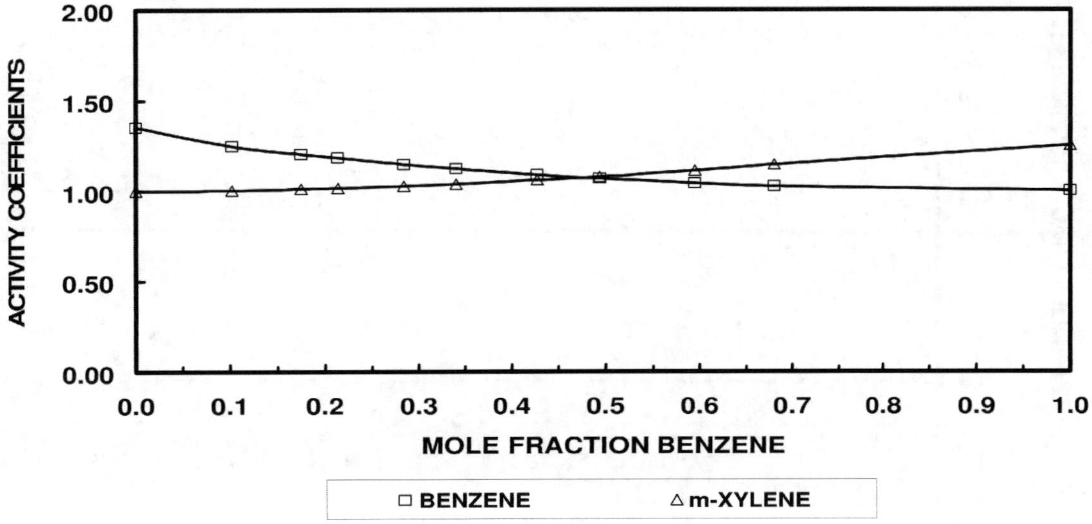

Figure B9. Activity Coefficients: Benzene–*m*-Xylene at 760 mmHg
Data from Reference **[37]**

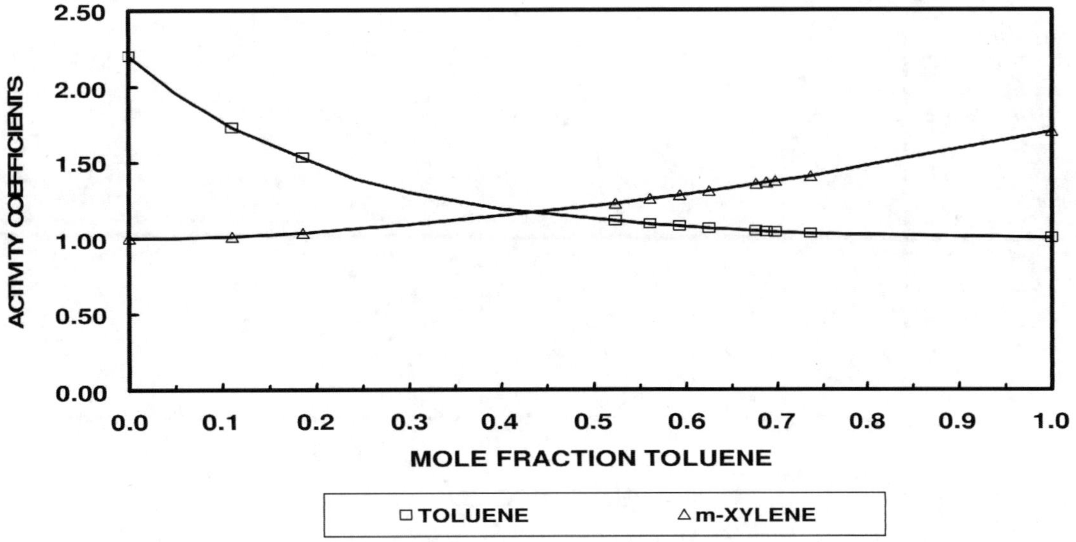

Figure B10. Activity Coefficients: Toluene–*m*-Xylene at 760 mmHg
Data from Reference **[37]**

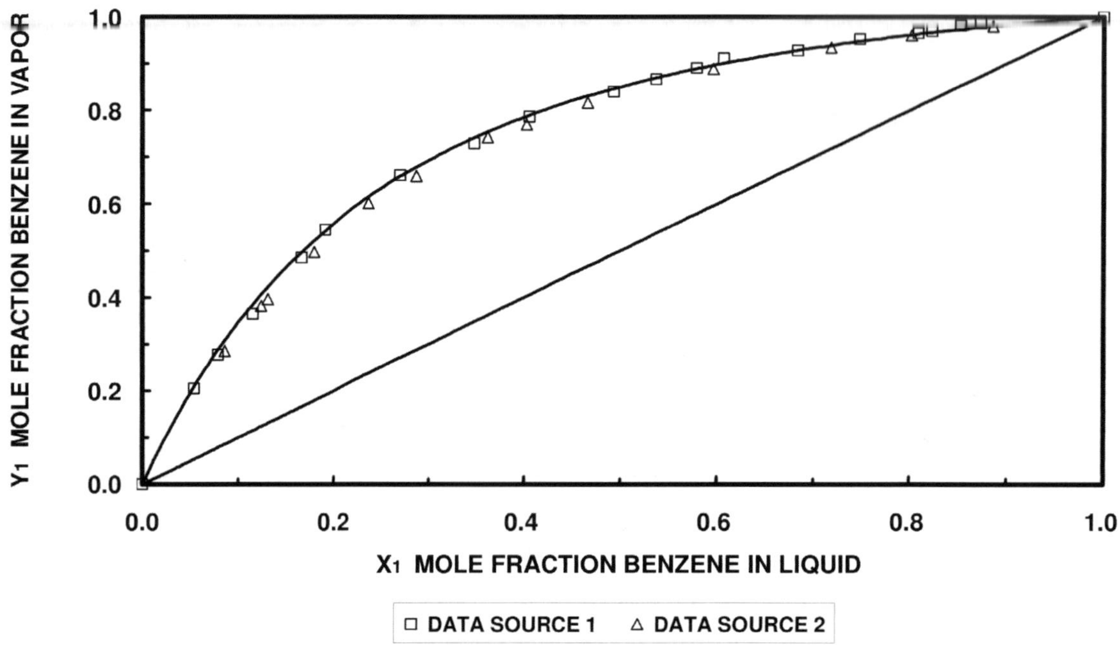

Figure B11. Vapor-Liquid Equilibrium Diagram Benzene–*p*-Xylene at 760 mmHg
Data from References **[38, 77]**

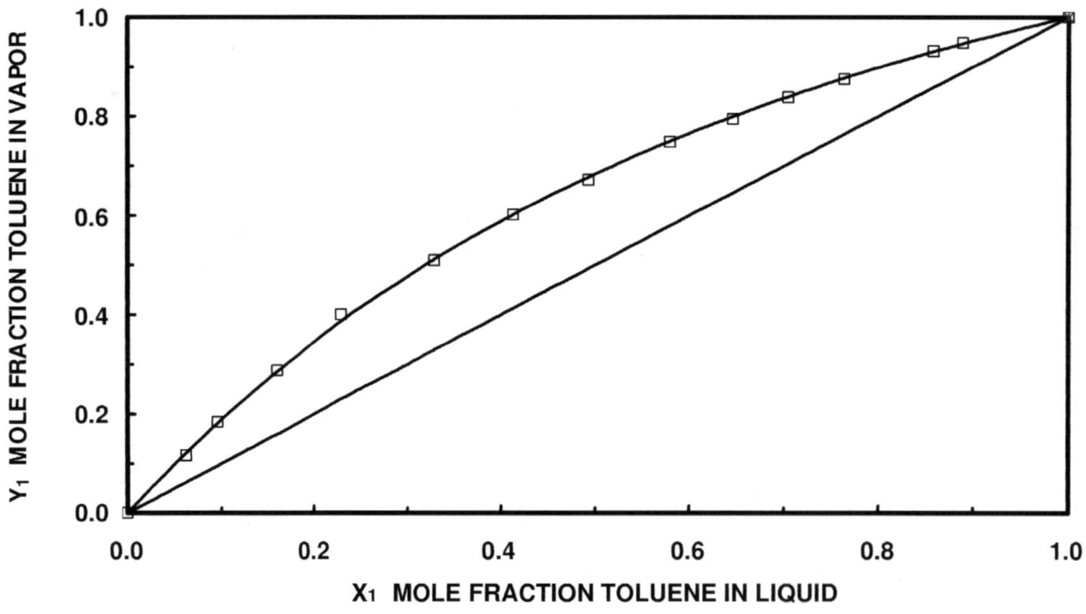

Figure B12. Vapor-Liquid Equilibrium Diagram Toluene–*p*-Xylene at 760 mmHg
Data from Reference **[38]**

The Benzene–Toluene–Ethylbenzene System at 760 mmHg Total Pressure

The **benzene–toluene–ethylbenzene** ternary stands a good chance of being ideal since the three constituent binaries are regarded as idea.[4,75,76] However, the limited number of ternary data points uncovered (ten in all) and the lack of agreement between the two independent investigations results in an inconclusive test case (Table B2). The y-value data from one investigation (the first six points in the table)[31] and ideal VLE calculations agree rather poorly, differing by up to 0.16 mole fraction; in the other (the last four points in the table),[32] all measured y-values deviate from ideal calculations by 0.01 m.f. or less. Except for Point 5, where the absolute value of the deviation in all y-values is well below 0.01, deviations in all y-values for Points 1-6 range up to 0.16. Deviations from ideal calculations for the last four points are about 0.01 or less. The liquid-phase compositions from Points 3 and 9 and Points 6 and 7 measured in the two investigations are reasonably similar, but the calculated vapor-phase compositions show superior agreement with data from the second study, suggesting that the quality of those data may be better and more closely follow the ideal model. Conclusion is that benzene–toluene–ethylbenzene is most likely ideal but that not enough ternary data have been found to confirm this assessment.

Table B2. Comparison of Ideal Ternary VLE Calculations with Experimental Data – Benzene(1)–Toluene(2)–Ethylbenzene at 760 mmHg Pressure (References **[31,32]**)

No.	T C	PT at T mm Hg	X1 meas.	X2 meas.	X3 meas.	Y1 meas.	Y1 calc.	Y2 meas.	Y2 calc.	Y3 meas.	Y3 calc.
1	125.75	755.082	0.0670	0.0890	0.8440	0.3340	0.2286	0.1630	0.1351	0.5030	0.6363
2	118.20	731.677	0.1130	0.1730	0.7140	0.4880	0.3326	0.2030	0.2218	0.3090	0.4456
3	110.70	747.771	**0.2020**	**0.2430**	**0.5550**	**0.5800**	**0.4830**	**0.2600**	**0.2475**	**0.1600**	**0.2695**
4	103.75	758.968	0.2950	0.3420	0.3630	0.6650	0.5804	0.2640	0.2805	0.0710	0.1391
5	94.80	783.779	0.5200	0.2780	0.2020	0.7760	0.7764	0.1600	0.1681	0.0640	0.0555
6	89.10	743.007	***0.6120***	***0.3010***	***0.0870***	***0.8550***	***0.8193***	***0.1300***	***0.1600***	***0.0150***	***0.0206***
7	90.62	752.982	***0.6000***	***0.2400***	***0.1600***	***0.8380***	***0.8282***	***0.1340***	***0.1323***	***0.0280***	***0.0395***
8	84.40	745.270	0.8000	0.1200	0.0800	0.9350	0.9296	0.0510	0.0545	0.0140	0.0159
9	113.62	771.546	**0.2000**	**0.1600**	**0.6400**	**0.4910**	**0.4988**	**0.1780**	**0.1715**	**0.3310**	**0.3297**
10	85.16	747.012	0.8000	0.0400	0.1600	0.9570	0.9487	0.0180	0.0186	0.0250	0.0327

The Benzene–Toluene–*o*-Xylene System at 772 mmHg Total Pressure

Minimal ternary data from a single source for the **benzene-toluene-o-xylene** ternary were found for this system at 772 mmHg,[33] along with binary data from the same laboratory for benzene–o-xylene at 761 mmHg[34] (Figure B3) and toluene–o-xylene at 770 mmHg (Figure B5). Ternary data are listed in Table B3.

Deviations from ideal temperatures average 1.7 °C and range from 0.4 °C to 2.7 °C. Except for Point No. 6, at the extremes of benzene and *o*-xylene investigated, maximum deviation in the absolute value of $y_{ideal} - y_{data}$ for all species is less than 0.006 mole fraction. At that point maximum deviation in y_1 is less than 0.01 and about 0.02 for y_2 and 0.03 for y_3. Certainly, the data nearer the center of a triangular ternary VLE diagram for this system appear to be ideal. In a sample problem dealing with a liquid composition of 50% benzene and

Table B3. Comparison of Ideal Ternary VLE Calculations with Experimental Data – Benzene(1)–Toluene(2)–*o*-Xylene(3) at 772 mmHg Pressure (Reference **[33]**)

No.	T C	PT at T mm Hg	X1 meas.	X2 meas.	X3 meas.	Y1 meas.	Y1 calc.	Y2 meas.	Y2 calc.	Y3 meas.	Y3 calc.
1	87.00	738.688	0.7311	0.0872	0.1817	0.9243	0.9259	0.0409	0.0435	0.0348	0.0306
2	90.10	760.296	0.6205	0.2681	0.1114	0.8323	0.8357	0.1423	0.1439	0.0254	0.0204
3	100.85	713.374	0.3914	0.1313	0.4773	0.7629	0.7582	0.1051	0.1051	0.1320	0.1368
4	104.20	736.194	0.2810	0.3656	0.3534	0.5710	0.5768	0.3170	0.3133	0.1120	0.1099
5	112.45	734.180	0.1134	0.5123	0.3743	0.2849	0.2886	0.5595	0.5584	0.1556	0.1530
6	129.60	727.661	0.0565	0.1214	0.8221	0.2271	0.2185	0.2341	0.2111	0.5388	0.5703

and 25% each of toluene and *o*-xylene, Treybal[2] assumes the benzene–toluene–*o*-xylene system at atmospheric pressure to be ideal. This point lies in the central region of a ternary VLE diagram, where such a system has the best chance of coming close to ideality.

Nevertheless, activity coefficients derived via Equation (18) from the binary data cited and plotted in Figure B4 (γ^{∞}'s = 1.145 and 1.866 from DECHEMA's fitted van Laar constants) and Figure B6 (γ^{∞}'s = 1.244 and 1.391, again from DECHEMA's fit) are clearly non-ideal and therefore inconsistent with an overall ideal ternary system. The activity coefficient of o-xylene above a benzene mole fraction of 0.7 seriously deviates from ideality. Activity coefficients in toluene–o-xylene solutions significantly deviate from ideality as well when the other constituent is in the majority. Activity coefficients from more recent data (not shown) have been found to exhibit large <u>negative</u> deviations from ideality.[36]

The Benzene–Toluene–*m*-Xylene System at 760 mmHg Total Pressure

The system **benzene–toluene–*m*-xylene** and its three constituent binaries have been studied in a single investigation at 760 mmHg,[37] but with fewer data entries than for benzene–toluene–*p*-xylene. Table B4 lists the ternary data and ideal calculated values.

Average y-value deviation is 0.015 (y_1), 0.022 (y_2), and 0.016 (y_3); maximum deviation is 0.0275 (y_1) 0.066 (y_2) and 0.093 (y_3), the maximum deviation for all y-values. For y_1, 61% of the deviations are less than 0.02, and 100% are less than 0.03; for y_2, 61% are less than 0.02, and 69% are less than 0.03; for y_3, 77% are less than 0.02 and 0.03, and 92% are less than 0.05, with 61%-77% less than 0.02 for all y-values. Temperature difference ranges from a minimum of 0.02 °C to a maximum of 3.0 °C, with an average of 1.0 °C and 61% less than 1.0 °C and 92% less than 2.0 °C.

Benzene–toluene binary activity coefficients computed using Equation (18) are virtually flat at 1.0 between x = 0 and x = 1.0 (Figure B2). Activity coefficients similarly calculated for the two binaries containing *m*-xylene indicate that both are non-ideal (Figures B9 and B10), more so at lower benzene or toluene concentrations (Figures B7 and B8, x-y curves). The intercepts, infinite dilution activity coefficients (γ^{∞}'s), extrapolated by "eyeball" equal 1.25 and 1.35 in benzene–*m*-xylene solutions and 2.2 and 1.7 in toluene–*m*-xylene mixtures.

Conclusion is that this ternary system is not completely ideal for all compositions.

Table B4. Comparison of Ideal Ternary VLE Calculations with Experimental Data – Benzene(1)–Toluene(2)–*m*-Xylene(3) at 760 mmHg Pressure (Reference **[37]**)

No.	T C	PT at T mm Hg	X1 meas.	X2 meas.	X3 meas.	Y1 meas.	Y1 calc.	Y2 meas.	Y2 calc.	Y3 meas.	Y3 calc.
1	**114.98**	719.662	**0.1360**	**0.2520**	**0.6120**	0.3600	0.3761	0.3590	0.3007	0.2810	0.3232
2	**116.21**	720.769	**0.1530**	**0.1470**	**0.7000**	0.4630	0.4355	0.2470	0.1812	0.2900	0.3833
3	**110.20**	723.127	**0.2170**	**0.1980**	**0.5850**	0.5130	0.5298	0.2530	0.2056	0.2340	0.2647
4	**105.17**	731.731	**0.2980**	**0.2010**	**0.5010**	0.6110	0.6313	0.2130	0.1783	0.1760	0.1904
5	**98.66**	697.583	**0.3750**	**0.2240**	**0.4010**	0.6750	0.7000	0.1970	0.1715	0.1280	0.1285
6	**98.22**	740.971	**0.4130**	**0.2620**	**0.3250**	0.6940	0.7171	0.2050	0.1863	0.1010	0.0966
7	**95.77**	748.204	**0.4620**	**0.2930**	**0.2450**	0.7220	0.7425	0.2050	0.1913	0.0730	0.0663
8	**93.47**	751.810	**0.5060**	**0.3330**	**0.1610**	0.7430	0.7587	0.2110	0.2013	0.0460	0.0400
9	**92.87**	751.578	**0.5250**	**0.3150**	**0.1600**	0.7580	0.7742	0.1970	0.1869	0.0450	0.0389
10	**90.98**	753.809	**0.5650**	**0.3460**	**0.0890**	0.7760	0.7872	0.2000	0.1927	0.0240	0.0202
11	**80.08**	759.532	**1.0000**	**0.0000**	**0.0000**	1.0000	1.0000	0.0000	0.0000	0.0000	0.0000
12	**110.58**	759.021	**0.0000**	**1.0000**	**0.0000**	0.0000	0.0000	1.0000	1.0000	0.0000	0.0000
13	**139.14**	762.009	**0.0000**	**0.0000**	**1.0000**	0.0000	0.0000	0.0000	0.0000	1.0000	1.0000

The Benzene–Toluene–p-Xylene System at 760 mmHg Total Pressure

The classic "ideal" benzene and toluene binary solution forms a nearly ideal ternary with para-xylene. The benzene–toluene–p-xylene ternary and its three constituent binaries have been studied at atmospheric pressure.[38] The tabulated binary x-y data are plotted here in Figures B1, B11, and B12. The benzene–toluene x-y data of Figure B1 would fit in nicely with other investigators' data depicted in Figure 1. The x-y data for benzene–toluene and its companion binaries follow the curves for Raoult's Law. Antoine Constants employed are given in Table B1. Activity coefficients computed from Equation (18) closely approximate 1.0.

Ternary data from this and another study[39] are listed in Table B5, along with results calculated here assuming ideal conditions. Values of vapor-phase mole fractions (y_1 benzene, y_2 toluene, and y_3 p-xylene) were calculated at the stated temperatures using calculated total pressure [Equations (1)-(4) and (6)-(8)]. Those calculated pressures (P_T) are listed in Table B5. A minor change to a slightly different temperature (not shown) is required to adjust calculated total pressure to 760 mmHg (by trial and error). The resulting y-values calculated before and after the temperature adjustment are virtually identical, and calculated y-values listed in Table B5 correspond to the original temperatures. The first 26 entries in Table B2 summarize the data from Ref. **[38]**. Average deviation in the absolute value of ($y_{ideal}-y_{data}$) ranges from 0.002 to 0.005 (0.005 for y_1, 0.003 for y_2, and 0.002+ for y_3). Maximum deviation is 0.0135 for y_1, 0.0081 for y_2, and 0.01 for y_3. In summary, 81-96% of the deviations are less than 0.01 for y_1 and y_3, and 100% less than 0.01 for y_2. Temperature difference ranges from a minimum of 0 °C to a maximum of 0.83 °C, with an average of 0.27 °C, with 85% less than 0.5 °C and 87% less than 0.7 °C. These minor deviations from the data for both x-y and temperature values compare favorably with the original investigators' more complicated calculations using experimentally determined activity coefficients.

Table B5. Comparison of Ideal Ternary VLE Calculations with Experimental Data – Benzene(1)–Toluene(2)–p-Xylene(3) at 760 mmHg Pressure (References [38], [39])

#	T °C meas.	P_T at T mm Hg	X_1 meas.	X_2 meas.	X_3 meas.	Y_1 meas.	Y_1 calc.	Y_2 meas.	Y_2 calc.	Y_3 meas.	Y_3 calc.
1	80.10	760.000	1.0000	0.0000	0.0000	1.0000	1.0000	0.0000	0.0000	0.0000	0.0000
2	110.65	760.533	0.0000	1.0000	0.0000	0.0000	0.0000	1.0000	1.0000	0.0000	0.0000
3	138.94	772.055	0.0000	0.0000	1.0000	0.0000	0.0000	0.0000	0.0000	1.0000	1.0000
4	122.96	767.363	0.0694	0.2451	0.6855	0.2247	0.2183	0.3407	0.3403	0.4346	0.4414
5	116.56	774.181	0.1353	0.2889	0.5758	0.3665	0.3616	0.3313	0.3347	0.3022	0.3037
6	114.01	754.771	0.1647	0.2604	0.5749	0.4346	0.4240	0.2803	0.2884	0.2851	0.2876
7	110.75	777.560	0.2297	0.2457	0.5246	0.5172	0.5288	0.2426	0.2410	0.2402	0.2302
8	106.59	769.391	0.2947	0.2238	0.4815	0.6027	0.6162	0.2028	0.1968	0.1945	0.1870
9	104.47	773.630	0.3125	0.2978	0.3897	0.6058	0.6147	0.2490	0.2448	0.1452	0.1405
10	101.12	766.636	0.3610	0.3211	0.3179	0.6453	0.6555	0.2475	0.2410	0.1072	0.1035
11	100.71	761.759	0.4023	0.1886	0.4091	0.7180	0.7270	0.1487	0.1407	0.1333	0.1322
12	97.55	761.802	0.4391	0.2817	0.2792	0.7159	0.7280	0.1965	0.1909	0.0876	0.0811
13	95.19	753.412	0.4890	0.2562	0.2548	0.7608	0.7679	0.1674	0.1631	0.0718	0.0690
14	91.59	766.387	0.5876	0.2589	0.1535	0.8136	0.8194	0.1488	0.1446	0.0376	0.0360
15	89.54	761.066	0.6447	0.2233	0.1320	0.8540	0.8535	0.1169	0.1176	0.0291	0.0290
16	88.03	762.660	0.6968	0.1919	0.1113	0.8792	0.8809	0.0963	0.0960	0.0245	0.0231
17	87.57	751.578	0.7265	0.0814	0.1921	0.9168	0.9196	0.0421	0.0407	0.0411	0.0397
18	86.60	759.455	0.7415	0.1642	0.0943	0.8991	0.9027	0.0802	0.0787	0.0207	0.0186
19	85.95	759.685	0.7653	0.1489	0.0858	0.9095	0.9136	0.0727	0.0698	0.0178	0.0166
20	85.69	751.495	0.7857	0.0639	0.1504	0.9457	0.9409	0.0261	0.0300	0.0282	0.0290
21	84.38	753.397	0.8318	0.0512	0.1170	0.9599	0.9556	0.0177	0.0230	0.0224	0.0215
22	83.56	754.360	0.8617	0.0419	0.0964	0.9652	0.9646	0.0169	0.0183	0.0179	0.0171
23	82.55	765.710	0.9161	0.0240	0.0599	0.9828	0.9799	0.0068	0.0100	0.0104	0.0101
24	81.24	763.867	0.9619	0.0117	0.0264	0.9930	0.9911	0.0022	0.0047	0.0048	0.0042
25	80.73	765.180	0.9845	0.0026	0.0129	0.9962	0.9970	0.0017	0.0010	0.0021	0.0020
26	80.50	763.286	0.9896	0.0036	0.0068	0.9979	0.9975	0.0004	0.0014	0.0017	0.0011
27	121.46	767.949	0.0910	0.2290	0.6800	0.2700	0.2760	0.3070	0.3053	0.4230	0.4187
28	123.87	795.228	0.1100	0.1240	0.7660	0.3020	0.3411	0.1690	0.1702	0.5290	0.4888
29	113.76	776.780	0.1100	0.5070	0.3830	0.2410	0.2734	0.5650	0.5418	0.1940	0.1848
30	113.76	771.204	0.1190	0.4640	0.4170	0.2980	0.2979	0.5090	0.4994	0.1930	0.2026
31	111.02	773.048	0.1230	0.5880	0.2890	0.2790	0.2868	0.5830	0.5846	0.1380	0.1286
32	114.10	775.931	0.1370	0.3970	0.4660	0.3440	0.3438	0.4230	0.4288	0.2330	0.2274
33	117.77	769.470	0.1380	0.2190	0.6430	0.3540	0.3822	0.2830	0.2639	0.3630	0.3539
34	112.75	736.353	0.1860	0.2040	0.6100	0.4840	0.4756	0.2190	0.2236	0.2970	0.3008
35	103.87	776.365	0.2360	0.6090	0.1550	0.4410	0.4553	0.5020	0.4901	0.0570	0.0546
36	100.15	774.415	0.3520	0.4430	0.2050	0.6090	0.6163	0.3220	0.3197	0.0690	0.0640
37	98.42	774.167	0.3980	0.4040	0.1980	0.6690	0.6651	0.2600	0.2767	0.0710	0.0583
38	99.67	769.074	0.4210	0.2170	0.3620	0.7430	0.7327	0.1510	0.1554	0.1060	0.1119
39	99.36	789.797	0.4220	0.3020	0.2760	0.6840	0.7092	0.2070	0.2086	0.1090	0.0822
40	97.62	756.985	0.4570	0.1970	0.3460	0.7750	0.7640	0.1260	0.1346	0.0990	0.1014
41	99.13	794.681	0.4730	0.1550	0.3720	0.7610	0.7850	0.1150	0.1057	0.1240	0.1093
42	99.07	865.177	0.5160	0.2370	0.2470	0.7720	0.7853	0.1450	0.1481	0.0830	0.0665
43	91.16	759.844	0.6120	0.1840	0.2040	0.8370	0.8502	0.1100	0.1022	0.0530	0.0475
44	87.98	766.218	0.6990	0.2050	0.0960	0.8610	0.8783	0.1210	0.1019	0.0180	0.0198

Note: Measured values in first 26 entries are from Ref. (38); remainder of data are from Ref. (39).

The last 18 entries in the table represent the data from the other investigation.[39] Deviations for the second study are greater but tolerable. Average deviation is greater: 0.015 mole fraction (m.f.) for y_1 (benzene), <0.01 m.f. for y_2 (toluene), and 0.01 m.f. for y_3 (*p*-xylene). Maximum deviation is 0.04 (y_1 and y_3) and 0.02 (y_2), with 89% less than 0.03 for y_1 and about 90% less than 0.02 for y_2 and y_3. Seventy-two percent of the deviations are less than 0.02 for y_1, 94% < 0.02 for y_2, and 89% < 0.02 for y_3; 89% are less than 0.03 for y_1. Minimum deviation in y-values is 0.0001 for y_1, <0.0012 for y_2, and 0.0018 for y_3. About 90% of the absolute differences are less than 0.03 for y_1 and less than 0.02 for y_2 and y_3. Temperature difference ranges from a minimum of 0.14 °C to a maximum of 4.5 °C, with 56% less than 0.7 °C and an average of 1.0 °C.

These two data sets echo the point that reproduction of experimental data by means of ideal-system calculations depends both upon how closely any real system approximates ideality and the accuracy of the experimental data to which the calculated values are being compared. Choice of a vapor-pressure relationship also affects the calculations somewhat. Binary data are plotted in Figures B1, B11, and B12. As shown in the figures, the x-y data for benzene–toluene and the other binaries follow the curves for Raoult's Law. As noted earlier, the benzene–toluene x-y data of Figure B1 would fit in nicely with the other data in Figure 1. Binary activity coefficients for benzene–toluene computed from Equation (18) are virtually flat at 1.0 between x = 0 and x = 1.0 (Figure B2) Activity coefficients for the other binaries scatter around 1.0 over the range of the data, the greatest deviation occurring for toluene in the benzene–toluene system at the lowest toluene concentration.

Conclusion is that the **benzene–toluene–*p*-xylene** system is nearly ideal and comparable to the alcohol systems described above.

APPENDIX C

Binary VLE Methanol–Ethanol–1-Propanol with Isomeric Butanols at 1 Atm

There are four isomers of butanol: the straight-chain 1-butanol and three others with various degrees of branching – *sec*- or 2-butanol, *iso*-butanol or 2-methyl-1-propanol, and *tert*-butanol or 2-methyl-2-propanol. Only *iso*-butanol and 1-butanol appear likely to produce a reasonably ideal quaternary solution with methanol–ethanol–1-propanol because they are the only ones that form "ideal" binary solutions with all three components, as explained below.

According to Perry's 3rd and 4th eds.,[4] a solution of methanol and *iso*-butanol[78] displays ideal behavior. Binary solutions of *iso*butanol with ethanol[79] and 1-propanol[80] follow Raoult's Law curves as well (Figure C1). Activity coefficients are uniformly close to 1.0. Fitted γ^{∞}'s for ethanol and isobutanol are 1.06 and 1.01[79] and 1.05 and 0.98[80] for 1-propanol and isobutanol. Hence, adding isobutanol to the methanol–ethanol–1-propanol ternary stands a good chance of resulting in a close-to-ideal quaternary mixture. Antoine constants are listed in Table A1.

In Figure C2, 1-butanol mixtures with methanol,[81] ethanol,[82-83] or 1-propanol[84] also appear to be close to ideal. These too stand a good chance of resulting in a reasonably ideal quaternary. Their binary activity coefficients scatter near 1.0, with the greatest deviations occurring for the methanol–1-butanol system, where γ^{∞} for 1-butanol is somewhat high at 1.15. Data for ethanol–1-butanol come from two sources in agreement,[82-83] with data from three other investigations[18,85-86] (not shown) scattering around the ideal x-y curve.

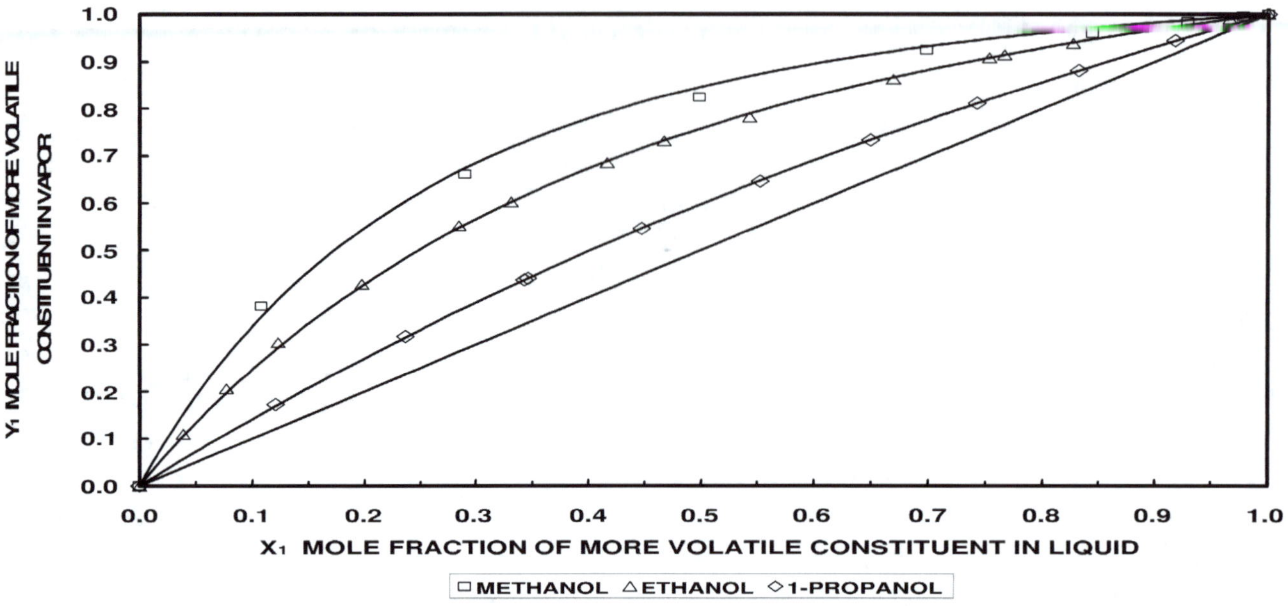

Figure C1. VLE x-y Diagrams for Isobutanol with Methanol, Ethanol, and 1-Propanol at 1 Atm Data from References **[78-80]**

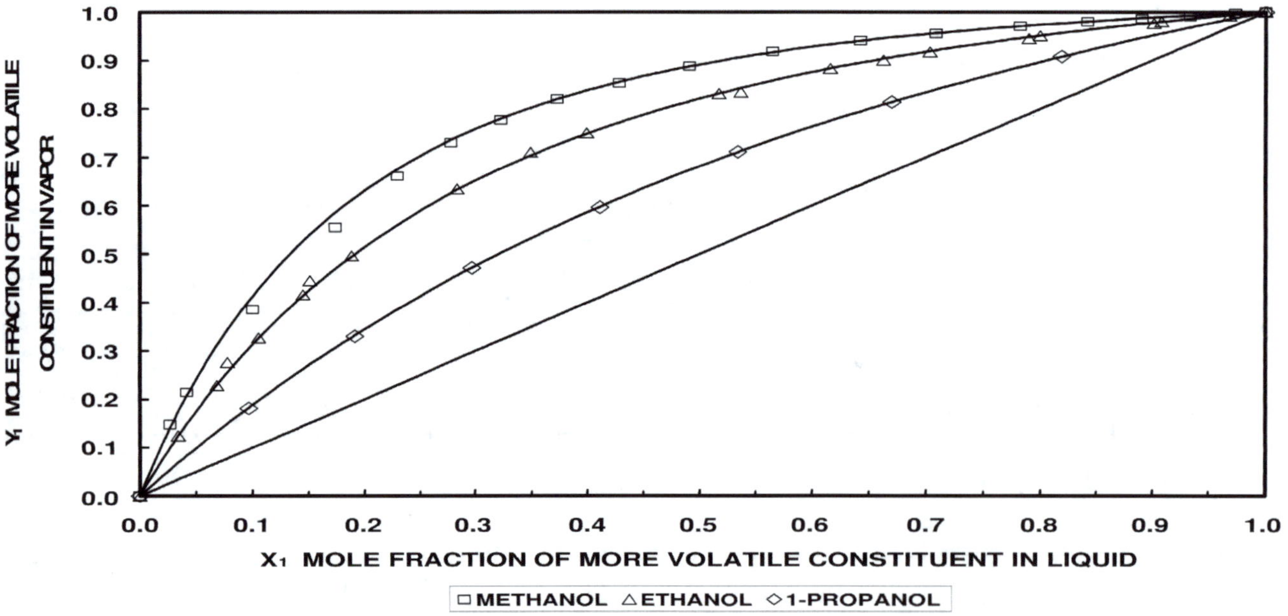

Figure C2. VLE x-y Diagrams for 1-Butanol with Methanol, Ethanol, and 1-Propanol at 1 Atm Data from References **[81-84]**

The other butanol isomers are less likely to produce ideal quaternary solutions since the γ^{∞}'s for the methanol binaries with 2-butanol[87] and *tert*-butanol[88-91] range from 0.75 to 0.85. However, no quaternary data containing any of the butanol isomers with methanol–ethanol–1-propanol could be found for verification.

400182 Distillation Tray Fundamentals

Abstract:

Distillation column internals, such as trays or packings, are used as phase contacting devices. The most common types of distillation trays are sieve trays, valve trays, bubble cap trays, and dual flow trays. Of these, sieve and valve trays are the most common.

An understanding of the fundamentals of trays is the key to successful design, operation, and troubleshooting of trayed distillation columns. This paper provides an overview of distillation tray design and operations for the practicing engineer. A brief history of trays and trayed columns will be presented. New generations of high performance trays, including augmented crossflow, counter current flow and cocurrent flow trays, will be discussed. The performance differences between trays and packings will be summarized. Column designs that provide high energy-efficiencies will be briefly discussed.

Extensive video footage of tray operations will be shown, from commercial size columns with various distillation systems ranging from deep vacuum to high pressure

Copyright © American Institute of Chemical Engineers. All rights reserved.

400183 Distillation Packing Fundamentals

Abstract:

Distillation packings are widely used in industrial distillation columns, particularly for low-pressure-drop applications. Distillation packings are generally classified as random packings, structured packings, and grids.

A brief history of packings, and basic performance characteristics, such as HETP, pressure drop, and capacity, and liquid distribution quality are reviewed. Guidelines on the performance differences between random and structured packings will be discussed.

Packed columns are equipped with liquid distributors to ensure uniform distribution of the liquid over the packed bed. Good liquid distributors are very important for effective packed column performance. An understanding of the fundamentals of packings and liquid distributors is the key to successful design, operation, and troubleshooting of packed distillation columns.

In this paper, packing physical characteristics, installation, and applications are discussed. Video footage prepared by FRI will be shown to illustrate the basics of packings and distributors.

Copyright © American Institute of Chemical Engineers. All rights reserved.

401746 Industrial Perspective on Licensing and Credentialing

Abstract:

In the USA large and small industrial organizations provide many engineering services, and they design, build and commission facilities in North America and globally. Standards and practices for continuing professional competency (CPC) vary for all aspects of the profession including engineering skills, ethics and social responsibility. ABET has increasing taken a role of adding these types of requirements to undergraduate chemical engineering programs. The recent policy statement by AIChE and ASME on the Industrial Exemption could be considered a compromise among industry and society and the engineers. This presentation will give one view of a licensed PE who has worked in organizations as small as two engineers to the largest ones in the world doing research to ECP to plant operations. Often, the hurdles to an engineer's obtaining a license are the exams to show engineering skills competency.

Copyright © American Institute of Chemical Engineers. All rights reserved.

401691 Texas Professional Engineers Practice Update and Ethics

Abstract:

Updates on Board rule changes within the last year, general updates on initiatives the Board is working on, information on enforcement statistics, and the engineering Code of Conduct.

Copyright © American Institute of Chemical Engineers. All rights reserved.

402127 Instrumentation, Automation and Control Applications in a Pilot Scale Supercritical Biodiesel Process

Abstract:

The supercritical biodiesel process uses temperature and pressure to drive both the alcohol and the feedstock beyond their critical points so that the reaction occurs in one phase. The critical point for methanol is 1415 psia and 480 °F. These reaction conditions allow the methanol to lose its polar nature and solvates both the triglyceride and free fatty acid. This method is able to use very high free fatty acid (FFA) feedstocks as there is no sodium hydroxide saponification reaction that competes with the transesterification with this method. High FFA feedstocks are commonly found available as waste streams that have no other economic purpose. The use of ethanol instead of methanol at supercritical conditions presents what some may term a more complete "green" solution since ethanol can be fermented from renewable feedstocks whereas most methanol is currently produced from the hydrolysis of natural gas. The greatest advantage with supercritical process is that the lowest value feedstocks may be used and the conversion has very few detrimental byproducts. As with all transesterification, water is a concern and must be controlled carefully.

Copyright © American Institute of Chemical Engineers. All rights reserved.

Size Control Valves for Lab-Scale Laminar Flow

Paul Martin, P.Eng.
Grant Girouard
Zeton Inc.

Existing equations used to size control valves are inadequate for laboratory- and pilot-scale work. Use the simple equation introduced in this article to accurately size needle control valves for laminar and transitional flow.

Sizing valves for laboratory- or pilot-scale use must account for flow in the laminar and transitional regimes. Conventional equations for calculating the flow coefficient (C_V) — the essential parameter for sizing control valves — work only for turbulent flow (Re > 10,000).

This article provides an overview of the available equation to calculate C_V and discusses its applicability to turbulent flow. It then offers a new way to size valves for flow in laminar and transitional regimes. The discussion focuses on the needle valve (Figure 1), which is the most accurate among the various types of control valves and is most often used for lab- and pilot-scale work.

Laminar and transitional flow

The flow coefficient (C_V) is the volumetric flowrate (gpm) of water that can pass through a valve with a 1-psi pressure drop across the valve at standard temperature. Valve manufacturers provide flow capacities expressed in terms of C_V for their valves of varying sizes and types.

The conventional equation used to calculate C_V is:

$$C_V = Q\left(\sqrt{\frac{SG}{P_1 - P_2}}\right) \tag{1}$$

where Q is the volumetric flowrate (gpm), SG is the specific gravity of the fluid relative to water (dimensionless), P_1 is the inlet pressure (psi), and P_2 is the outlet pressure (psi). This equation is valid only for turbulent flow.

In the small valves used in the lab or pilot plant, however, the flow of liquids is often not turbulent, despite what many may believe.

* This article is based on a paper presented at the AIChE Annual Meeting, Nov. 3–8, 2013, San Francisco, CA.

Consider the flow of a fluid through a needle control valve sized for a lab or pilot plant. The valve's trim, which consists of the seat and the plug, defines the flow path. (Figure 2 shows three types of trims commonly found in needle control valves.) The annulus formed between the seat hole (on the order of 0.060 in. dia.) and the needle or plug (on the order of perhaps 0.0595 in. dia.) is very small,

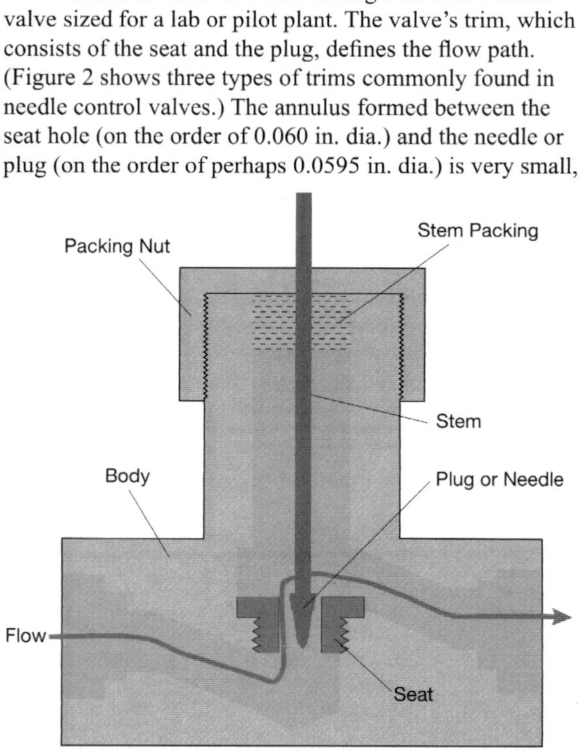

▲ **Figure 1.** A typical needle control valve consists of a body, stem and plug, seat, and packing. The body encases the valve parts and includes the fluid flow path and the inlet and outlet connections. An actuator moves the stem up and down through the packing in the body. The plug (or needle) is a machined part on the end of the stem that fits into the seat of the valve. The internal parts of the valve are collectively referred to as the valve's trim.

Copyright © 2014 American Institute of Chemical Engineers (AIChE)

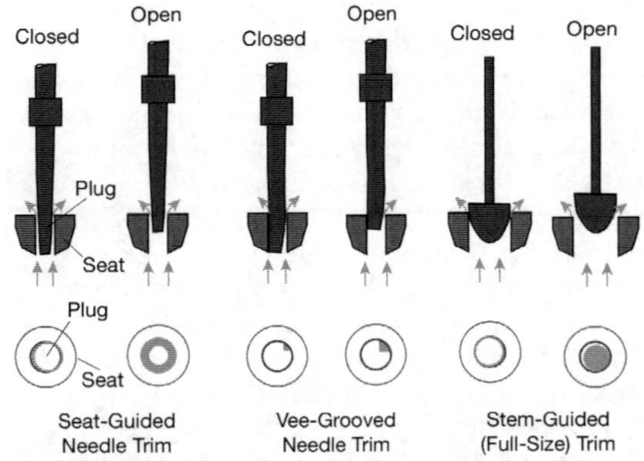

▲ **Figure 2.** Three common types of needle valve trims are seat-guided, vee-grooved, and stem-guided trims. The trim type determines the fluid flow path through the valve. In a seat-guided valve (left), the flow path is a circular annulus that varies in both length and inner diameter with the travel of the valve, and the plug is never completely removed from the seat, even when the valve is 100% open. In a vee-grooved valve (center), vee-shaped grooves are milled into the plug and the fluid flows along these grooves. In the stem-guided trim (right), which is also referred to as full-size trim, opening the valve raises the plug and creates a circular hole that the fluid passes through when the valve is 100% open.

so the velocity of the fluid moving through this valve will be very high. But high fluid velocity alone does not mean that the flow is turbulent ($Re > 10{,}000$), because the Reynolds number depends on more than just the velocity. In these small needle valves, the surface area (or wetted perimeter) of the annulus is large enough to stabilize laminar flow, even at very high velocities. Thus, Eq. 1 is not applicable to pilot-scale needle valves.

Applying Eq. 1 to a system operating at laminar or transitional flow conditions will result in valves that are significantly undersized, sometimes by many trim sizes.

Methods to correct for laminar flow

Rely on vendor expertise. Some manufacturers of lab- and pilot-scale control valves are aware of the consequences of laminar flow on sizing and have developed their own methods to correct for this phenomenon. However, we have found that vendors' sizing predictions for viscous liquids do not match measurements; frequently, the predictions and the measurements can differ by 200–500%. Therefore, we do not recommend relying solely on vendor sizing for laminar flow conditions.

Use the ISA method. The International Society of Automation (ISA) has developed a rather complex and cumbersome method to size valves in the laminar and transitional flow regimes *(1)*. This iterative method is often simplified to nomographs that can be used to determine an effective valve Reynolds number and the associated valve sizing correction factor (F_R). F_R is a "fudge factor" and is expressed as:

$$F_R = \frac{C_{VL}}{C_{VT}} \qquad (2)$$

where C_{VL} is the flow coefficient for laminar or transitional flow (gpm/psi$^{0.5}$) and C_{VT} is the flow coefficient for turbulent flow (gpm/psi$^{0.5}$). F_R approaches 1 as flow approaches fully developed turbulent conditions, and becomes significantly less than 1 as flow becomes laminar.

For laminar conditions, ISA defines F_R as:

$$F_R = aRe^b \qquad (3)$$

where a and b are constants that depend on whether the goal is to size a valve for a particular flowrate and pressure drop, to calculate the flowrate given the C_{VT} and the pressure drop, or to predict the pressure drop required to obtain a given flowrate through a valve of known C_{VT}. For transitional flow, F_R simply transitions smoothly between the two equations $F_R = aRe^b$ and $F_R = 1$, and is generally given in tabular or graphical format.

Unfortunately, determining the valve-trim Reynolds number (*Re* in the annulus between the plug and the seat) in valve trims of the size used in lab and pilot plant equipment ($C_V < 1$) is not straightforward. The ISA method does not clearly specify how to determine or select the appropriate variables (velocity, diameter, etc.) to use in the equations. And, valve manufacturers consider the geometric dimensions of their trims a proprietary part of their design and seldom publish that information.

Apply Page's equation. George W. Page described a simpler and more elegant method of sizing valves in the laminar and transitional flow regimes *(2)*. This method models a control valve in the transitional flow regime as two valves in series: one valve in purely laminar flow (representing flow through the body piping), and one in turbulent flow (representing flow through the throttling orifice). Applying the simple equations in Crane *(3)* for each of the two conditions, Page arrives at a new equation for the laminar/transitional flow sizing correction factor, F_R:

$$F_R = \frac{1}{\sqrt{\left(1+\left(\frac{\nu L}{44{,}142Q}\right)\left(\frac{C_{VT}}{D_O^2}\right)^2\right)}} \qquad (4)$$

where ν is the fluid kinematic viscosity (cSt), L is the length of pipe between the upstream and downstream pressure-measurement taps (in.), and D_O is the diameter of a circle that has the same area as the valve opening (in.):

$$D_O = \sqrt{\frac{4A}{\pi}} \qquad (5)$$

where A is the area of the annulus between the plug and the seat (in.^{2}).

Copyright © 2014 American Institute of Chemical Engineers (AIChE)

Page demonstrated good agreement between the results of Eq. 4 and experimental data. However, this equation is not easy to use because D_O is not always known. D_O is calculated from the valve geometry, and is equal to the valve size only for full-sized globe-type trims.

For a lab- or pilot-scale valve, frictional loss in the inlet and outlet tubing is not the controlling resistance to flow and seldom contributes detectably to the overall pressure drop, so L is meaningless in the laminar flow situation. Thus, Eq. 4 is not valid for nonturbulent conditions in reduced-port valves.

Use Zeton's K-factor method. Rearranging Page's equation (Eq. 4) reveals that L/D_O^2 is an adjustable parameter with little practical physical meaning. Indeed, it is essentially a parameter of fit for the Fanning/Darcy equation *(4)*.

When flow is purely laminar, Page's equation (Eq. 4) *(2)* reduces to:

$$F_R = 210\left(\frac{D_O^2}{C_{VT}\sqrt{\dfrac{Q}{\nu L}}}\right) \tag{6}$$

Upon rearrangement of this equation, it appears that a new parameter, K, can be constructed to account for the contributions of the lumped parameter L/D_O^2, the fluid density (ρ, kg/m^3), and a unit conversion factor. K can be obtained from a plot of flow versus $\Delta P/\mu$, where ΔP is the pressure difference across the valve (psi) and μ is dynamic viscosity (cP). This plot is a straight line through the origin for a valve in fully laminar flow (Figure 3). Accordingly, the value of K for any particular valve can be determined through a single experiment with a Newtonian fluid viscous enough and at a flowrate low enough to ensure fully laminar flow through the valve trim. (The flow must be fully laminar such that the flowrate varies linearly with $\Delta P/\mu$.)

Thus, for fully laminar flow:

$$Q = \frac{\Delta P}{K\mu} \tag{7}$$

Page's equation then reduces to:

$$F_R = \frac{1}{\sqrt{1+\dfrac{\mu K C_{VT}^2}{Q}}} \tag{8}$$

Separate experiments for each trim size are required to determine the value of K, which varies with the C_{VT} of the valve.

Zeton's new laminar flow sizing parameter

An equation relating K and C_{VT} could reduce the number of experiments needed to determine K for a series of trims that have similar geometry but different C_{VT} values. We hypothesized that families of trims of similar geometry would need only a single parameter to describe their laminar flow performance relative to that observed in turbulent flow, since the same underlying geometric factors define pressure drop for a particular trim type under any flow regime.

The ISA *(1, 5)* provides the relationship $F_R = aRe^b$, with the exponent b set to 0.5, 0.67, or 1, depending on the intended use. We tested the relationship for $b = 0.5$.

Consider a valve of a particular geometry with a particular fluid flowing through it at a constant valve opening (%). The Reynolds number, $Re = DV\rho/\mu$ (where D is the diameter of the opening in meters and V is the velocity in m/sec) varies directly with the volumetric flowrate. Taking the definitions of K (Eq. 7) and F_R (Eq. 3), and then grouping all of the proportionality constants together as a_v, we obtain the following relationship:

$$C_{VT} = \frac{a_v}{\sqrt{K}} \tag{9}$$

where a_v and K are both lumped parameters of best fit (without a clear physical interpretation). Substituting the laminar flow geometric parameter a_v into Page's equation (Eq. 8) gives the following simple relationship:

$$F_R = \frac{1}{\sqrt{1+\dfrac{\mu a_v^2}{Q\times SG}}} \tag{10}$$

Next, the equations that define C_V (Eq. 1) and F_R (Eq. 2) can be combined with Eq. 10 to derive a new sizing equation:

$$C_{VT\text{-}Req} = Q\sqrt{\frac{SG}{\Delta P}}\left(\sqrt{1+\frac{\mu a_v^2}{Q\times SG}}\right) \tag{11}$$

Solving for Q yields a somewhat more complex equation:

$$Q = \frac{-\mu a_v^2 + \sqrt{\left(\mu a_v^2\right)^2 + 4(SG\times C_{VT}^2\times\Delta P)}}{2(SG)} \tag{12}$$

Equation 12 reduces to Eq. 1, which is the definition of C_V for turbulent flow, when $a_v = 0$ or when the product μa_v^2 is sufficiently small. A high a_v value for a particular trim geometry indicates that the trim has a tendency to stabilize laminar flow to a greater extent than trims with a lower a_v value, perhaps due to the higher ratio of wetted perimeter to cross-sectional area in the flow path.

This model uses a single parameter to correct for the laminar flow behavior of a series of geometrically similar trims, across a wide range of C_{VT} values. For two popular trim brands, we found that a_v values of 0.135–0.14 provide an excellent fit.

This simple, one-parameter model given in Eqs. 11 and 12 may be cautiously extended to use with other seat-

Copyright © 2014 American Institute of Chemical Engineers (AIChE)

guided annular needle-type valves. In this case, the model equations should not be trusted completely and should be verified with experiments. However, it will surely give better results than either the turbulent sizing equation, or the ISA sizing equations without values for F_d and D validated by real testing on viscous liquids.

Equations 11 and 12 with a conservative value of a_v (*e.g.*, 0.18) can be used for the approximate sizing of annular needle-type trims in the absence of other data. Equation 11 with $a_v = 0.18$ can also be used as a test of whether laminar-flow correction is necessary. If the calculated F_R is significantly less than 1, it would be prudent to either perform validation experiments with your particular trim and fluid, or to have your valve supplier do so for you.

Model verification for seat-guided needle trims

The model (Eqs. 11 and 12) was verified for a pneumatically actuated needle control valve (Brand A). The valve's trim, which is a type widely used in lab- and pilot-scale work, has a proprietary design, the dimensions of which are not published. In general, however, the trim involves a small-bore circular seat with a tapered needle plug that is never completely removed from the seat even when the valve is fully open (*i.e.*, the plug is seat guided).

The testing apparatus consisted of a reservoir of liquid that could be pressurized with air or nitrogen, a filter to remove fine particulate matter, an accurate differential-pressure transmitter with tubing tees mounted immediately upstream and downstream of the valve body to connect the pressure-measurement impulse lines, a bucket or vial, and an accurate scale. Data on weight and time were collected and used with the liquid density to calculate volumetric flowrate.

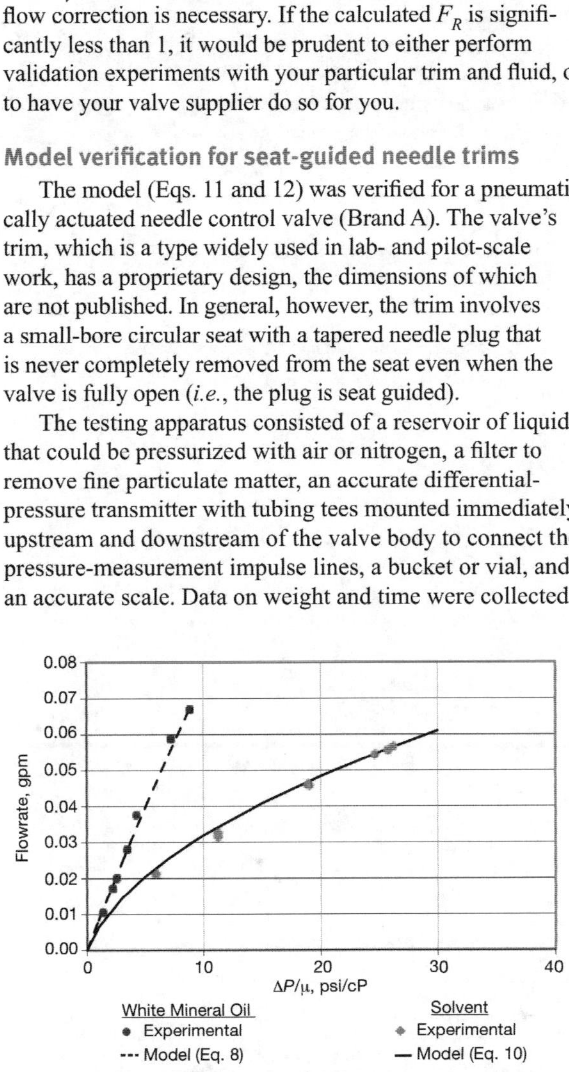

▲ **Figure 3.** The volumetric flowrate as a function of $\Delta P/\mu$ is plotted for two liquids. The model of Eq. 8 fits well for Liquid 2 (mineral oil) in purely laminar flow (F_R under these conditions ranges between 0.25 and 0.6), while the more-general model of Eq. 10 does a good job of predicting the experimental data for transitional to turbulent flow for the much-less-viscous Liquid 1 (solvent) (F_R ranges from 0.93 to about 1).

Data were collected over a range of applied pressures (5–500 psig) for several trim sizes. Brand A's manufacturer sells valves with trim sizes designated by letters ranging from A (largest) to P and micro-flow trims designated as P1 (largest of the micro trims) to P14. The verification tests evaluated trim sizes M, P1, P3, and P5.

Gas flow measurements were taken to verify the actual C_{VT} of the valves. A rotameter and thermal mass flowmeter were used to measure the flowrate of bottled nitrogen, with measurements made in choked flow (*i.e.*, $P_1 > 2P_2$).

Figure 3 presents the results of the first set of tests, which used two fluids — Liquid 1 (a kerosene-like solvent with $\mu = 0.8$ cP and $\rho = 788$ kg/m^3 at 20°C) and Liquid 2 (a paraffin white mineral oil with $\mu \sim 29$ cP and $\rho = 846$ kg/m^3 at 20°C) — flowing through a 0.25-in. Brand A control valve with an M trim size. The vendor specifies this valve's C_{VT} to be 0.01 nominal; the measured C_{VT} based on nitrogen flow measurements is 0.012.

As shown in the Q vs. $\Delta P/\mu$ plot in Figure 3, the model of Eq. 8 is in excellent agreement with the experimental data for Liquid 2 in purely laminar flow in this trim, with F_R under these conditions ranging between 0.25 and 0.6. The more-general model of Eq. 10 does a good job of predicting the experimental data for the much-less-viscous Liquid 1 (F_R ranging from 0.93 to about 1) in transitional to turbulent flow.

The next set of tests used more-viscous fluids, Liquid 3 ($\mu \sim 90$ cP) and Liquid 4 ($\mu \sim 296$ cP), and Brand A valves with the smaller trims of P1 ($C_{VT} \sim 0.002$), P3 ($C_{VT} \sim 0.001$), and P5 ($C_{VT} \sim 0.0004$). The unique K values obtained for each trim size were in excellent agreement with Eqs. 8 and 10.

Figure 4 is a plot of $K^{-0.5}$ vs. measured C_{VT} for all of the valve trims of Brand A that were tested, including the M trim at varying stem positions. The slope of this plot is the laminar flow sizing coefficient, a_v, which for this brand and

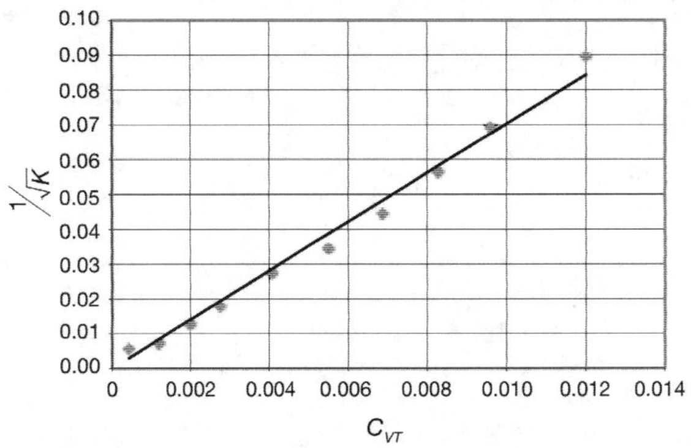

▲ **Figure 4.** $K^{-0.5}$ vs. measured C_{VT} for all of the Brand A valve trims. The slope of this plot is the laminar-flow sizing coefficient, a_v.

Copyright © 2014 American Institute of Chemical Engineers (AIChE)

these trims had a value of about 0.135. The graph demonstrates the excellent fit of the new model (Eqs. 11 and 12) to the entire set of measured data for the entire family of trims.

Subsequent testing used a different brand of control valves with similar seat-guided needle-type trims in a similar trim size range (C_{VT} = 0.01–0.0001). In these tests, a_v had similar values, in the range of 0.12–0.14. The value of a_v did not change much with trim travel, similar to what was observed for Brand A.

Low-flow control valves manufactured by other vendors with similar seat-guided needle-type valve trims were found to have a_v values of similar magnitude. An experiment on a single valve from a third manufacturer also demonstrated that the a_v value was constant as the valve's open area changed.

Vee-grooved needle trims

The new equations were also demonstrated for a vee-grooved needle valve. This valve has a trim consisting of a cylindrical plug with one or more triangular grooves of varying depth machined into the plug. The triangular groove trim design has a smaller ratio of wetted perimeter to cross-sectional area than a trim with a purely annular cross section, which translates into a higher Reynolds number at a given flowrate and viscosity than a circular annular trim of the same C_{VT}. Thus, this type of valve would be expected to have a smaller a_v value.

Using the same setup as for the seat-guided valves, a_v values were measured for the vee-grooved needle valve in the range of 0.05–0.065. The a_v values decreased as the valve's open area decreased, falling to 0.02 in some of the trims. Even at a_v values in the range of 0.02–0.06, the F_R values for these valves were significantly less than one with liquids of even modest viscosity. Assuming that flow in these valves is turbulent, therefore, will lead to the selection of valves far too small for the duty.

Stem-guided trims

Based on our observations, we would not expect a_v to remain constant with varying stem position for valves with full-size trims. These valves are fully stem guided rather than seat guided, and the trim becomes an unrestricted circular orifice when the valve is fully open. Such valves include some of the fine-metering manual needle valves offered by several tube-fitting manufacturers. Individual sizing experiments using the K factor previously discussed would be required for accurate sizing of this type of valve in laminar flow. CEP

Nomenclature

A = area of the annulus between the plug and the seat of the valve, in.^{2}
a_v = trim family valve-sizing factor for laminar flow, (gpm/cP)$^{0.5}$
C_{VL} = flow coefficient for laminar and transitional flow, gpm/psi$^{0.5}$
C_{VT} = flow coefficient for turbulent flow, gpm/psi$^{0.5}$
D_O = diameter of a circle with the same area as the valve opening, in.
F_R = laminar flow adjustment ratio (dimensionless)
K = single trim valve-sizing factor for laminar flow, psi/gpm-cP
L = length of pipe between upstream and downstream pressure measurement taps, in.
P_1 = valve inlet pressure, psi
P_2 = valve outlet pressure, psi
ΔP = pressure difference across valve, psi
Q = volumetric flowrate, gal/min
Re = Reynolds number (dimensionless)

Greek Letters

ρ = density, kg/m^3
μ = dynamic viscosity, cP
υ = fluid kinematic viscosity, cSt

Acknowledgments

The authors wish to thank Kevin Nold, Min Kwan Kim, Johnathan Caguilat, Kenneth Edwards, and Sun Young Gong for their painstaking work in collecting the data on which this article is based.

Literature Cited

1. **International Society of Automation,** "Flow Equations for Sizing Control Valves," ISA/ANSI 75.01, Research Triangle Park, NC (1975).
2. **Page, G.,** "Simplified Valve Sizing for Laminar Flows," *Chem. Eng.*, **105** (11), pp. 195–198 (Oct. 1998).
3. **Crane Co.,** "Flow of Fluids Through Valves, Fittings, and Pipes," Technical Paper No. 410, Crane, New York, NY (1982).
4. **Sakiadis, B. C.,** "Fluid and Particle Mechanics," in Green, D. W., and R. H. Perry, eds., "Perry's Chemical Engineers' Handbook," 6th ed., Section 5, p. 5-24, McGraw-Hill, New York, NY (1984).
5. **Baumann, H. D.,** "A Unifying Method for Sizing Throttling Valves under Laminar or Transitional Flow Conditions," *Journal of Fluids Engineering*, **115** (1), pp. 166–169 (Mar. 1993).

PAUL MARTIN, P.Eng., is a senior project manager and engineer at Zeton Inc. in Burlington, Ontario, Canada (Phone: (905) 632-3123; Email: pmartin@zeton.com). He has 17 years of experience designing pilot- and demonstration-scale plants for the chemical process industries. He received his BASc and MASc from the Univ. of Waterloo, Ontario, both in chemical engineering. He is a member of the AIChE Process Development Div. and a licensed P.Eng. in Ontario.

GRANT GIROUARD is a project engineer at Zeton Inc. in Burlington, Ontario, Canada (Phone: (905) 632-3123; Email: ggirouard@zeton.com). He received a BASc from the Univ. of Waterloo, Ontario, in chemical engineering.

Copyright © 2014 American Institute of Chemical Engineers (AIChE)

397527 Using Dimensional Analysis to Scale Fixed-Bed Adiabatic Reactors

Abstract:

The most common reactor in catalyzed chemical processes is the fixed-bed reactor. The chemical processing industry (CPI) prefers this reactor for catalyzed processes because it has a relatively light capital requirement, catalyst is readily located --- most of the time, and it is inexpensive to operate. The CPI generally operates fixed-bed reactors adiabatically.

Fixed-bed reactors are deceptively simple at first blush. However, they are not easy to scale, either up or down. Strictly speaking, fixed-bed reactors do not meet the requirements for geometric similarity; thus, they invalidate the model theory used for their design. Also, most fixed-bed reactor scaleups or scaledowns do not meet the requirement of dynamic similarity, which again invalidates the model theory used for their design.

This presentation discusses the Dimensional Analysis of fixed-bed reactors. Since Dimensional Analysis provides the foundation for Model Theory, we will be discussing the foundation for adiabatic fixed-bed reactor design.

Copyright © American Institute of Chemical Engineers. All rights reserved.

403088 Energy-Water-Food: Maui and the World

Abstract:

This session is part of the effort to raise awareness of the Nexus, to present results of specific case studies which serve to illustrate the issues, to present new tools and methodologies and to identify research needs and opportunities. We welcome your contributing to our developing an effective response to this important challenge.

Copyright © American Institute of Chemical Engineers. All rights reserved.

401333 A Process Methodology for Assessing Sustainability

Abstract:

Assessing the sustainability impact of any system - factory, business, institution, supply chain, industry, city, province, country - is essential to setting a policy for sustainability. This assessment can be made by using a set of indicators, but selecting indicators is not straightforward. The Process Analysis Method considers the system as a group of processes causing triple-bottom line impacts. The indicators are then related to these impacts. The Method provides a set of sustainability indicators and metrics tailored to the particular system, but which will be similar for similar systems, facilitating comparison and benchmarking. The method has been applied in a variety of Case Studies: manufacturing operation, transport system, river basin management, community water purification, which demonstarte its flexibility and usefulness.

Copyright © American Institute of Chemical Engineers. All rights reserved.

402196 Research Needs and Opportunities to Address the Nexus

Abstract:

Open discussion on issues around water-food-energy Nexus, including providing supply of energy for developing world, protection of water resources, providing food for expanding population, addressing climate change and sustainability issues.

Copyright © American Institute of Chemical Engineers. All rights reserved.

A Tutorial on Model Based Loop Tuning

James Beall
Principal Process Control Consultant

15 AIChE
Spring Meeting, Austin, TX

EMERSON.
Process Management

James Beall

- Principal Process Control Consultant, Emerson Process Management
- 33 Years experience in process control
  - 14 Years Emerson Process Management
  - 19 Years at Eastman Chemical Company
- Chairman ISA 75.25 Committee - Control Valve Performance Testing
- James.Beall@emerson.com 903-235-7935

15 AIChE
Spring Meeting, Austin, TX

EMERSON.
Process Management

"PID = Proportional, Integral, Derivative"

- **P = proportional** - output contribution is "proportional" to difference between Set Point and PV (**error**). Therefore, output contribution is "gain" times error.
- **I = Integral** - output contribution is "integration" of the **error**. Therefore, always tries to make PV = Set Point (eliminates offset).
- **D = Derivative** - output contribution is based on the derivative (rate of change) of the **error** or more often the **PV**

15 AIChE Spring Meeting, Austin, TX

EMERSON. Process Management

Caution!!!

- Controllers have different units for their tuning constants. For this presentation:
  - Proportional action will be expressed in **gain** (%out/%PV)
  - Integral action will be expressed as integral time, **Tr,** in **seconds/repeat**
  - Derivative in units of **seconds**
- Controllers have different **Forms** (Parallel, Series, Standard) of the PID algorithm. This can significantly affect the actual tuning values to achieve the desired response. The tuning rules in this presentation are for Standard and Series PID Form. Call for conversion to Parallel Form

15 AIChE Spring Meeting, Austin, TX

EMERSON. Process Management

Tuning Methods

- First "calculation" tuning method by Ziegler & Nichols (1942)
  - Ultimate Gain/Period
  - Reaction Curve
- "Little black books"
- Many people still do not use any method preferring to "tune-by-feel"
- Default tuning (gain=1.0, Reset=1 min)
- Net result is each loop tuned independently
  - process dynamics not coordinated

15 AIChE
Spring Meeting, Austin, TX

EMERSON.
Process Management

Tuning Procedure

1. **Determine basic process type**
   - Self-regulating
   - Integrating
2. **Determine process dynamics for the "approximate" model**
3. **Choose desired closed loop response time, "Lambda"**
4. **Calculate tuning constants**

15 AIChE
Spring Meeting, Austin, TX

EMERSON.
Process Management

1. Identify Process: Self Regulating Process

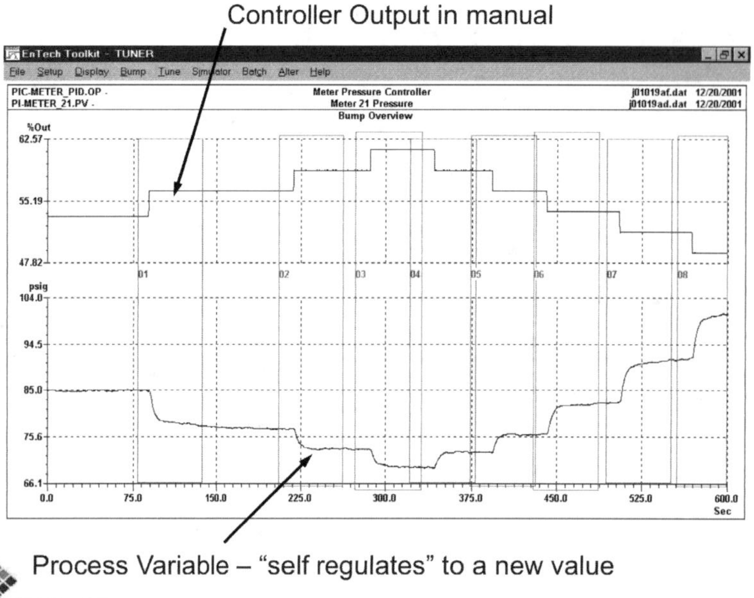

15 AIChE Spring Meeting, Austin, TX

EMERSON. Process Management

2. Determine Process Dynamics with a Manual Output Step Test (example 1st Order SR)

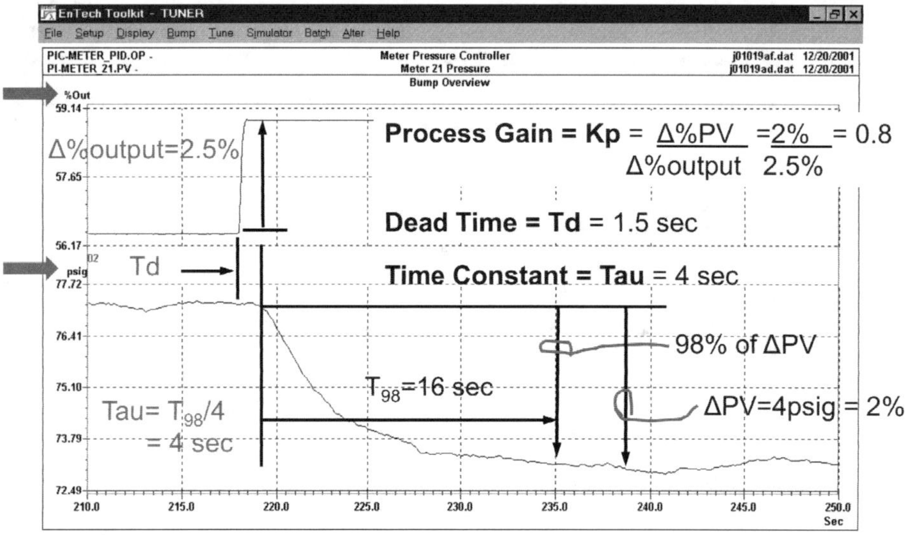

15 AIChE Spring Meeting, Austin, TX

Must use %OUT and %PV
Out_Scale=0-100%; PV_Scale=0-200 psig

EMERSON. Process Management

3. Choose Lambda – Self Regulating

- Lambda is the closed loop, aka Auto, response time to a Setpoint change (T_{98}=4*Lambda)
- Can be used to coordinate the speed of loops (e.g. cascade loops)
- Can be used to stop interaction between loops
- For self regulating processes, a STARTING point for Lambda is 3*(Larger of Dead Time or Time Constant)
- This allows Kp AND Dead Time to double and tuning will remain stable
- Lambda can be smaller (more aggressive) but loop may go unstable as conditions change

15 AIChE Spring Meeting, Austin, TX

EMERSON. Process Management

4. Calculate Tuning – Self Regulating Example – 1st Order SR

- Recall Kp=0.8, Dead Time = 1.5 sec.; Tau= 4 sec.
- Choose Lambda (λ), the "closed loop time constant"
  - Starting point = 3*(larger of Dead Time or Time Constant)
  - Lambda = 3*(4 sec.) = 12 sec.
- Tr = Reset Time = Tau = 4 sec.
- Kc = Controller Gain = $\dfrac{Tr}{Kp\ (\lambda + Td)} = \dfrac{4\ \text{sec.}}{0.8(12+1.5\ \text{sec})} = 0.37$

- NOTE – Tr stays the same, only Kc changes with Lambda

15 AIChE Spring Meeting, Austin, TX

EMERSON. Process Management

Lambda Tuning Example – 1st Order SR

- Note impact of different Lambda
- Integral Time Tr = Tau = 4 seconds
- Kc = Tr / Kp (λ + Td)
  - Kp = 0.8
  - Td= 1.5 sec.

Lambda	Gain	Integral Time
12 sec	0.37	4 sec.
8 sec	0.53	4 sec.
4 sec	0.91	4 sec.
1.5 sec	1.67	4 sec.

Faster

Higher Gain

15 AIChE Spring Meeting, Austin, TX

EMERSON. Process Management

Response – Lambda = 12 sec , 8 sec

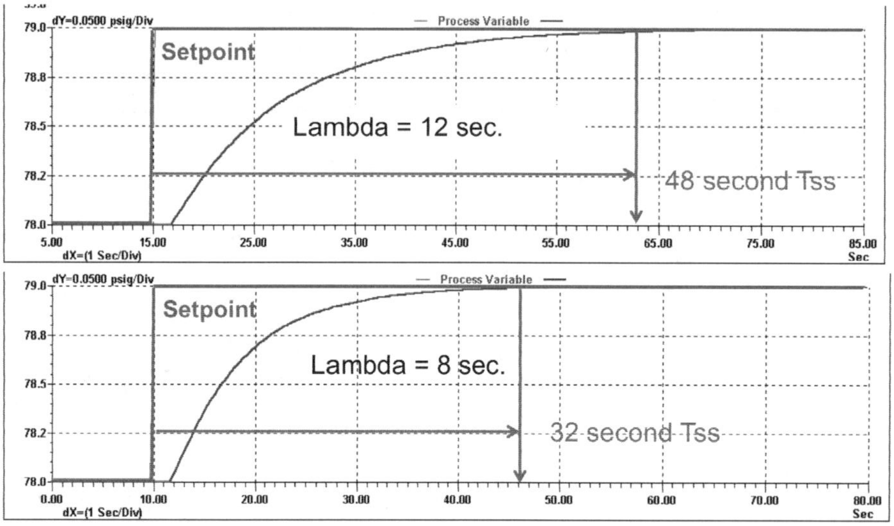

15 AIChE Spring Meeting, Austin, TX

EMERSON. Process Management

Response – Lambda = 4 sec , 1.5 sec

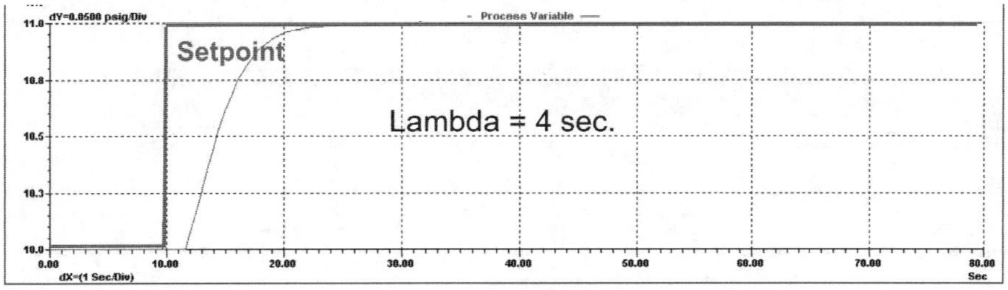

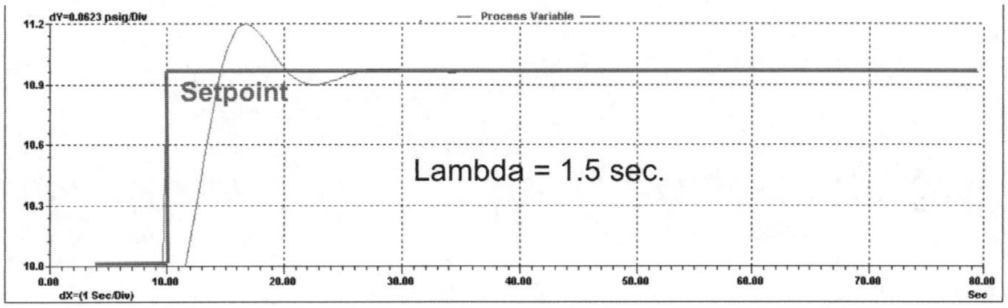

15 AIChE
Spring Meeting, Austin, TX

EMERSON.
Process Management

Lambda Tuning - Is it slow? Compared to what?

Typical "Tune by Feel"

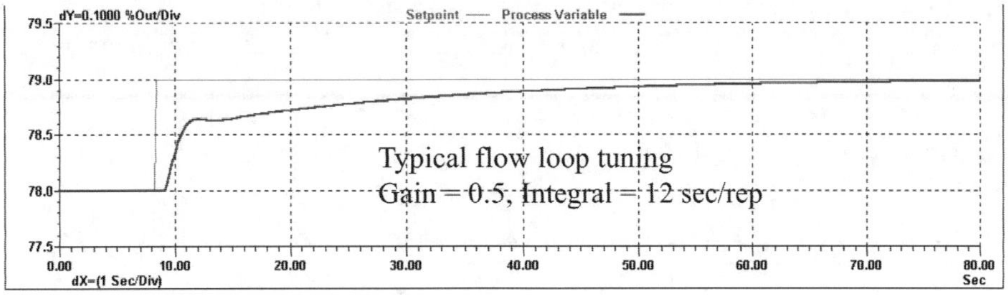

Lambda = 6.6 seconds (normal robustness)

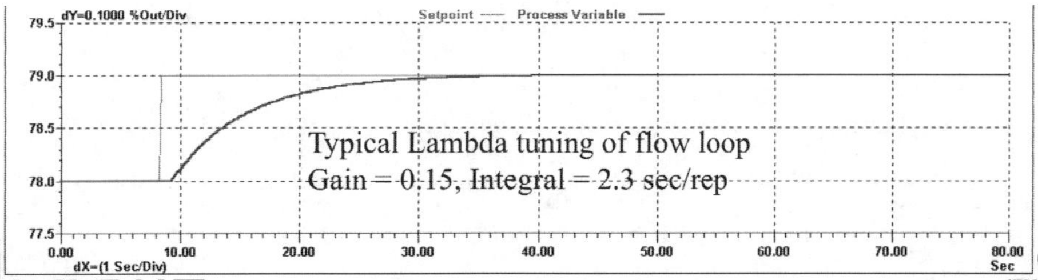

15
Spring Meeting, Austin, TX

EMERSON.
Process Management

1. Identify Process: Integrating Process

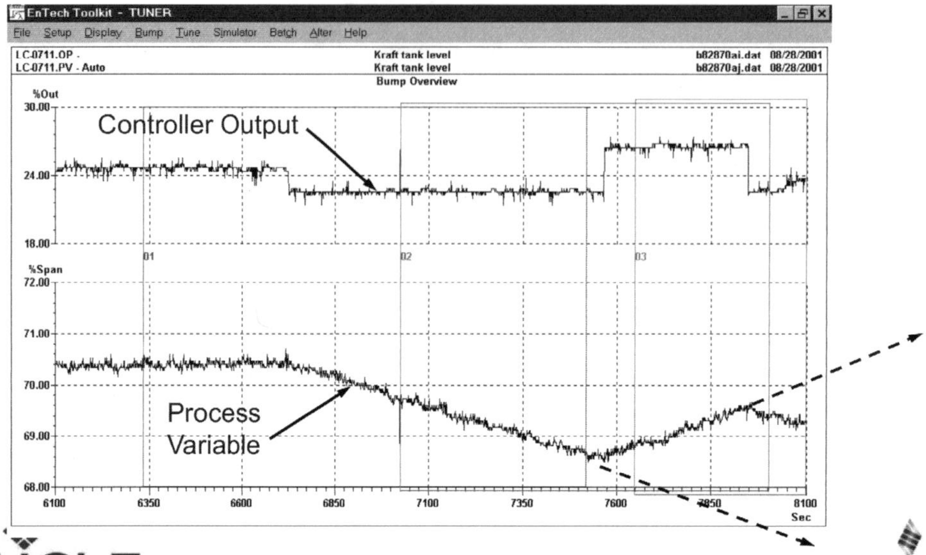

15 AIChE Spring Meeting, Austin, TX

EMERSON. Process Management

2. Determine Process Dynamics with a Manual Output Step Test

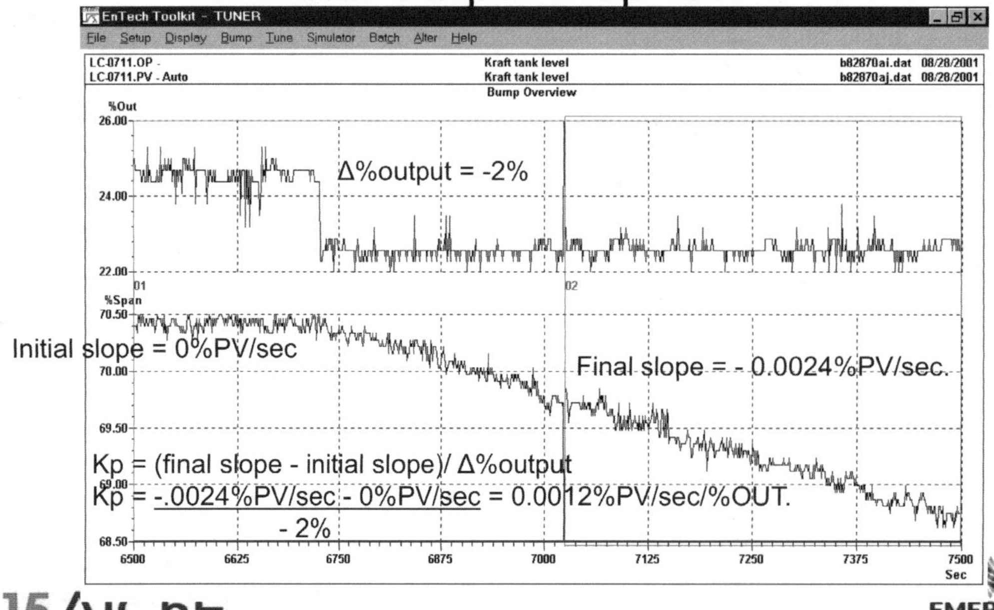

15 AIChE Spring Meeting, Austin, TX

EMERSON. Process Management

3. Choose Lambda (Integrating)

- Recall for SR, T_{98} for SP Change = 4* Lambda
- But for Integrating Processes, SP change:

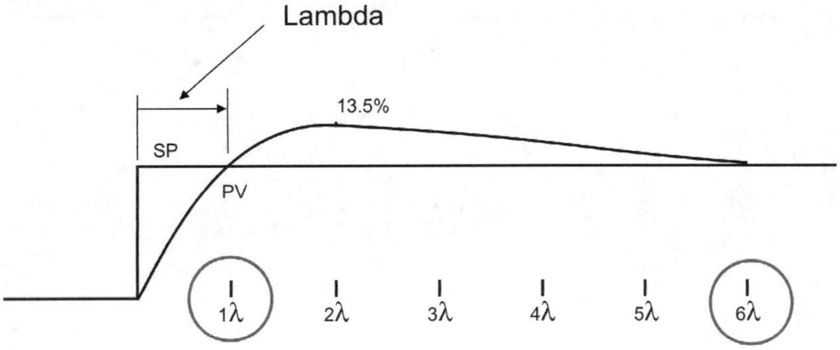

15 AIChE Spring Meeting, Austin, TX

EMERSON. Process Management

3. Choose Lambda (Integrating) - Load Disturbance Response

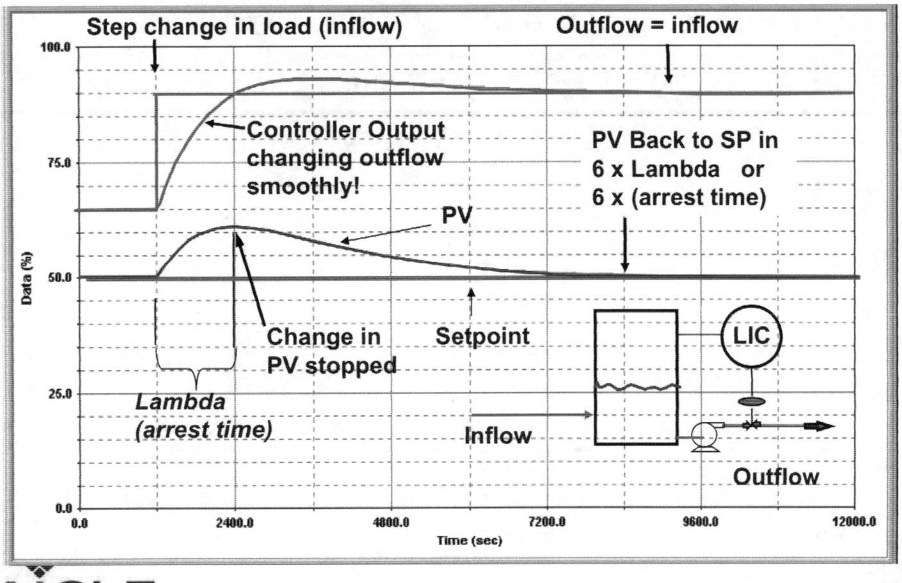

15 AIChE Spring Meeting, Austin, TX

EMERSON. Process Management

4. Calculate Tuning – Integrating Process

- THIS RULE DOES NOT INCLUDE DEAD TIME OR A LAG. Contact the author for applications with significant dead time or lag.
- Choose Lambda (λ) (more on this later!)
- $Tr = (2^* \lambda)$
- $Kc = \dfrac{Tr}{Kp(\lambda)^2}$

- NOTE – Gain AND Reset Time change with Lambda

15 AIChE
Spring Meeting, Austin, TX

EMERSON.
Process Management

4. Calculate Tuning – Integrating Process

- Example – Choose Lambda = 10 min. = 600 sec.
- $Tr = (2^* \lambda) = (2^*600\ sec) = 1200\ sec/rep$
- $Kc = \dfrac{Tr}{Kp(\lambda)^2} = 1200\ sec/((.0012/sec)^*(600sec)^{2)}$
- $Kc = 2.78$

Lambda	Gain	Integral Time
600 sec	2.78	1200 sec/rep
300 sec	5.56	600 sec/rep

Faster | Higher Gain | Faster (shorter) Reset Time

15 AIChE
Spring Meeting, Austin, TX

EMERSON.
Process Management

Lambda = 10 min. for Integrating Process

Kc = 2.78, Tr = 1200 sec./rep

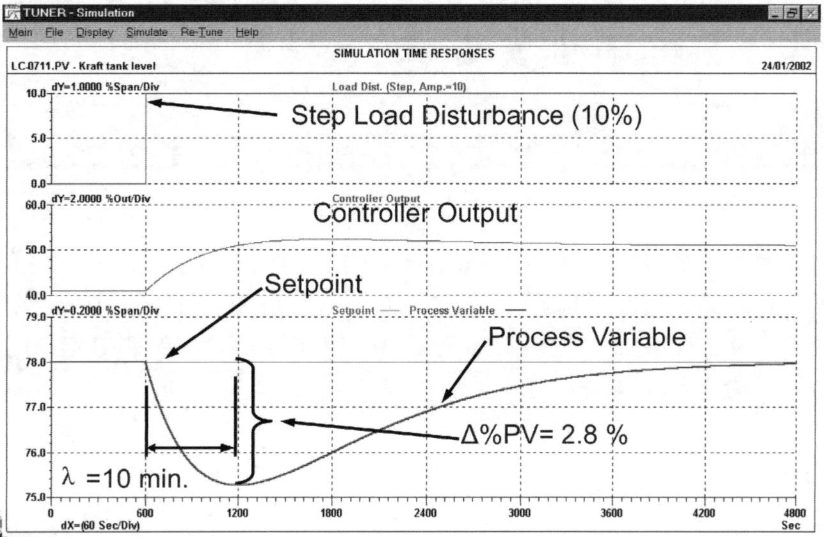

15 AIChE Spring Meeting, Austin, TX

EMERSON. Process Management

Lambda = 5 min. for Integrating Process

Kc = 5.56, Tr = 600 sec./rep

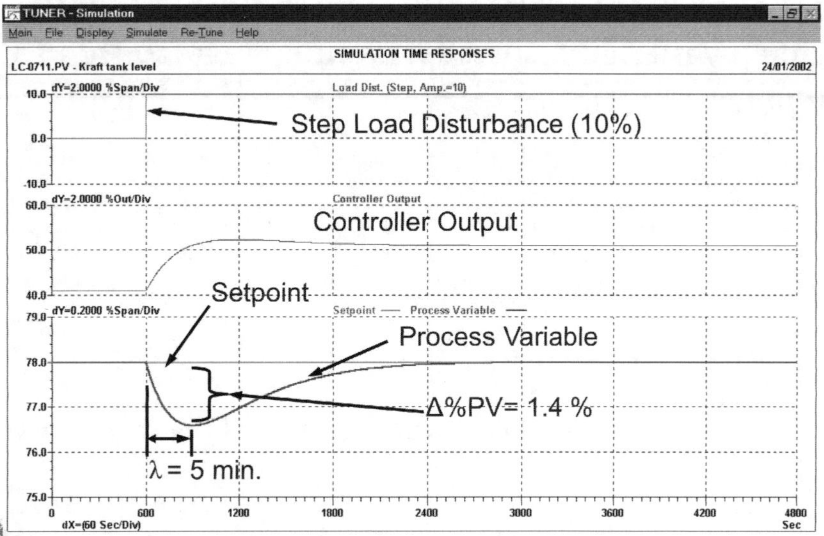

15 AIChE Spring Meeting, Austin, TX

EMERSON. Process Management

Attenuate Variability with Control/Equipment

- "Capacity" in the process can be used to attenuate or absorb variability
- Primary source of process capacity is level control
- To utilize level control as a capacity tune the controller only "fast" enough to hold the PV within the allowable level deviation (ALD) for a maximum load change

15 AIChE Spring Meeting, Austin, TX

EMERSON. Process Management

Utilize Level Control as Variability Sinks

- Choose the arrest time (Lambda) "slow" enough to provide a variability sink yet maintain level within the allowable variation
- Lambda $\alpha \quad \dfrac{\textbf{ALV}}{\textbf{Kp * MLD}}$
  - ALV = Allowable Level Variation
  - Kp = Integrating process gain
  - MLD = Maximum Load Disturbance (in % of controller output)

15 AIChE Spring Meeting, Austin, TX

EMERSON. Process Management

Example of Lambda Calculation

- Previous Example Kp= 0.0012/sec
- Let ALV = +- 15%
- Let MLD = 25%
- Lambda = 1333 seconds = 22.2 minutes
- **Tr = 2666 seconds; Kc =1.25**

15 AIChE
Spring Meeting, Austin, TX

EMERSON. Process Management

Example of Lambda Calculation

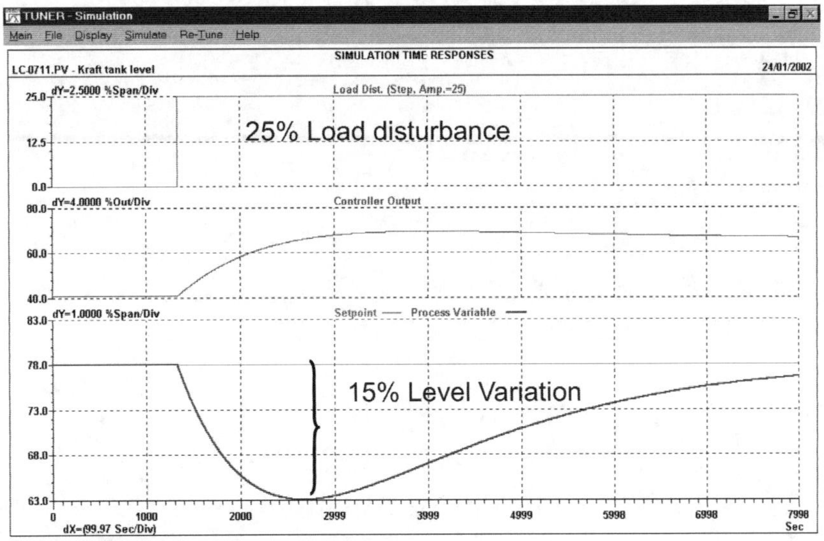

Process: Integrator
Kp=0.00120 %Span/Sec/%Out, Td=10.0 Sec
Controller: Generic Standard PID Ctrlr, Standard PID, Ctrl Int.=1 Sec
KC=1.236 Gain, TR=2676 Sec/Rep

15 AIChE
Spring Meeting, Austin, TX

EMERSON. Process Management

Results – Integrating Process Tuning

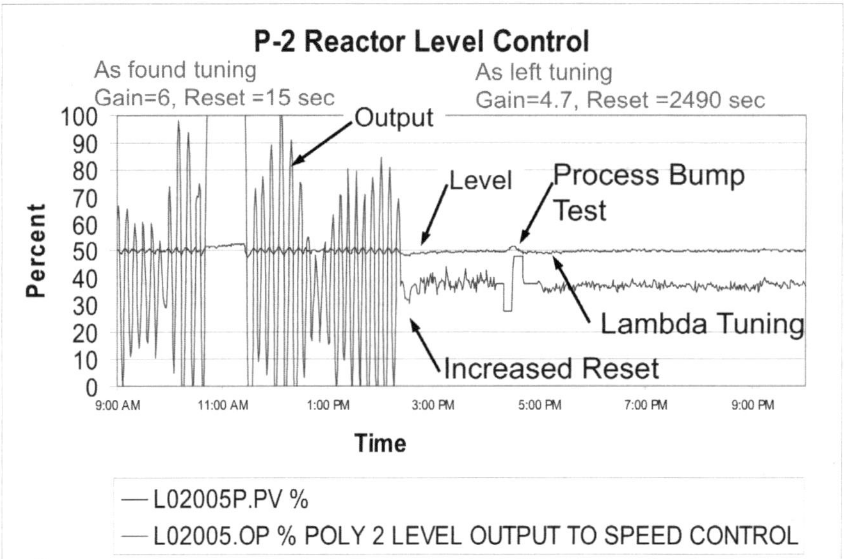

15 AIChE Spring Meeting, Austin, TX

EMERSON. Process Management

Process Dynamics – *Model Based Tuning May Include Derivative

- Pure Gain
- 1st Order
- *2nd Order, Over Damped
- *2nd Order, Under Damped
- 2nd Order, Lead
- 2nd Order, Lead with Overshoot
- 2nd Order, Non-Minimum Phase
- Integrator
- *Integrator, 1st Order Lag
- Integrator, 1st Order Lead
- Integrator, Non-Minimum Phase

15 AIChE Spring Meeting, Austin, TX

EMERSON. Process Management

Summary Tuning Procedure

1. **Determine basic process type**
   - Self-regulating
   - Integrating
2. **Determine process dynamics**
3. **Choose desired closed loop response time, "Lambda"**
4. **Calculate tuning constants**

- NOTE – For better load regulation, self-regulation loops with long lags can be tuned as in integrator

15 AIChE Spring Meeting, Austin, TX

EMERSON. Process Management

Questions?

James.Beall@Emerson.com

903-235-7935

15 AIChE Spring Meeting, Austin, TX

EMERSON. Process Management

400333 Risk Analysis of Wired Versus Wireless Transmission of Process Measurements

Abstract:

There is no doubt that wireless technology lowers the cost barriers to remote assessment of equipment condition. As such, wireless technology is relied upon for more and more functionality in the process industries. We are seeing interest in the application of wireless devices in functions typically devoted to wired applications. This session will explore some fundamental questions for consideration when evaluating transmission methods of measurements. The tools for risk analysis of safety systems will be utilized to analyze potential faults versus the benefits for wired and wireless networks of process measurements.

Copyright © American Institute of Chemical Engineers. All rights reserved.

400394 Common Plant Asset Failures

Abstract:

Review some of the common asset types in a facility (pumps, blowers, heat exchangers) and common failure modes for those assets.

Copyright © American Institute of Chemical Engineers. All rights reserved.

401286 Economic Optimization and Hazard Analysis of a Mono-Chlorobenzene Process

Abstract:

The process of chlorinating benzene to produce mono-chlorobenzene was modeled using five series/parallel reactions to include the dichloro and trichlorobenzene byproducts. A target of 10,000 lb/hr of 99.5% pure mono-chlorobenzene was specified. Up to 1000 lb/hr of dichlorobenzene (ortho or para) could be sold. Hydrochloric acid maybe sold as a byproduct at a concentration of 35 wt% depending upon impurities. All trichlorobenzene had to be disposed of as hazardous waste. The system was modeled using the process simulator ChemCAD and priced using cost equations from Turton, Bailie, Whiting and Shaeiwitz (2012). Both topological and parametric optimizations were considered. The recycle streams increased the complexity of the simulator significantly, but were required to accurately model the concentration build-up in the system. The impact of reactor residence time, separator recycle purity, and feed ratio on the economics will be presented. Process hazards will be identified and suggestions to mitigate hazards will be presented.

Copyright © American Institute of Chemical Engineers. All rights reserved.

398764 Design of Propane Dehydrogenation Plant

Abstract:

After ethylene, propylene is the most important raw material in the chemical processing industry. The recent shift in ethylene production from cracking to dehydrogenation has brought with it a decline in production of propylene, which is generally a side product of cracking. However, due to the abundance of low cost natural gas from shale gas, feedstock such as propane is now readily available for dehydrogenation into propylene. Our goal is to generate a preliminary design for a propane dehydrogenation (PDH) plant that is capable of producing 500kta of propylene with a minimum purity of 99.5wt% with a stream factor of 91%. The plant will also produce hydrogen as a side product with a minimum purity of 99.9wt%. The plant feed contains 97wt% propane with 1.5wt% ethane, 1.5wt% butane, and 30ppm sulfur impurities. The preliminary design will include heat and mass balances, equipment sizing, utility summary, and capital and operating cost estimates. Additionally we will discuss site location and we will prepare an economic evaluation of the site taking into consideration the local health, safety, and environmental regulations.

Copyright © American Institute of Chemical Engineers. All rights reserved.

400114 Introduction to Rate Based Mass Transfer Models

Abstract:

Rate based mass transfer models are built on five key elements –

- Mass and energy balances around individual phases on a tray or in a packed segment
- Thermodynamic phase equilibrium

- Equilibrium across interfaces
- Kinetic effects on mass transfer rates
- Mass and heat transfer rate models for transport across interfaces.

In addition, the five key elements have interwoven relationships with chemical, physical, phase and transport properties based on the chemical species being used and the actual mass transfer hardware. In sharp contrast, an equilibrium stage model uses only mass and energy balances around an entire ideal stage plus thermodynamic phase equilibrium and use empirical correction factors to account for efficiencies and mass transfer hardware. For applications such as absorption, where the mass transfer efficiencies are poor (5-25%) equilibrium stage based models prove to be grossly non-predictive, while rate based models are mechanistic, and predictive. This presentation will focus on the characteristics of true mass transfer rate based models, contrast them with the ideal stages and introduce briefly the ProTreat ® process simulator.

Copyright © American Institute of Chemical Engineers. All rights reserved.

396291 Fundamentals of Liquid-Liquid Extraction

Abstract:

Distillation has been the process industry workhorse for separating components in liquid mixtures. However, there are a broad range of applications in which liquid-liquid extraction (LLE) should be the method of choice. Unfortunately, many chemical engineers are unfamiliar with the technology and how it can be applied to separation processes. Practical aspects of the technology are not widely known, partly because the design of extraction equipment and systems has been in the hands of equipment suppliers. But familiarity with the basics of LLE can enable a process engineer to fully take advantage of the attractions this technique offer, and to interact effectively with those supplies. This paper will discuss the fundamental aspects of LLE and an overview of when and how it should be applied as a separation process. Calculation techniques will be reviewed, including generation and use of equilibrium data and applying this data to process design. Also various aspects of LLE process development from concept formulation, to bench-top testing, to pilot plant testing will be addressed. Simple scale-up correlations will be presented for the design of specific types of LLE columns, based upon results from pilot plant testing. By means of flowsheets, several extraction processes will be incorporated into the presentation.

Copyright © American Institute of Chemical Engineers. All rights reserved.

399807 Radioisotope Applications for Obtaining Detailed Process

Abstract:

In all chemical plants and refineries there generally is not enough instrumentation in the right places in our processes. During the design phase, decisions were made between knowing more about the process, which means buying more instruments, and saving capital funds. When a problem appears with the operation of a vessel, there are always gaps in the knowledge of what is happening due to a lack of instruments in the places where knowledge is not routinely required. Process engineers and operations personnel need to know what is happening so they can take corrective action. When knowledge is missing, they often take actions that make matters worse.

For distillation and separation processes Chemical Engineers have historically relied upon plant process measurements such as flow rates, temperatures, and pressures and model or simulation results for data with which to solve troubleshooting or re-design problems. An array of on-line diagnostic services are available that can provide additional data offering real-time information on how pieces of process equipment are actually operating.

The most common application is Gamma Scanning. This test is primarily applied, but not necessarily limited to, distillation or separation columns. Gamma scans provide a density profile of the internal process of operating distillation columns and other process vessels. The density profile can be used to diagnose the hydraulic operating conditions of mass transfer devices such as damage to internals, flooding, degree of entrainment or weeping, liquid levels on trays and distributors, liquid distribution through packed beds, etc. The presenter will show case studies where scanning revealed vital process information that helped solve an operating problem, or helped make a revamp/re-design successful.

Additional derivatives from gamma scanning of detecting flow distribution patterns will be demonstrated. The CAT-Scan technique is a specialized horizontal gamma scan used to generate a topographic profile of the internal cross-sectional density of process equipment. This profile is useful for the detailed study of liquid flow distribution through packed columns as well as having applications beyond distillation columns. A hybrid application involving tracers and scanning on fixed bed catalytic reactors will also be discussed.

Another technique involving radioactive materials is the Tracer Test, in which a radioactive tracer is introduced into the process for investigating the process flows. External radiation detectors are placed at strategically chosen locations outside vessels and piping to monitor the tracer flow through the process. Either liquid, gas or a solid tracer can be utilized depending on process conditions and the fluid to be studied. This tracer technique can identify liquid or gas maldistribution in fixed or fluidized bed reactors. The residence time through a tank or reactor can be measured and compared with the calculated residence time. Observed differences can relate directly to the degree of mixing or channeling inside the reactor. These tracer techniques have also been used in identifying leakage of heat exchangers and valves.

Copyright © American Institute of Chemical Engineers. All rights reserved.

397222 We Learn and Share but We Don't Get Better! Time for Huaa

Abstract:

Any organisation can improve by leveraging its knowledge and lessons learned. These learnings or knowledge can be **external** or, in a global enterprise, **hidden** within your own company where lessons learned are difficult to share beyond their initial impact. There is so much data, information and knowledge (opportunities) coming at your company that can make you stronger or make you look like your not paying attention. Not paying attention or managing the available knowledge poorly, can **sink your ship**... nobody wants that!

When lessons are learned at a local level, large companies struggle to spread learnings across the organization. Even when they share in a passive manner, organizational value is not always realized. When sharing is not enough to ensure we don't repeat missteps organizationally, you need a process like HUAA (Heard, Understood, Acknowledged and Actioned). Holding sr. managers accountable is the only way to ensure that painful learnings are leveraged for improved process safety performance.

Copyright © American Institute of Chemical Engineers. All rights reserved.

Young, Inexperienced, and Learning:
How Process Safety Hit Home While Attending a Conference

Ashley M. Weckwerth
Burns & McDonnell Engineering Co., Inc.
9400 Ward Parkway, Kansas City, MO 64114
aweckwerth@burnsmcd.com

Prepared for Presentation at
American Institute of Chemical Engineers
2015 Spring Meeting
11th Global Congress on Process Safety
Austin, Texas
April 27-29, 2015

UNPUBLISHED

AIChE shall not be responsible for statements or opinions contained
in papers or printed in its publications

Young, Inexperienced, and Learning: How Process Safety Hit Home While Attending a Conference

Ashley M. Weckwerth
Burns & McDonnell Engineering Co., Inc.
9400 Ward Parkway, Kansas City, MO 64114
aweckwerth@burnsmcd.com

Keywords: Process Safety, Lessons Learned, Conference, GCPS

Abstract

Many professionals believe you will only understand process safety if you have seen incidents firsthand. Some professionals believe firsthand experience is more valuable than reading or hearing about the experience. I believe they are right. Does that mean if these incidents haven't happened to you that you haven't had process safety "hit home"? Does that mean you will never gain the experience unless an incident arises? I disagree.

I have never been a plant engineer or an operator. I am under twenty-five years old and have worked for an engineering firm ever since I graduated from college. Even with this background, I can tell you the exact moment when process safety hit home for me (and I didn't even step foot inside a chemical facility). That moment occurred when I attended my first professional conference. At the 9th Annual GCPS, I was listening to a speaker discuss the 2012 Chevron Refinery Fire. During the presentation, I first learned about the chemical safety board and its website. I was so interested in the CSB that later that night I watched the CSB video animation for the 2005 BP Texas City incident. To say the least, I was stunned. How did I never hear about this 2005 incident in college or since graduation? Was I living under a rock?

Until that day and that moment, I did not truly realize that the instrumentation and control systems that engineers design every day could impact someone's safety. Yes, my employer emphasizes safety on a daily basis. A safety moment is always shared before any meeting starts, and every Wednesday we huddle as a group and discuss a safety topic. Also, a safety email is sent out weekly and we identify opportunities to improve safety in design through LOPA and HAZOP. However, because I had never seen an incident of this magnitude until this point, this was when process safety really hit home. This video brought home the professional responsibility that I have, that we all have, as engineers to consider the broad impact of decisions we make. This was an awakening for me.

Knowledge can be obtained by listening, reading, and understanding history from the "senior engineers", the experienced co-workers and peers. The younger generation doesn't need to repeat incidents to fill the senior engineers' shoes. We can learn from the senior engineers, but not in the same way they necessarily learned, through trial and error. Through collaboration and knowledge sharing, we can learn from the past. The saying "history repeats itself" does not have to exist in the world of process safety, so let's get rid of it.

The future of process safety is in the hands of four essential parties: the younger generation, the senior engineers, the employers, and the technology. The younger generation has to be ready to learn without "living through" all the experiences. The senior engineers have to be ready to teach effectively. The employers have to be ready to support the resources and tools for communication between the generations.

Another hindrance to learning by shared experience, I have learned, is fear of litigation. In the earlier days of small firms, knowledge was shared over the water cooler, at lunches with co-workers and other small settings. But as firms get larger and larger, with offices spread all over the country and senior engineers retiring, how do we share institutional knowledge? Some companies teach "lessons learned" or post the same on intranet sites to pass on the mistakes made, and successes achieved. But lawyers are fearful that such documented mistakes might be misused by an aggressive plaintiff's lawyer, blowing them up on a large screen for a jury to see how bad the company has been. The result is that many firms don't teach lessons learned, or do so in a limited and protected manner that does not disseminate the information to those who need it. One solution that I've recently learned about is "peer review privilege" laws passed in a few states to encourage teaching these lessons so that history does not repeat itself.

Finally, technology needs to be used to keep the communication strategy effective for those individuals trying to teach and learn in the future. All parties have to work together to learn from the past and not repeat it in the future.

I am so privileged that my employer chose me to attend the conference that year, where I had my "Ah-Ha" moment. Luckily, through support from my employer and from coworkers, I was able to attend the conference that year and learn about the CSB. This guided me to use technology to seek the 2005 BP Texas City animation videos which effectively communicated the situation.

I did not experience that incident firsthand; however, I know that video was when process safety hit home for me. That day is when I truly realized why my employer emphasizes safety on a daily basis and how my work could impact someone's life any time they worked in a facility I designed.

1. Introduction and Background

Many professionals believe you will only understand process safety if you have seen incidents firsthand. Some professionals believe firsthand experience is more valuable than reading or hearing about the experience. I believe they are right. Does that mean if these incidents haven't happened to you that you haven't had process safety "hit home"? Does that mean you will never gain the experience unless an incident arises? I disagree.

I have never been a plant engineer or an operator. I am under twenty-five years old and have worked for an engineering firm since I graduated from college. Even with this background, I can tell you exactly when process safety hit home for me and I didn't even step foot inside a chemical facility.

This paper describes the moment process safety hit home for me. It describes the event as well as the approach that will help the younger generations have its "Ah-Ha" moment early on in their careers rather than having to learn from their own mistakes. With the younger generation, senior engineers, and employers all working together, history does not have to repeat itself.

2. Process Safety Hitting Home

2.1 The Event

In 2013, I attended the 9th Global Congress on Process Safety. I had only been in the industry for 9 months, but was excited about the opportunity. Throughout the conference, I had generally stayed in the PSM^2 (Process Safety Management Mentoring) Track. This track definitely catered to what I needed, which was an introduction to Process Safety. Being a young professional, this track was extremely beneficial and opened my eyes to the importance of process safety. However, my "Ah-Ha" moment didn't happen until the last day of the conference. After hearing all the speakers and presentations, I had learned so much, but the last day was when my experience came full circle. On that day there was a joint session with all the tracks together. This session was "Case Histories and Lessons Learned". Multiple case histories were discussed, but the one that stood out to me was the CSB investigation on the Chevron Richmond Refinery Fire. I had heard of "CSB" before but never really knew what they did, or that I could access the information online at www.CSB.gov. This incident had me riveted during the presentation. I couldn't believe this event happened while I was working in industry and hadn't heard about it. After the session, I decided that I wanted to visit the CSB website to learn more and see other incidents. That night I stumbled upon the BP Texas City refinery explosion. Throughout the conference, I had heard the Texas City incident mentioned but was unaware of what actually happened. The incident happened March 23, 2005. At that time, I was a freshman in high school. I had no intentions of being an engineer and hadn't even taken my first chemistry class yet. I also didn't follow the news closely and was usually unaware of anything that was outside a 50 mile radius from my parent's home. Therefore, throughout the conference whenever Texas City was mentioned I just tuned it out, unaware of the significance of that incident.

Some background on the BP Texas City explosion would be helpful because this might also be your "moment" process safety hits home. The CSB accident description states: "At approximately 1:20 p.m. on March 23, 2005, a series of explosions occurred at the BP Texas City refinery during the restarting of a hydrocarbon isomerization unit. Fifteen workers were killed and 180 others were injured. Many of the victims were in or around work trailers located near an atmospheric vent stack. The explosions occurred when a distillation tower flooded with hydrocarbons and was overpressurized, causing a geyser-like release from the vent stack". - http://www.csb.gov/bp-america-refinery-explosion/

Figure 1. CSB's Final Investigation Report – Impacts of the Disaster

The animation of the CSB video describes the incident. Here is a summary of what CSB states in the video.

During the restart of the isomerization unit, as the operators feed the tower, the level transmitter failed and key alarms did not warn operators of the process conditions in the raffinate splitter tower making them unaware of the unsafe situation. This caused the actual liquid level to be higher than what they read in the control room. Poor communication and planning contributed to the incident next. As the day operator arrived for his shift (for his 30th day in a row – 12 hr shift), he read the log book. However, the log book was unclear. The day shift then added more feed being unaware how much liquid was already in the tower. The operators were unsure on what had already been completed in routing the product due to contradictory information; so, they left the product valve closed and continued with the procedure. Next they lit burners in the raffinate furnace to heat up the feed as the procedure stated. This heated the feed and served as a reboiler. Another personnel issue then happened, the day supervisor left for the day to attend a family emergency and no one replaced him. This left the operator to run the three refinery units by himself; the second operator position was removed due to budget cuts several years earlier (in 1999). The feed continued to fill the tower, reaching a height of 98 feet by noon (15x normal level). The reason they kept filling was that the liquid level indicator was saying it was only at 8.4 feet. Poor design to the HMI screen didn't show the operators the flows in and out of the tower on the same screen. There was also no calculation or indication of the total liquid in the tower. The operators then noticed that the pressure was too high, so the operators opened the manual chain valve to vent gases to the emergency relief system. The operators thought they should turn off the burners and also opened up the product valve to storage to remove some liquid from the bottom of the tower. However, this continued to heat the tower because prior heating had created a lot of hot liquid at the bottom of the tower which then caused vapor bubbles to rise to the top cold liquid to heat it up. Also, as the hot raffinate was leaving the bottom of the column it was transferring additional heat to the feed through the heat exchanger. So, the tower continued to get extremely hot and full even with the burners off. The liquid started to boil and swell. When the tower was completely full, the overhead vapor line filled with liquid. This caused the three pressure relief valves to open and send the liquid to the blowdown drum. The liquid level alarm on the blowdown drum failed to go off. This lead to the point where the flammable liquid was spilling out of the blowdown vent stack described as a "geyser". The vapor cloud began covering the area including the trailers where workers had been in a meeting nearby. Two workers were in an idling truck near the unit. The diesel engine began to "race", so the workers fled and were unable to shut off the engine. This caused the truck to backfire and ignite the vapor cloud. Explosions and fires began throughout the unit.

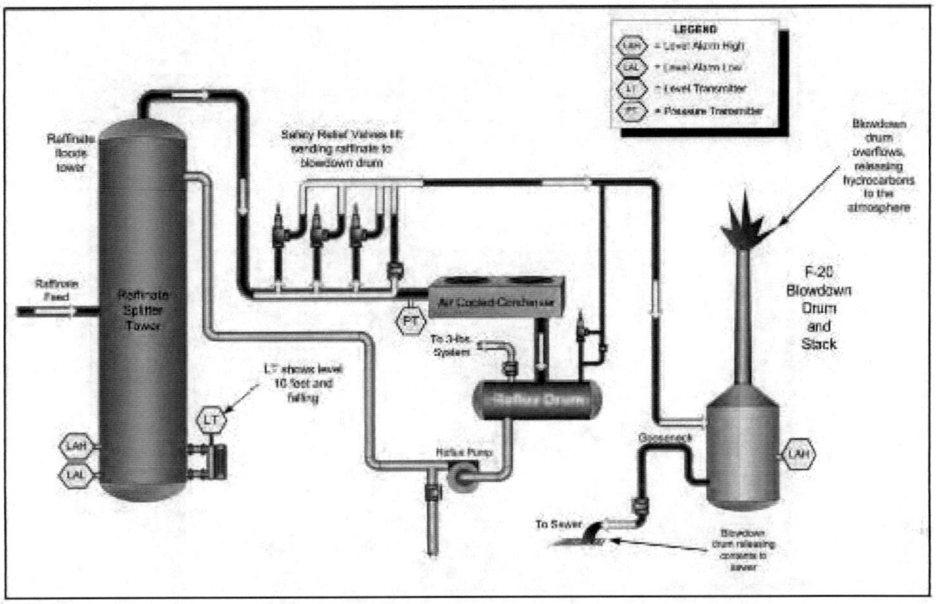

Figure 2. CSB's Final Investigation Report – Tower overfills and blowdown drum releases hydrocarbons

As the photo indicates, this was a major accident. After I read the accident description, I saw that CSB had an animation for this report. After watching the fifty-five minute animation, it was clear that the faulty instrumentation, poor maintenance, and ineffective process safety leadership, culture and knowledge, led to the catastrophic accident. After finishing the animation, I realized that I could be a part of an incident like this. My name (or yours) could be associated with faulty design or equipment. In the early stages of the Texas City incident, CSB concluded that the liquid was actually at 13 feet at the time the operator stopped the feed, thinking it was only at 9 feet. The reason the operator didn't know the true level was because the level indicator only read the bottom 9 feet of the vessel. It couldn't read any higher.

I work primarily in the instrumentation design field, meaning my work could impact others and their lives! This moment was when Process Safety hit home for me.

2.2 *The Aftermath*

It doesn't matter that I don't work on the plant design on a daily basis. It doesn't matter that I am just one person. My work still impacts lives. Safety needs to be on my radar at all times from design to start-up. My employer emphasized our impact on safety in design every day before I reached my "Ah-Ha" moment, so why was I just not realizing it? It is because I had not put myself in a real life catastrophe before. I had not seen how small failures or mistakes lining up could cause a major accident. At the GCPS, we discuss the "Swiss cheese model" and how hazards can line up, which cause the accident. In figure 3, you see an example of this method. This is what happened at Texas City, which is why all details and changes need to be taken seriously.

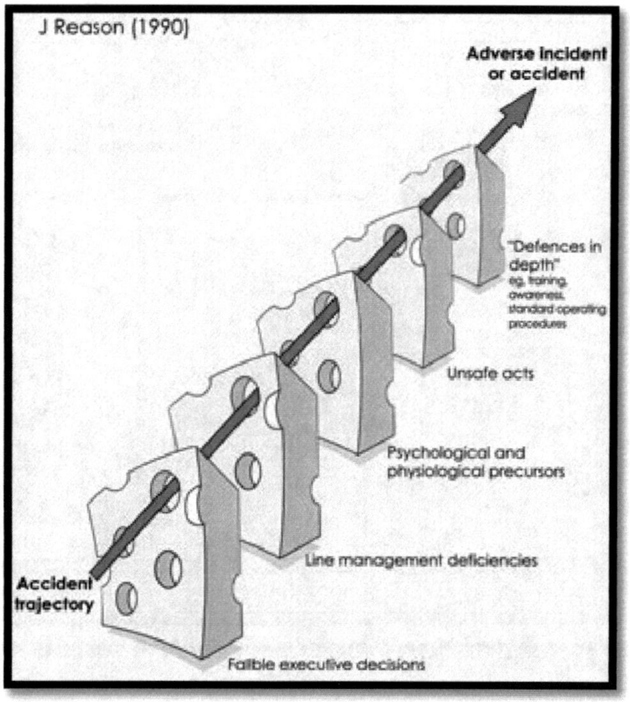

Figure 3. The Swiss Cheese error management model (*The Pharmaceutical Journal*).

Until that day and that moment, I had not realized the importance of my career path. We hear in college that engineers are licensed to protect public health, safety and welfare, but what does that really mean to a young EIT? Even though my employer emphasizes safety continuously (the word "Safety" is shown on my desktop background), I had personally not taken it as seriously as I should before my "Ah-Ha" moment. I did not know about this Texas City incident (nor other major incidents) and its magnitude because of my age and inexperience. I did not realize fully that the instrumentation and control systems I help design every day could impact someone's safety. Working in a nice office setting, the biggest hazards I face daily at my desk are paper cuts. I was naïve and invincible until that moment, when suddenly I became vulnerable.

My vulnerability did not come from living the experience myself or experiencing a tragedy first-hand. I didn't learn from my own mistakes. I learned from others. Knowledge can be obtained by listening, reading, and understanding history from the "senior engineers", the experienced. The younger generation doesn't need to repeat incidents to fill the senior engineers' shoes. We can learn from the senior engineers, but not in the same way they necessarily learned. Through collaboration and knowledge sharing, we can learn from the past. The saying "history repeats itself" does not have to exist in the world of process safety, so let's get rid of it.

3. The Approach

The future of process safety is in the hands of four essential parties: the younger generation, the senior engineers, the employers, and the technology. The younger generation has to be ready to learn without "living through" all the experiences. Because of what has happened already in history, we have to realize that we are vulnerable - not learn after we make a mistake ourselves. The senior engineers have to be willing to teach and mentor effectively. The employers (and their lawyers) have to be ready to support the resources and tools for communication between the generations, to pass on "institutional knowledge" to the "up and comers". Technology these days allows companies to create tools that can impact the communication between people and generations to fill any gaps. Below are the roles of all four parties.

3.1 The Younger Generation

"They're talented, educated, tolerant, sociable, collaborative, and oriented toward achievement." – Paul Hersey (Goldsmith, 2010). This is the generation I relate to – the younger generation. We have multiple names: the millennials, generation Y, the "ME" generation, and so on. For the sake of this paper, let's say those in the workforce between 22-35 years old fall into this category. So, the average work experience for this generation lies between new graduate and about 13 years post-graduation (assuming a bachelor's degree at age 22). The millennials are graduating college and entering the workforce with little knowledge of past engineering failures, but for a few episodes taught in college (e.g. the Tacoma-Narrows Bridge Collapse, the Hyatt Skywalk Collapse, and the Twin Towers in New York). With the amount of things to teach students in the engineering curriculum in four years increasing, process safety is not covered as in depth. For example, in my curriculum, Process Safety was a chapter in my senior-year design book. Safety was not a full semester course, only a mere chapter in a book. Of course, safety is discussed in most ChemE classes; however, the history of past events and lessons learned is usually not specifically discussed.

As a recent ChemE grad, May 2012, I personally did not have a curriculum that emphasized process safety on an industry level. Therefore, it is left to the senior engineers and the employers to take the importance of process safety into their own hands and to fill in the gaps missed in formal education. I believe this is a better approach than teaching it at the university because of the variety of jobs ChemE grads go into. The senior engineers and employers can better mentor and teach the specific applications of process safety in their specific industry rather than the universities having to teach all the varied industries that ChemEs enter. For example, my friend who works in a lab does not deal with distillation columns or large scale processes, so specific examples of refineries and past lessons learned might not apply to her work.

The younger generation needs to realize that we are not invincible and just because we are new hires doesn't mean the things we work on won't impact someone's safety. We need to take our engineering role seriously and realize that bad things can still happen to good people. We need to be thorough in our work and listen to the senior engineers (even if some don't know how to use Social Media or unhide Excel spreadsheet columns).

3.2 The Senior Engineers

The senior engineers are the experienced ones, our mentors. They are the ones getting ready to retire now or shortly down the road. They have "been there and done that." They have lived through the accidents and have seen the evolution of process safety and understand how it became what it is today. Now instead of just leaving the workplace with their wealth of knowledge they have to make sure it stays in the industry. This is where senior engineers need to mentor and assist the younger generation while they still can, to download that head full of knowledge into our somewhat empty heads. Instead of having 20-30 years to relearn all what the senior engineers know, we need them to start mentoring to cut the learning time in half. Don't forget to "Consider it a blessing that Generation Yers want to do everything better and faster and they are confident in their ability to do so.... As a leader, you need to MENTOR them. They want to add to your company, not own it. Do not be afraid to give feedback, positive or negative" (Goldsmith, 2010).

With how busy the industries are right now it can be difficult to find time to mentor. Who has time to train and teach? It doesn't seem like a priority right now to the companies or some senior engineers, but it should be. A big topic in today's workforce is the generation gap. We are losing a LOT of baby boomers and don't have the people to fill those positions with experience. Employers are worried about losing their knowledge and their subject-matter experts. And their lawyers are concerned about how this information is shared. Training and mentoring is needed now; it is a priority.

3.3 The Employers

So what can the employers do to help? They can support and encourage.

It will take time for the senior engineers to mentor and teach the younger generation. This time is money to an employer, non-billable in most instances from a consulting perspective. It is an investment that employers will have to make, however, if they want to excel. Shared knowledge will help employees feel more needed and want to work harder for their employers. Many leadership books will state similar things about training like: "Developing people within an organization should be viewed as an investment. Then it becomes easy to understand the importance of maintaining this key resource so that it keeps performing at optimum level. If we know the cost of replacing an employee, then it is easy to conclude that getting the most from employees just makes good business sense" (Gulati). A company should encourage those in upper level positions to mentor the younger crowd. Employers should consider a mentor program, giving each younger person a senior engineer mentor. A mentor can share his or her many years of experience to the younger employee.

Employers should support the younger generation. Invest in them. Invest in Process Safety. Send the younger employees to conferences. Yes, it costs money. (Keep in mind, the cost of attending training and a conference is cheaper than if that younger employee causes a major incident or a shutdown). Yes, it is a risk to spend that kind of money on new hires. Most employers know that a lot of younger professionals tend to move around in the industry rather than spend their entire careers for one employer. It is a risk to invest in the younger generation, because we don't know if they will still be with the company in a year, but it is worth the risk in my humble opinion. A common characteristic or perception for the younger professionals is: "No sense of Loyalty." As a FORTUNE reporter stated about Generation Yers, "When it comes to loyalty, the companies they work for are last on their list – behind their families, their friends, their communities, their coworkers, and, of course, themselves" (Hira, 2007). If employers would invest in the younger crowd, they would feel more connected and engaged in the company. The new hires would see that their future would be growing and that their employers saw potential in them and wanted to take them to the next level. Sending them out to conferences shows that employers are willing to pay them to learn and pay them to grow. And after they attend conferences, employers should have them share what they learned with the other new hires. Pay it forward.

For example, the company I work for, Burns & McDonnell, paid for me to attend the 9th Global Congress on Process Safety in 2013. My employer knows the importance of process safety in the work we do. They make it their priority to communicate the importance of safety. They invest in their new hires to attend training on safety so that we can learn from others in the industry. It was because my employer invested in me that I know the importance of process safety in my career and my every day work more than I did. Also, I was able to attend the GCPS the next year; and when I came back to work, I ended up making a presentation to share with my entire department. In my presentation, I brought up the Chevron Richmond Fire animation and showed my colleagues CSB website. Not that it surprised me, I found out that many of the young professionals I work with didn't know about CSB and about the Chevron Richmond Fire or BP Texas City. It was just as new to them as it was to me the year before.

This is why employers need to invest in the younger generation: to encourage the senior engineers to mentor and to support the new employees in attending conferences and other off-site training opportunities.

3.4 The Technology

Technology ties all the three groups together. Technology can be used to improve the communication between senior engineers and the younger generation as well as give employers the tools to maintain the knowledge and distribute.

"As a leader, you need to provide access to technology.... Capitalize on their [Gen Yers] expertise and involve them in the process of choosing and implementing technology." (Goldsmith, 2010)

There are other types of technology tools that can be developed and used depending on your company's needs. From a process safety standpoint, MOC (Management of Change) logs and other change logs are used to make sure everyone is on the same page and that further research or evaluation doesn't need to be done. Using technology to improve these areas of communication will help share knowledge to fill the generation gap. This improved knowledge should help history NOT repeat itself.

3.5 The Peer Review Privilege

One hindrance to learning by shared experience, I have learned, is fear of litigation. In the old days of small firms, knowledge was shared over the water cooler, at lunches with co-workers and other small settings. But as firms get larger and larger, with offices spread all over the world, and senior engineers retiring, how do we share institutional knowledge? Some companies teach "lessons learned" or post the same on intranet sites to pass on the mistakes made, and successes achieved. But lawyers are fearful that such documented mistakes might be misused by an aggressive plaintiff's lawyer, blowing them up on a large screen for a jury to see how bad the company is. The result is that many firms don't teach lessons learned, or do so in a limited and protected manner that does not disseminate the information to those who need it. One solution that I've recently learned about is "peer review privilege" laws passed in a few states to encourage teaching these lessons so that history does not repeat itself.

In the law, courts protect certain communications (called a "privilege") in order to promote candid discussions. Examples are communications between a doctor or therapist and patient, a pastor or priest and congregation member, a lawyer and client, even a husband and wife. The privilege prohibits people from being forced in court to disclose what was said in these settings. For many years, however, a peer-to-peer communication was fair game in the courtroom. A doctor who discussed a failed surgery with another doctor was not protected. So, in the 1980's in what was called "tort reform," the medical profession got laws passed in nearly all states and even in Congress to create a "peer review privilege." As one court explained: "The peer review privilege is designed to encourage health care professionals to engage in candid, critical analysis of their peers' performance *by shielding participants from liability for their comments during peer review* and by ensuring that their disclosure cannot be compelled." State ex rel. Lester E. Cox Medical Centers v. Darnold, 944 S.W.2d 213 (Mo.1997). But what about engineers? If we hold "lessons learned" sessions among our peers, reveal what went well, and what didn't, can we be forced to disclose this in court? That real fear prevents many companies from teaching or posting lessons learned.

The process works well for healthcare providers. In a 2011 Article from the *American Journal of Medical Quality* it was reported that: "Clinical peer review has long served *as the primary process* by which physicians evaluate each other's performance with the aim of improving the quality and safety of patient care." (Edwards, 2011). Taking a tip from the doctors, in 2012 engineers and architects in Missouri lobbied for and passed SB 667 which enacted the design-professional peer review privilege law, the first in the nation. Under this law, lessons learned can be taught within a company that is not admissible in court, and outside peer reviewers are immune from liability. Kansas followed Missouri's lead in 2014 by passing HB 2246 with an even broader version of peer review privilege for design professionals. Several other states are considering such laws, but not fast enough. In order to learn from our mistakes, we need to be able to share them with our peers, without fear that some lawyer will misuse this information.

4. Conclusion

I did not experience BP Texas City incident firsthand; however, I know that was when process safety hit home for me. It was a major incident in the United States that we can all learn from. That day I learned about Texas City was when I realized fully my work could impact someone's life. Just like my employer, we need employers to invest in their younger generation. Most companies have young, inexperienced, and eager-to-learn employees just like me. If the senior engineers would help mentor, the laws would encourage lessons learned, then the employers would invest, and the technology would help with the communication, I believe that process safety can be taught. It can be effectively taught and passed on instead of learned firsthand before a major incident. We don't need to relive history. Let's invest now in our people, our laws and technology to avoid the high costs of incidents tomorrow.

5. References

[1] Goldsmith, M. B. (2010). *AMA Handbook of Leadership - 14.3 Characteristics of Generation Y.* AMACOM - Book Division of American Management Association.

[2] Gulati, R. (n.d.). *Maintenance and Reliability Best Practices - 10.3 Understanding the Generation Gap.* Industrial Press.

[3] Hira, N. A. (2007, May 15). *Attracting the Twentysomething Worker.* Retrieved from Fortune: http://archive.fortune.com/magazines/fortune/fortune_archive/2007/05/28/100033934/index.htm

[4] U.S. Chemical Safety and Hazard Investigation Board. (2007, March 20). Investigation Report – Refinery Explosion and Fire – BP Texas City, Texas. Retrieved from CSB: http://www.csb.gov/assets/1/19/CSBFinalReportBP.pdf

[5] Edwards, M.T. (2011). *The Objective Impact of Clinical Peer Review on Hospital Quality and Safety.* The American Journal of Medical Quality, http://ajmq.sagepub.com

[6] *The Pharmaceutical Journal*, Vol. 282, p743 | URI: 10967647

The Road between Reality and Philosophy (a.k.a HSE Success within an Engineering Organization)

Rao Akula
WorleyParsons
125 West Huntington Drive, Arcadia CA 91007, USA
Rao.Akula@worleyparsons.com

Prepared for Presentation at
American Institute of Chemical Engineers
2015 Spring Meeting
11th Global Congress on Process Safety
Austin, Texas
April 28, 2015

UNPUBLISHED

AIChE shall not be responsible for statements or opinions contained in papers or printed in its publications

The Road between Reality and Philosophy (a.k.a HSE Success within an Engineering Organization)

Rao Akula
WorleyParsons
125 West Huntington Drive, Arcadia CA 91007, USA
Rao.Akula@worleyparsons.com

Keywords: WorleyParsons, Zero Harm, Safety Culture, Group Integrity, Reality and Philosophy

Abstract

Compared to the days when the industrial revolution started, safety as a discipline has come a long way. The advancements in the civilization and technology - including the concepts of democracy, law enforcement, freedom of expression, and more recently the internet based social media have taught us what we know today – regarding how to value human life, dignity, and the common good for a society. We humans use the never ending process of introspection, conversation, and continuous improvement to raise the bar constantly. Today, we all accept that no one needs to get hurt at work; and we have enough knowledge, skills and abilities to prevent injuries at the workplace. But then, injuries and accidents continue to occur. What's wrong with this picture? Is an injury free workplace as hard to realize as the world peace? I don't think so.

A long time ago, I had this epiphany that men would naturally turn into philosophers when they cross the age 50. Also, I had a spontaneous thought few years ago, during the celebration of the 50th birthday of one of my supervisors, that age 50 is a landmark, with some more significance. It is (approximately) the time when the person's need for the world (which decreases as one gets older) becomes less than the world's need for that person (which increases as one gets older, assuming that most people get wiser as they grow older, and the world needs their knowledge!). This transformation helps one become more secure with increased level of confidence in one's own knowledge. Also, our ability to articulate the definition of the problems and their potential solutions gets better with time.

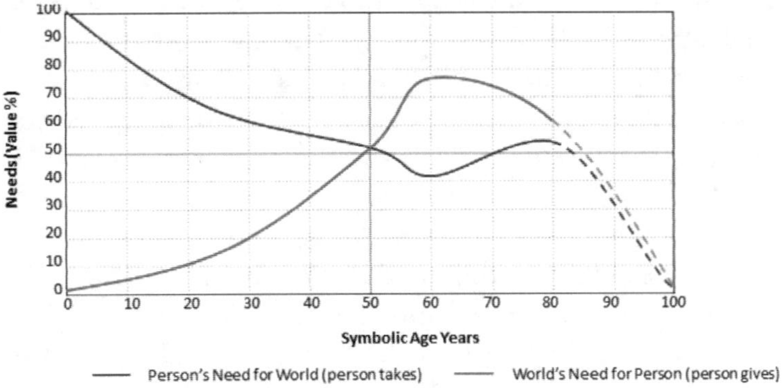

Figure 1. (Symbolic) Age vs. Needs

Having spent 37 years in a Chemical Engineering career (and over 50), I now see myself as a philosopher, and feel happy to share what I know. *I must explain that the age mentioned in the Figure 1 is symbolic, not absolute. No age discrimination is intended in this matter, since competence and age are not necessarily related.*

The purpose of this paper is to share my philosophy of Health Safety and Environmental (HSE) success (a.k.a., an injury free workplace) within an Engineering Organization executing hydrocarbon processing projects (my area of experience – I have no plans to address anything beyond that!), as we know today, and what it can be (or should be).

1. The Philosophy of HSE Success

I think almost every manager in the engineering and contracting business is familiar with the acronym CTR (Cost, Time, Resources) used in planning for projects and executing them. As a Program HSE Manager at an Engineering Organization, I had the opportunity to support multiple projects for several years. I still remember my first week at this job, when I was confronted by one of our Project Managers, with a series of "reasonable" questions - "what do you do for my project?", "what does HSE do?", "why is HSE required on my Project?" and so on, leading to an "urgent implication" that unless I explain what I have done for his project, he does not want to/have to approve my time charges. That conversation may not sound like much of anything and in fact, similar questions are faced by many personnel in engineering/contracting companies on a regular basis). Therefore it is important that I include my subsequent conversation with my supervisor (our Vice President), to make my point. My supervisor stated something like "you were hired to support all my projects; at the end of your performance year, I would like to see that you did. It is your responsibility to work with each of my project managers (and me) to help the Program with all the ongoing projects, and our CTRs with respect to HSE". Having met these expectations, I offer the following thoughts **for HSE success within in an engineering organization**:

1.1 What does HSE do?

It is tricky to separate what an individual does from what a group is doing, what is billable and what is overhead, in the name of HSE. Therefore, you can get startled by questions like "what does HSE do?" - When I faced that question I was wondering whether the question was referring to me, as an individual or to a project or program I was assigned to, or to the HSE function of the company that I just joined. In an engineering company, the HSE function ensures full legal compliance to local laws, corporate and client assurance requirements. Many companies, and some of the project teams within almost all the companies, are not very serious in walking the talk on HSE, especially in the matter of *HSE culture* (This is the single biggest challenge in any global company trying to establish a HSE culture of its own). Any group of professionals, no matter how good they are, will have trouble landing on consensus with respect to allotment of HSE man-hours in CTRs. What we do as a group (considering there are many groups of various sizes in a company), or as a company, in the name of HSE is hard to capture in CTRs – for reimbursement by our clients or otherwise. A mature organization looks at these needs with equal criticality – whether the HSE task is reimbursable or not i.e., for both the two sacred, but separate HSE responsibilities of any engineering company: Corporate HSE and HSE in Design. I think it would be fair to say Corporate HSE is generally driven by law and HSE in design is driven by the need for Zero Harm to people, assets, and environment (in aspirations as well as reality). In smart and successful companies these objectives blend well together and the gap between the aspirations and reality can become very small (if not disappear – hence the aspiration of zero harm).

1.2 Corporate HSE and HSE in Design

In the United States (and in many of the developed countries) Corporate HSE is critical for the company's ability to do business. In other words, if your company does not meet the required minimum HSE standards, it goes out of business. However, many clients and engineering organizations themselves are demanding HSE in Design as well. The skill set of a successful HSE Manager with responsibilities in both Corporate HSE and HSE in Design, must include competence, integrity, communication and relationship skills. Building a respectable reputation in HSE in Design requires good understanding of the technical subject matter in the context of the Project (as well as the organizational capabilities & limitations), along with ability to design/alter roles and responsibilities of multiple highly talented technical "professionals" depending on the context. This is because of the common belief that organizations try to inculcate – "HSE is everyone's business", and who is responsible for what, can be clouded by the context. Other subtle issues of understanding or knowing and acknowledging who knows what (*and doesn't know*), and ability to predict what the overall outcome is, can be a challenge.

1.3 HSE and Process Safety

Some people, and even some companies, think that HSE and Process Safety are two different functions. They would explain to you that HSE is for people safety, and Process Safety is a complex and very exclusive area for highly skilled specialists who understand engineering, and have deep technical knowledge about how to keep hydrocarbons where they belong – contained, exactly as designed (and as meant to be). I respect and agree with the complexity and engineering/technical knowledge part; having worked in design and operations of complex refineries for several years, I even believed that I may have that type of "process safety" knowledge. What I learned over the years is that the exclusivity part does not make sense. In fact, in the context of success, the operative word should be *inclusive*. All the established disciplines of the engineering organization/function (chemical, mechanical, electrical, instrumentations & controls, civil, structural) have central aspects of process safety in them. Therefore, a separate process safety discipline, albeit a good idea, has not become very wide-spread in the engineering industry. It may have a better place and significance in operating companies. Another source of support for the exclusivity perceptions of process safety from HSE may come from the desire to separate the process safety engineers from people with low HSE competence (or level of HSE knowledge) who may be looking after personal safety matters only. I think process safety engineers are still part of an HSE organization, and also belong as much in an engineering function. A competent HSE Manager should be good at both – not just one. He should be able to grasp all possible risks that come up during "execution" of projects - which can cut across multiple disciplines, and may have process safety as well as people safety issues.

1.4 Construction HSE

The project execution part with the highest HSE exposure is construction. Some engineering companies consciously stay away from construction responsibilities just to be more competitive with respect to their HSE performance and injuries record. I think the philosophy of HSE success in construction is no different – a highly active and responsive HSE culture, with the work practices driven by team integrity, competence, and reward system for job well done. HSE Compliance enforcements and budgets practices should not compromise the HSE culture while there are gaps between aspirations and reality of HSE performance.

1.5 HSE in Programs with MSAs

Large Operating Companies have recognized that they get diminishing returns on their efforts/costs for the contractor selection process when they want help with projects with decreasing levels of Total Installed Cost/effort. Therefore, Master Services Agreements (MSAs) are very common for engaging a Program Contractor, who can handle a number of small projects with minimal (or no) necessity to bid for the work on a project by project basis. HSE management in such programs requires a (slightly unique/different) combination of CTRs and HSE culture – separate CTRs for the Projects and Program, a HSE culture that is comparable to (if not better than) the customer's HSE culture, *and* aspects to sustain the long-term relationship between the client and contractor. A good MSA contractor should be able to add value to the customer's own HSE process/needs. In this context, having *control* processes for HSE quality, and *assurance* (self-verification as well as independent) processes for HSE quality on behalf of the client may be put in place with appropriate administrative controls. Also, there may be opportunities to set up common teams that can help address knowledge management, lessons learned programs, joint representations at industrial conferences, mentoring, and long-term development of personnel, resources, technology, identification of new projects, and tools benefiting both sides.

1.6 When do we need a full-time HSE Manager in a Project?

These days, many small projects do not have full time HSE Managers. The Project's HSE needs are either managed by the Project Manager or assigned to some other manager. Sometimes it is even worse (and demoralizing for the group), when there is a lip service for the HSE aspirations with empty slogans like HSE is everyone's responsibility – and there are no real resources/efforts to address the HSE gaps between aspirations and reality. However, it is very common that large Projects do have a full time HSE Manager.

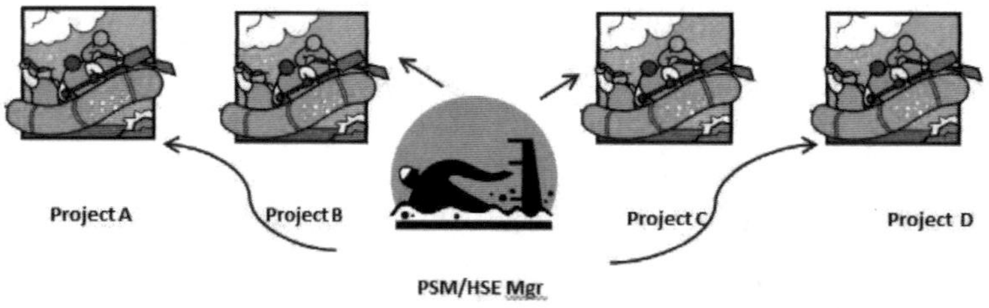

Figure 2. The Challenge of Part-time HSE Resources

Within the culture of engineering and contracting companies, the answer to the question (of a need for full time HSE Manager) typically would be – it depends on the context of the Project – how big or complex the project is, how many safety in design issues exist, what is the development stage of the project (you typically need more of HSE as we get closer to completion), etc.

In general, many engineering companies will say if the client thinks they need a full time HSE Manager, they will be happy to hire one. In fact, I have seen that such a decision (whether to have a full-time HSE Manager for a Project or not) depended more often than not on the Client preferences rather than the engineering company practices.

What does it mean? HSE means a lot of things to lot of people, but when it comes to the world of projects, HSE requirements are much less understood (than what could be best for the projects); Since money drives "HSE" actions more often than promises, many small projects choose not to have a full time HSE Manager.

1.7 How should we estimate requirements for HSE resources?

When it comes to determining requirements for HSE resources (people and tools), it is not a good idea to use CTRs for analyzing level of effort from one HSE person (or a couple of HSE folks – to ensure depth and availability of the service). The basis for such decision(s) should be simple – we need a fulltime (competent) HSE person (or a couple), or not. When we need multiple HSE personnel (like in large/complex projects and organizations), CTRs may be a good start – but there should be a lot more that should be considered besides CTRs. A lot of the effort will have to be driven by a *competent* HSE Manager, who can use the CTRs, and the entire context based information to come up with a "what does good look like" definition of HSE needs.

1.8 Limitations of CTRs

What limits the use of CTRs in this context is that we are supposed to use CTRs for deliverables. A whole lot of what HSE (and few other similar groups - think risk management, and the like) does is tasks or "level of effort". CTRs may not capture HSE tasks that all the members of the teams perform to sustain a highly active HSE culture. While program management encourages those tasks, project teams tend to push back on what is not in CTRs, and complain that their billable activities suffer from the "need" for positive HSE culture. As we try to "account for" (and minimize – perhaps in the context of re-organizations, and the passion to reduce the "overhead" budgets since HSE is *perceived* to be an overhead function) HSE needs with CTRs, the time needed for doing the HSE tasks (without reducing the number of tasks) can go down, and HSE budgets can be reduced without much thought. As a consequence the quality of HSE service can suffer. For HSE, we cannot (and should not) have a low quality service.

1.9 What HSE Activities are not billable to a Project?

I think HSE Assurance and Audit activities should not be billable to a project unless exclusively agreed with/requested by the client. Albert Einstein said that "the significant problems we face cannot be solved at the same level of thinking we were at when we created them". Therefore, HSE budgets planning for assurance should be done at an organizational level higher than the one you are trying to protect. Also, HSE culture related activities, personal protective equipment, and activities like completing Corporate Required HSE Training, should not be billable to a project – unless explicitly agreed to with the Clients in the contracts. In general, only scope-specific HSE in Design activities, reimbursable HSE deliverables and tasks (including self-assurance measures required by clients) should be billable and defined by the organizational/group level tasked to deliver them.

1.10 What matters most, and the importance of HSE competence

I cannot emphasize enough, the importance of HSE competence in HSE Managers. Those who lack in HSE competence/ knowledge tend to compensate it by their passion for the HSE culture. This can help generate a large volume of conversation within their group. However, it is very likely to lose focus on what matters most while the project teams are chasing issues that rank lower in the priority. I believe that a close analysis of all past major safety incidents would substantiate my point. I think humans in groups have a tendency to work with "on-off" control type of response to safety incidents (similar to air conditioning thermostats at homes). In other words, attention to safe behaviors and procedures increases dramatically following an incident, but such a hyper attentive culture is difficult to sustain. There comes a natural reduction to attention over time, until the next incident. The story never ends. Think of all major disasters and safety incidents in the industry. Every incident (and its post-incident analysis) shows at least some contribution of HSE incompetence leading to the incident. What is sad is that we can almost be sure, there is going to be another disaster/safety incident – we just don't know when and where. I happen to think that it won't happen on the watch of a competent HSE Manager. I think a competent HSE Manager is exceptionally responsive (personally) and helps maintain a highly responsive HSE culture. I also think that all smart successful engineering companies should include personnel with plenty of knowledge/experience in HSE, Process Safety and Operations – if their aspiration includes adding value to energy companies with hydrocarbon processing operations.

1.11 HSE Culture and group integrity

Arguably, a voluminous HSE conversation is an indicator of a good HSE culture. According to my philosophy, integrity of any person connects with constant validation and consistency among thoughts, words, and actions – especially in important matters. The same applies to professionally successful groups working towards common goals, including some in HSE. I also believe that I should not worry about things upon which I do not have any control – meaning that I try to minimize those wishes/expectations for which there can be no corresponding action *and* refrain from those actions that do not correspond to good thoughts. Similarly, for successful groups, the trick of maintaining of a good HSE culture lies in their ability to sustain the volume of (honest and informed) conversation, and a group-wide genuine passion (and consistency between actions and the inputs/thoughts/conversations/ behaviors observed) to minimize gaps between group HSE aspirations and reality.

1.12 The sweet spot of success in HSE

a.k.a. of "OneWay to Zero Harm" becoming a reality, in all Projects of an engineering organization – is therefore difficult, if not impossible. A wise friend of mine said once – if you leave a group of very smart engineers in a room, close the doors, and ask them to reach consensus on a complex design including all options and for all possible scenarios – it may never happen. That is where a competent Project Manager can make a difference in the outcomes, to keep the project within reasonable timelines and budgets. The sweet spot of zero harm mixes up individual success with that of a group, and eventually many groups or projects in the whole organization. It crosses the lines of fixed/defined "roles and responsibilities", what is reimbursable (or not), who gets the credit (or should get) and so on. It includes a mature HSE culture – with many ingredients - healthy policies on tracking good metrics, competence developed/nurtured by diversity & inclusion, and creativity. I sincerely believe it is possible and achievable – it is not just a slogan/rhetoric - when there is a pragmatic consensus among the project teams, guided *by two competent mutually respectful managers (for every one of the company's projects) – one passionately managing the business (P&L), another managing HSE with similar passion and both are trusted and treasured by their upper management.*

2. Summary

An injury free workplace is achievable. It takes HSE competence to properly assimilate and fulfill the HSE needs and success of an engineering organization executing hydrocarbon processing projects. Do not use CTRs (alone) for sizing up HSE requirements – either in a project or in an organization. Do not make HSE Management a part-time service (even) at any physical location of an engineering organization. In the business of small projects, a full-time HSE person may sometimes sound excessive to the passionate business manager. Let the Location business (P&L) managers negotiate additional tasks/responsibilities with the Location HSE manager, depending on the context of the Location's priorities and business needs. When we think about it, a full-time Location HSE manager may alleviate some of the issues on the HSE culture companies advertise to have and (try to) live by. Zero Harm in reality is feasible, when HSE culture is part of what needs to be nurtured, and highly competent and well respected P&L Managers and HSE Managers work as partners with mutual respect and passion.

About the Author:

Rao Akula is a California Registered Professional Chemical Engineer, with a total of 37 years work experience in three countries. Rao is currently a Senior Manager for HSE and Risk Management in WorleyParsons *Improve* business line – Americas Region, Arcadia Operations. Rao has a Master's degree in Computer Information Systems, and a Bachelor's degree in Chemical Engineering. Rao joined WorleyParsons in 2007 and has worked in a wide variety of settings in the industry - Management, HSE, Process Safety, Environmental Permitting and Compliance, Advanced Controls, Process Design; Government, Operating Refineries, Engineering & Consulting. Rao's past employers include BP, Long Beach, California, State of California Department of Toxic Substances Control, Glendale & Sacramento California, Brown & Root, Alhambra, California, Kuwait National Petroleum Company, Mina Abdulla Refinery, Kuwait, and Engineers India, New Delhi.

402856 Mercury Compliance While Avoiding Halogen-Driven Corrosion

Abstract:

EPA's Mercury and Air Toxic Standards (MATS) requiring stringent emission controls for coal-fired electric generating unit (EGU) power plants will be enforced beginning April 2015. Powdered activated carbons (PACs) have been demonstrated as the leading technology for achieving compliance with the stringent mercury control levels required by MATS. The PACs are sourced as halogenated or non-halogenated materials. In some cases halogens are added to the coal or flue gas stream separately from the PAC. As more of these systems are coming online and accumulating operating hours, issues of plant equipment corrosion are becoming known and work is being done to understand the nature and extent of the problems. Solutions can include operational changes and/or changes to the materials injected. We will discuss the design and application of advanced PACs to minimize or avoid corrosion issues and retain effective mercury capture and co-benefit integration with the EGU's air quality control system (AQCS).

Copyright © American Institute of Chemical Engineers. All rights reserved.

401729 BACT in Air Pollution Control

Abstract:

Best Available Control Technology for air pollution is mandated by many regulatory agencies for projects above certain thresholds. The Prevention of Significant Deterioration (PSD) permitting guidance recommends selection of control technology using a top-down process. This session will cover the structure of the top-down selection process and basic cost effectiveness calculations using USEPA methods.

Copyright © American Institute of Chemical Engineers. All rights reserved.

399709 Air Pollution Aftertreatment Control for Natural Gas Applications

Abstract:

Natural gas is an important part of the American infrastructure and economy, growing in significance with the rise of domestic shale gas. In 2013, 31% of the natural gas consumed in the United States was for electric power generation in 2,154 power plants. Regulated emissions from natural gas combustion include NOx, CO and VOC's (volatile organic compounds). These pollutants can be controlled by the use of a post-combustion catalyst installed downstream of the engine prior to the stack. For the power industry to provide electricity economically while maintaining low pollutant emissions, plant operators and equipment vendors face a variety of technological challenges which require innovative solutions.

NOx emissions are controlled by a Selective Catalytic Reduction (SCR) system. In a typical SCR system, gaseous ammonia (NH_3) is injected into the exhaust stream upstream of the SCR catalyst. In the presence of oxygen, NO and NO_2 react with NH_3 over the SCR catalyst to form N_2 and H_2O. Liquid ammonia or aqueous ammonia must be heated and injected into the unit in a precise amount in order to limit unreacted ammonia, also known as ammonia slip.

There are a number of issues currently facing the NOx control market for natural gas-fired applications. In order to improve air quality, regulators periodically reduce the emissions limits for new units, increasing the performance requirements of the technology, while users are trying to limit the cost and risks associated with operating these systems. Stricter limits require more catalyst and/or new designs from suppliers. As emissions limits become lower, the ammonia efficiency of the SCR system must be improved so ammonia injection is optimized and ammonia slip is minimized. Issues with ammonia quality are also common and can further increase operating costs by poisoning the catalyst and reducing the catalyst life cycle. As the emissions limits have been reduced, the emissions control system places more and more restrictions on the power producers and limits their operating flexibility to meet demand. To counter this, emissions control system designers are required to provide systems that operate outside the preferred operating window for the catalysts. Some large units require the catalyst to operate at higher temperatures. While on the other hand, faster start-up times and stricter start-up limits require the catalyst to operate at lower temperatures. Both of these issues are hardware and catalyst issues that require the catalyst supplier and system suppliers to work together to develop the right product.

Oxidation catalysts are used to oxidize CO and VOC's. The issues facing the oxidation catalyst are similar to the NOx control catalyst. The limits for CO and VOC's emissions are being reduced and operators are seeking more flexibility in terms of operating parameters. The big issue in terms of operating temperature has to do with poisoning of the catalyst by sulfur. At lower temperatures sulfur blocks the active sites of the oxidation catalyst limiting activity. At higher temperatures the sulfur desorbs, which is why it is a temperature-dependent issue. There are methods to remove the sulfur from the catalyst, but as long as it is present in the fuel the issue will return. Finding formulations that operate at lower temperatures is an important area of development currently.

Copyright © American Institute of Chemical Engineers. All rights reserved.

401809 2007 Valero McKee Refinery Fire

Abstract:

The Valero McKee Refinery experienced a significant fire in February 2007 after a propane line burst during winter weather conditions. The refinery was shut down and evacuated until the situation was stabilized. This presentation discusses lessons learned from this event.

Key Lessons:

Cold weather caused accumulated water in a stagnant section of piping to freeze, the ice expanded and broke a section of pipe. Subsequent increasing ambient temperatures allowed the ice to melt and a release of propane to occur. Important lessons from the event include the following.

1. Challenges associated with winter operations in refineries.

2. Criticality of understanding and managing parallel processes equipment in operating units.

3. Reviewing process units for change of service over time.

Copyright © American Institute of Chemical Engineers. All rights reserved.

401810 1987 Celanese Pampa Butane Vapor Cloud Explosion

Abstract:

On Saturday November 14, 1987, the presenter personally witnessed a catastrophic vapor cloud explosion at the Celanese Chemicals facility in Pampa, Texas resulting from the release and subsequent ignition of vaporized butane from a ruptured reactor feed line. The explosion resulted in three fatalities, dozens of injuries and several hundred million dollars in property damage.

This presentation will discuss the details of the incident, including a video containing original news clips and subsequent interviews, as well as key learnings that shaped the organization's approach to process safety management. Much of this information continues to be as relevant today as it was then.

The Pampa event was recognized across Celanese as a process safety wake-up call and led to the development and implementation of comprehensive company-wide process safety management programs, which continued to be expanded and fine-tuned upon and following promulgation of the 1992 OSHA Process Safety Management Standard and the 1996 EPA Risk Management Program Rule.

From the start, until his retirement in 2005 following 35 years of service to the company, the presenter was directly involved in virtually all key aspects of Celanese's PSM programs. He subsequently formed his own consulting firm and continues to advise clients on a range of process safety matters. He chaired the CCPS committee responsible for publication of *Guidelines for Risk Based Process Safety* and has been actively involved in the CCPS Vision 2020 project, among others. He holds a Bachelor of Science degree in Chemical Engineering from Texas Tech University.

Copyright © American Institute of Chemical Engineers. All rights reserved.

"Let Me Tell You................"

The impact of Eastman's Aniline Plant Explosion on Process Safety Awareness

Peter N. Lodal
Technical Fellow
Plant Protection Technical Services
Eastman Chemical Company, Kingsport Operations
Kingsport, TN

Copyright 2015, Eastman Chemical Company
All Rights Reserved

Prepared for Presentation at
American Institute of Chemical Engineers
2015 Spring Meeting
11th Global Congress on Process Safety
Austin, Texas
April 27-29, 2015

UNPUBLISHED

AIChE shall not be responsible for statements or opinions contained
in papers or printed in its publications

"Let Me Tell You..............."

The impact of Eastman's Aniline Plant Explosion on Process Safety Awareness

Peter N. Lodal
Technical Fellow
Plant Protection Technical Services
Eastman Chemical Company, Kingsport Operations
Kingsport, TN

Keywords: Lessons Learned, Explosion, Culture, Incident Investigation

Abstract

A single presentation covering one person's first-hand experience with a significant process safety incident, delivered at the beginning of OSHA's PSM regulation still provides contemporary lessons, even though the actual incident occurred more than 50 years ago

1. Introduction

"Let me tell you about the day I got serious about process safety. It was the day I was blown through my office wall, not knowing whether I would live or not when I hit the ground."

This was the introduction we received from our then Vice President of Safety, Mike Piazza, in early 1992 as the then-new OSHA PSM standard was being promulgated. Mike had gathered all of us together--safety and process design personnel for an after-lunch presentation on the new regulation, and why we needed to treat it with utmost seriousness. I was working in Process Engineering at the time, doing equipment design work—not dedicated process safety work as I do now.

After his introduction, Mike proceeded to tell us of his first-hand experience with the aniline explosion, and showed us a number of photographs (several of which are included here as figures 1-4) of the damage done to our facility by an incident that ultimately claimed 16 lives. This was the first time most of us had ever seen these photos, as the aniline plant explosion was not widely discussed within Eastman before this time.

Seeing extensive damage of buildings and parts of the plant that we all regularly passed by and worked in had a dramatic effect on the audience. Mike's presentation lasted 90 minutes, and was delivered on old fashioned, pre-PowerPoint acetate overheads. None of us had any problem staying awake, even immediately after lunch. The impact of this presentation is still felt today, as Eastman regularly shares the events and learnings from this incident with employees, through the Process Safety Boot Camp, group safety presentations, and an on-line video on our intranet system.

2. The Incident

I will not go into great detail on the specifics of the incident, as that has been well covered elsewhere (1). The essence of the event can be shown in the "before and after" photos in Figures 1 and 2--a massive, 6 ton TNT (equivalent) detonation of a nitrobenzene-nitric acid-water mixture in a distillation column where nitrobenzene, created as the first step in a two-step reaction process, was to be azeotropically distilled overhead in preparation for the second reaction step (hydrogenation to aniline).

The resulting explosion caused 16 fatalities, more than 45 overnight hospitalizations, and at least 250 medical treatment cases, in addition to the substantial property damage and business interruption costs.

While much of the evidence was destroyed in the explosion, one critical piece of evidence, a sample of the nitrobenzene purge stream, which provided the composition in the base of the nitrobenzene column did survive, and from that one piece of hard data, the most plausible (but by no means the only) failure scenario was developed:

1. The plant started during day shift on October 3, 1960, after an extended period of down-time for maintenance and repair.
2. Startup of the new, clean reboiler, using the same steam flow and temperature control setpoints used prior to the shutdown, resulted excessive vapor flow, causing undetected damage to column trays in the bottom section of the column. Most likely, this damage was a dislodging of tray panel sections due to excessive pressure drop caused by the higher than expected vapor flow.
3. Compromised separation ability of the nitrobenzene column due to the tray damage led to an unexpectedly high nitrobenzene concentration in the column base. Because this material was recycled upstream, back to the nitration reactor, the column underflow concentration of nitrobenzene continued to increase throughout the day.
4. Because of difficulties in getting the plant lined out, and what appeared to be erroneous temperature readings in the column, the foreman decided to put the nitrobenzene column on total reflux at about 3:15 PM on the afternoon of Tuesday, October 4.
5. By 4:43 PM that afternoon, the nitrobenzene concentration in the column underflow had increased to the point where it penetrated the single-phase, shock sensitive boundary (Figure 5), which was not detected in the chemical safety testing available at the time.
6. Shortly thereafter, an unknown disturbance (valve closing, pump starting, etc.) caused the mixture to detonate, resulting in a fireball of about 150 feet in diameter, with a TNT-equivalency estimated at 6 tons.
7. Neighboring warehouses were set afire by the fireball. Injuries resulted from a number of different sources; the fireball, the blast pressure wave, and significant shrapnel generated by the explosion.

This paper will focus on several of the process safety-related issues (not all of which are obvious, even now) which were uncovered as a part of our retroactive investigation.

3. Process Safety-Related Issues

3.1 Control Settings

The one piece of solid evidence (the underflow sample) indicated a concentration of 18% nitrobenzene where there should have been none. This, coupled with the observations of the operators at the time that the column just wasn't "behaving right", led to the post-incident inference of tray damage in the nitrobenzene column. People who actually operated this system stated that it ran very stably for the 2-3 years it was in operation. One common way of damaging or dislodging trays in a column is to force too much vapor through the column via excessive boil up.

The column's reboiler had just been replaced, and it is known that thermosiphon reboilers can transfer significantly more heat (6-10 times, depending upon the fluid being boiled and the heating medium temperature) until the fouling reaches steady state.

For those old enough to remember chemical operations controlled by pneumatic, panel-mounted instruments, operators often would put pencil marks on the panel to show where a particular loop "ran good". It is suspected that, since the new reboiler was "identical" to the old, the column was started by returning it to the settings where it "ran good". In doing so, an excessive amount of vapor was sent into the column on restart, causing tray damage and effectively reducing the number of separation stages and thus compromised the column's ability to remove nitrobenzene. Since nearly all the column underflow was recycled upstream to the nitration reactor, any nitrobenzene not removed in the column overhead would continue to build up in the reaction loop, until a critical concentration was reached. This critical concentration was reached at approximately 4:43 PM the following day, when the explosion occurred.

The Lesson to be Learned-- Be very cautious when returning a plant to service, especially during a "cold" restart. The uncritical use of previous "runs good" settings may result in an unexpected process response. Consideration should be given to what has been done to the plant between the last shutdown and the current startup, assessing what changes could impact performance, and adjusting procedures and instructions accordingly.

One additional question comes to mind at this point in the analysis—why didn't the operators recognize the unexpected behavior of this column as serious? More on this question in Issue #5 below.

3.2 Thermocouple Reading at 3:15 PM, Tuesday Afternoon

When the column was put on total reflux at 3:15 PM on the afternoon of October 4, a control system mechanic was called to the plant and asked to replace a thermocouple at tray 10 of a 30 tray column (the column was fed at tray 8, counting from the bottom up). The reason? It was reading 121° C when it should have been reading 101-102° C. The conclusion at the time? The thermocouple was bad, and therefore, it needed to be replaced.

In hindsight, the thermocouple was probably reading correctly. Staging analysis of the column with the trays below the feed compromised to ~5% efficiency indicates that the increase in nitrobenzene concentration would cause the 20° C rise in temperature.

The Lesson to be Learned---How often do we dismiss readings that are not normal as "erroneous"? While this is may be the case, shouldn't we first assume that the instrument is reading correctly, and then try to understand why the reading is not normal, using related measures and other information to perform a more comprehensive assessment. Shouldn't we ask "What are the consequences if this reading is accurate?" Doing so may lead to additional inquiry and help uncover otherwise unexpected issues. The process safety culture we are striving to establish should lead people to believe what their instruments are telling them, and only after a methodical assessment should the accuracy of the reading be called into question.

3.3 Chemical Safety Testing—Applicability of Results

Chemical safety testing in general and calorimetry in particular, were not nearly as refined in 1960 as they are today. Although two different types of safety testing were performed on various nitrobenzene-nitric acid-water mixtures, all testing was done at ambient temperature, not at the conditions of actual operation.

The Lesson to be Learned--In designing chemical safety testing experiments, test conditions must consider the full range of possible operating conditions for evaluation. Such conditions should include startup, shutdown, and emergency operations (including the possibility of fire exposure, among others). There are many process safety-related parameters that change dramatically with temperature, including:

- Reaction rate (both for primary, as well as competing side reactions)
- Solubility
- Flammability Limits

Additionally, emergency operation conditions (such as fire) should also be evaluated to see how the dynamics of the system change with changing conditions. Extreme caution should be used in extrapolating any results.

3.4 Change Management---what constitutes a change?

When presenting this incident in our on-site Process Safety Boot Camp class, the following question is generally asked of the class; "How many of you would consider replacement of the 316L stainless steel reboiler with one of identical design in Titanium Grade 2 a change to be managed?" Almost always, the answer is 100% of the class indicates that this is a modification that requires formal change management. The question is then asked; "How many would have considered it a change if the reboiler had been replaced like-for-like with stainless steel unit?" A few people will raise their hands, but not many. Finally, the question is asked; "How many would consider it a change had we simply cleaned the exchanger?" Almost no one raises their hand on this question. Yet, if our logic and assumptions as to the cause of this incident are correct, the ability of the "clean" heat exchanger to put 6-10 times the vapor into the system as the old "dirty" exchanger was most likely the initiating event for this incident.

The same logic can be applied to other "restore to like new" projects such as:

- Replacing a worn pump impeller with a new one of the same design
- Cleaning a cooling water strainer
- Replacing a worn seat on a block valve
- Changing the speed at which an isolation valve operates

The Lesson to be Learned--While few people would advocate using a formal change management system for every cleaning operation, it is critical to recognize that changes in the dynamic characteristics of a system (heat transfer, pumping rate, degree of agitation, etc.) can be just as important as changing the steady-state characteristics of the system (larger capacity lines, vessels, changes in material of construction, etc.) Operational control needs to be adjusted and monitored accordingly, and operating procedures should be modified and updated to address these possibilities.

3.5 When to call for a stand-down

Finally, why did the operators not recognize the tray damage on initial restart? The most likely explanation lies with the fact that "non-normal" operation is generally expected on startup. The tray damage was not noticed because the plant was in restart, and the expectation was that it would take some time to "line the plant out." Even after struggling for more than a day of "not right" behavior, the belief that "with a little more effort, the plant will be back up and running smoothly" is a common phenomenon, and even a source of pride amongst operational personnel.

The Lesson to be Learned-- When to call for a stand-down can be problematic. People in general take a great deal of pride in what they do, and to say "Stop, I don't understand what is happening" can be a difficult admission. Economic pressures to get the plant back up and running (and making money) can add to the general reluctance to stand down. On top of such pressures, the increased operational reliability of many plants makes start-up a more infrequent event, which may limit understanding of what is truly abnormal. One of the main goals of creating and sustaining a strong process safety culture is to empower people at all levels to be able to say "stop" without penalty when things are happening that do not make sense, or are not understood. Of all the issues that have been explored as a result of this incident, this is perhaps the most critical, and the most difficult to address universally.

4 Conclusions

There is a great deal to be learned by analyzing details from process safety incidents, even those that occurred decades ago. Close examination of details can reveal both technical and cultural issues that are relevant to today's operation, and the lessons to be learned through a heightened awareness of process safety issues can still yield significant, contemporary process safety benefits.

5 Acknowledgements

Many people over the years have contributed their insights to our common understanding of this incident, and its ramifications to current process safety thought. In particular, I would like to thank Dennis Hendershot, John Herber, Don Abrahamson and Louisa Nara, four of the Boot Camp instructors who have sat through my presentation on this incident multiple times, and who, through their insight, have pointed out some of the key learnings shared in this paper.

6. References

1. **"Distant Replay: What can reinvestigation of a 40-year-old incident tell you? A look at Eastman Chemical's 1960 Aniline Plant Explosion"**-- Process Safety Progress Volume 23, Issue 3, pages 221–228, September 2004

Figure 1

Figure 2

Looking southwest from Building 207
Building 122 in background

Figure 3

Figure 4

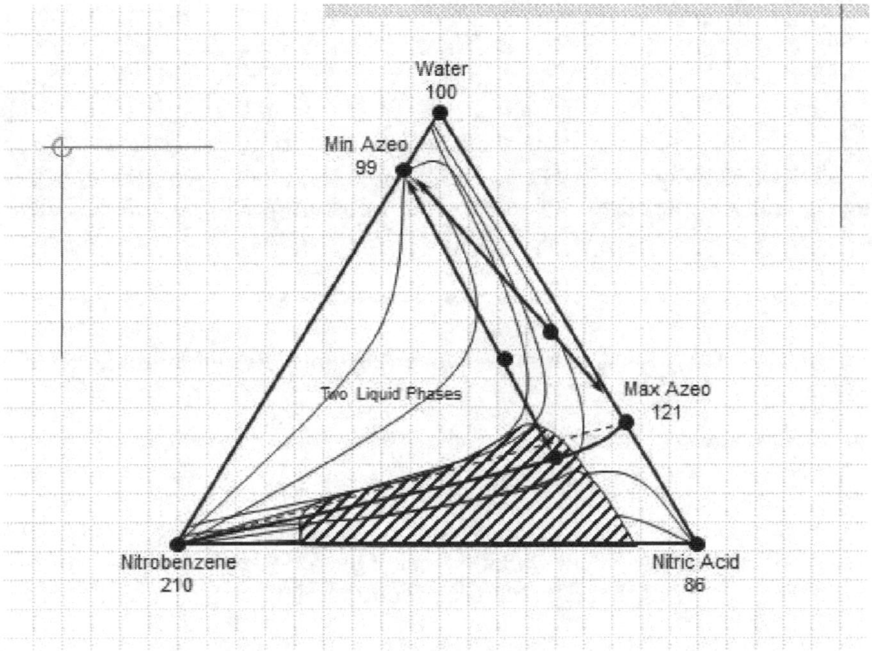
Water
100
Min Azeo
99
Two Liquid Phases
Max Azeo
121
Nitrobenzene
210
Nitric Acid
86

Figure 5

397207 Influence of Ice Cavity Geometry Changes on Burning Behavior of Large Scale Crude Oil Fires

Abstract:

Several Arctic council reports conclude that oil spills are the most significant threat to the Arctic ecosystem. Some studies have shown that *in-situ* burning (ISB) on open water can remove more than 90 % of the spilled oil, and it is the most promising technology for an efficient response to oil spills in Arctic. However, few studies have been undertaken on the influence of ice on the efficiency of oil burn. Prior studies have shown the coupling between the mass loss rate () and the geometry change of the ice cavity in pool fires smaller than 25 cm [1, 2]. In this study, a series of experiments are performed to develop an understanding of the influence of the coupling of ice cavity geometry changes and burning behavior of larger sized pool fires. The experiment involves Alaska North Slope (ANS) crude oil ignited in ice cavities of diameters (25-100 cm) with depth 6-25 cm. It is found that both burning rate and radiative heat flux are enhanced because of the cavity expansion. However, the overall efficiency of the burn is significantly lowered due to formation of lateral cavities. At some stage during the burning process, the heated oil tends to penetrate into the ice forming a pocket. This lateral cavity causes 10 - 35% of the crude oil to be trapped inside and hence un-recoverable. Further understanding of the reasons that may cause this unique formation and measures to prevent the same are discussed.

References

[1] P.W. Bellino, A.S. Rangwala, M.R. Flynn, A study of *in situ*burning of crude oil in an ice channel, Proceedings of the Combustion Institute, 34 (2013) 2539-2546.

[2] H.F. Farahani, X. Shi, A. Simeoni, A.S. Rangwala, A study on burning of crude oil in ice cavities, Proc. Combust. Inst., 35 (2014).

This study was funded by the Bureau of Safety and Environmental Enforcement, US Department of the Interior, Washington, D.C., under Contract Number E12PC00056. The contents do not necessarily reflect the views and policies of the BSEE, nor does mention of the trade names or commercial products constitute endorsement or recommendation for use.

Copyright © American Institute of Chemical Engineers. All rights reserved.

CFD Analysis of Supercritical Water Reactor for Flow Back and Produced Water Treatment

Swapnil Fegade, Abhijeet Shrawage, Jason Trembly

Ohio Coal Research Center

Ohio University, Athens OH 45701

Shale deposits have the potential of unlocking a vast reserve of fuel across the United States. However, the current high water usage of the horizontal drilling and hydraulic fracturing techniques used to develop shale resources needs to be addressed. OHIO is developing the Integrated Precipitative Supercritical Process (IPSC), with the aim of reusing flowback and produced water (F/P) from shale wells to reduce the water intensity of shale development activities. Of the components in the IPSC process, the Supercritical Water Reactor (SCWR) will be responsible in the treatment of the major salt constituents in F/P water. The purpose of this work was to utilize CFD analysis, using ANSYS Fluent (Fluent), to investigate the heat transfer and fluid flow in the SCWR, at different inlet temperatures (350° C and 390° C) and operating pressures (23 MPa and 26 MPa).

A user-defined material representing Supercritical Water (SCW) was developed in Fluent, which was later compared to existing experimental data from literature and implemented to evaluate SCWR operation. Temperature and density contours along with flow patterns within the SCWR are presented. The estimated TDS solubility in the lower half of the SCWR were estimated based upon SCW density. It was observed that at higher operating pressures, SCW could be heated to a higher temperature compared to pressures around the supercritical point (PSupercritical = 22.064 MPa). However, the minimum SCW density increases as the operating pressure increases, resulting in an increase of TDS solubility. As a result, the ideal conditions would need to be a balance between the inlet temperature of F/P water and the operating pressure of the system.

400845 Arsenic Removal from Aqueous Solution Using Nano-Structured Calcium Silicate

Abstract:

In this study, removal of arsenic from water was intended with using nano-structured calcium silicate as an adsorbent which was produced from rice hull ash.

For calcium silicate, two different production methods, which are calcium silicate produced in the presence of water and spacer (diethylene glycol), were performed. Arsenic removal efficincies were compared and nano-structured calcium silicate which was produced in the presence of spacer compound was choosen according to the experimental arsenic removal results.

The study covers the determination of the effect of time, pH, different arsenic concentrations and adsorbent quantities to obtain the optimum conditions for arsenic removal with nano-structured calcium silicate.

Copyright © American Institute of Chemical Engineers. All rights reserved.

401322 Optimal Strategies for NOx Reduction in Power Generation Using Optimization Approach

Abstract:

The power generation sector is considered as one of the major contributors to air pollution. An optimization model was formulated and written in a general format. The objective of this model is to select the best pollution control strategy for the power generation to reduce NO_x to specific level while meeting the electricity demand at minimum cost. The model is illustrated in a case study taken from Ontario Power Generation; Canada. For the NOx case when we consider two options (fuel balancing and switching), it was found that fuel switching is the best option of choice for reduction targets from 5 up to 30%.However, by considering all mitigation options, the result shows that applying SCR technology is the best option to reduce NOx emissions.

Copyright © American Institute of Chemical Engineers. All rights reserved.

15 AIChE
Spring Meeting, Austin, TX

HYDROPRIME® Modular Plants Provide Low Cost, Reliable Hydrogen

Kyle Finley
Hydro-Chem
125 Hickory Springs Industrial Dr.
Canton, GA 30115
kyle.finley@hydro-chem.com

Goutam Shahani
Linde Engineering

Nick Onelli
Linde Gas

Larry Lyda
Hydro-Chem

Steven Parente
Hydro-Chem

Copyright © 2015 Linde North America Inc. All rights reserved. No part of this publication may be reproduced or transmitted in any form or by any means, electronic or mechanical, including photocopying, recording, or by any information storage and retrieval system, without permission in writing from the Copyright owner.

Prepared for Presentation at
American Institute of Chemical Engineers
2015 Spring Meeting
11th Global Congress on Process Safety
Austin, Texas
April 27-30, 2015

UNPUBLISHED

AIChE shall not be responsible for statements or opinions contained in papers or printed in its publications

HYDROPRIME® Modular Plants Provide Low Cost, Reliable Hydrogen

Kyle Finley
Hydro-Chem
125 Hickory Springs Industrial Dr.
Canton, GA 30115
kyle.finley@hydro-chem.com

Goutam Shahani
Linde Engineering

Nick Onelli
Linde Gas

Larry Lyda
Hydro-Chem

Steven Parente
Hydro-Chem

Keywords: modular hydrogen generator, on site hydrogen, steam methane reforming

Abstract

This paper provides an overview of HYDROPRIME®, a unique steam methane reformer, employing an integrated heat recovery system, which can reduce the cost and improve the reliability for relatively small hydrogen consumers.

HYDROPRIME® plants offer many advantages over traditional supply modes such as electrolytic plants, conventional steam methane plants, and truck-delivered bulk hydrogen. These plants have a capacity of 0.15 – 0.3 MMSCFD (165 – 330 Nm^3/h), produce ultra-high purity hydrogen (99.999 %) at 200 psig (13.8 barg), reducing the need for product compression. HYDROPRIME® plants are highly heat integrated, which translates into low operating cost. These plants offer high reliability, and superior environmental and safety performance. HYDROPRIME® plants are fully automatic with fail-safe cutting edge process control technology, allowing unattended operation with remote start-up and monitoring. In addition, these plants are modular with an open skid design which provides easy site installation and excellent accessibility for maintenance.

HYDROPRIME® was developed to address an unmet need in the marketplace, by taking advantage of Linde's extensive experience in modular hydrogen plant design and fabrication.

HYDROPRIME® offers a flexible supply solution to the metals, glass, foods, chemicals, electronics and clean energy industries. The plants have demonstrated excellent results in actual commercial operation all over the world.

HYDROPRIME® is a registered trademark of Linde Aktiengesellschaft. All rights reserved.

1. Introduction

Hydrogen and nitrogen atmospheres are used in a variety of industries including metals, glass, foods chemicals, electronics and clean energy. Relatively small volumes of hydrogen (< 1 MMSCFD) are supplied in bulk by trailer or produced on-site by electrolysis, methanol and ammonia dissociation or steam methane reforming. While steam methane reforming is the dominant method of producing hydrogen at a relatively large scale (> 1 MMSCFD), this technology has not yet been widely adopted at a small scale due to cost and reliability considerations.

This paper will provide an overview of an innovative, hydrogen generator based on proven steam methane reforming technology. Linde's HYDROPRIME® plants are compact, efficient, and flexible. Ten units have been built and nine placed in commercial operation for an extended period of time, demonstrating their performance in a variety of applications all over the world. A picture of one such plant is presented below in Figure 1.

Figure 1: HYDROPRIME® plant

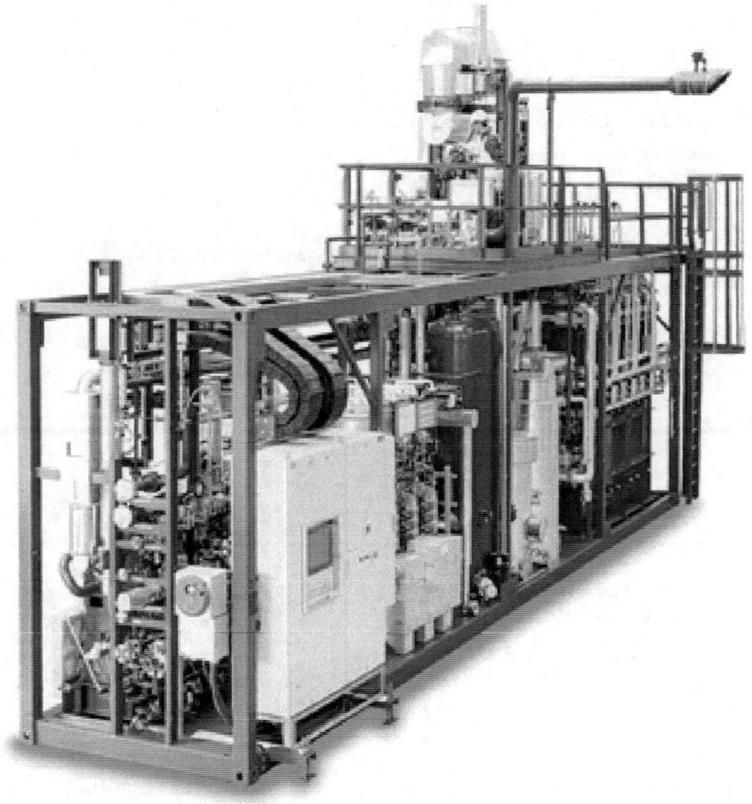

2. HYDROPRIME®

2.1 Technology

The predominant method of producing hydrogen on an industrial scale is steam methane reforming. HYDROPRIME® employs a unique heat integration concept, while using this proven technology. In this process, desulfurized natural gas feed is mixed with preheated water and fed to tubes filled with nickel catalyst. The following reactions take place at elevated temperature and pressure:

2.1.1 Reforming Reaction

$CH_4 + H_2O \leftrightarrow CO + 3H_2$ Endothermic (1)

2.1.2 Shift Reaction

$CO + H_2O \leftrightarrow CO_2 + H_2$ Exothermic (2)

Approximately 75% of the conversion to hydrogen takes place in reaction (1). Reaction (2) drives the equilibrium balance further to yield a hydrogen rich gas. Reaction (1) is reforming which is endothermic; reaction (2) is shift conversion which is exothermic. Both reactions (1) and (2) take place in the reformer. However, only reaction (2) takes place in the shift converter. The shift reaction employs promoted iron oxide catalyst. Both reactions are equilibrium limited, based on the outlet temperature and pressure. The reaction products are a mixture of H_2, CO, CO_2 and H_2O. A simplified block flow diagram of the SMR process is presented below in Figure 2.

Figure 2: Steam Methane Reforming (SMR)

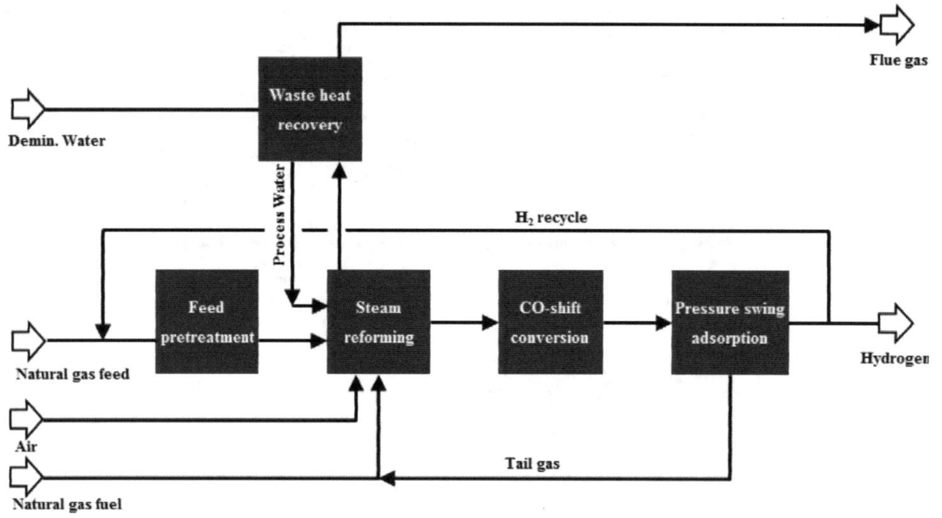

The overall reforming reactions are endothermic, requiring heat, which is supplied by the combustion of fuel. The tail gas from the PSA system meets most of the fuel requirement with the rest being supplied by natural gas.

The hydrogen rich stream is purified by pressure swing adsorption (PSA). The PSA is a physical process that depends on the selective, physical binding of gas molecules. Hydrogen, being non-polar and highly volatile, is essentially not adsorbed by the proprietary adsorbent material, which is a mixture of carbon molecular sieve and zeolites. The system operates on a repeated cycle having two basic steps: adsorption and regeneration. The regeneration consists of de-pressuring, purging and re-pressuring.

2.2 Design

The inputs to a HYDROPRIME® plant are natural gas, demineralized water, and electric power. The outputs are high purity, gaseous H2. There is no export steam. This plant produces H2 with the following specifications.

- Flow rate from 0.15 - 0.3 MMSCFD (165 -330 Nm^3/h)
- Purity of 99.999 +%
- Pressure of 200 psig (13.8 barg)

A picture of a partly assembled plant in transit is presented below in Figure 3.

Figure 3: HYDROPRIME® plant in transit

The major dimensions of the skid are listed below in Table 1. The plant is designed to be easily shipped all over the world. It can be placed in a flat-bed truck, thereby simplifying logistics and transportation.

Table 1. HYDROPRIME® Skid Dimensions

Aspect	Dimension, ft	Dimension, m
Length	46	14
Width	10	3
Height	14	4

The HYDROPRIME® plants are heat integrated for high thermal efficiency. The typical operating characteristic for a 0.3 MMSCFD (330 Nm^3/hr) plant are summarized below in

Table 2. These results have been achieved in actual commercial operation over an extended period of time.

Table 2. HYDROPRIME® Performance

Performance Parameter	Value
H_2 Capacity	
MMSCFD	0.3
Nm^3/hr	330
H_2 Purity, %	99.999
$CO + CO_2$, ppm	<1
Pressure	
Psig	20
Barg	14
Natural Gas /H_2 (HHV)	
BTU/scf	450
MM cal/Nm^3	4.4
Power, KW	60
Demineralized Water	
Lb/hr	1210
Kg/hr	550

HYDROPRIME® plants have a modular, open skid design which provides the following advantages:

- Simple and quick site installation
- Outdoor installation
- Excellent accessibility for maintenance
- Small foot print

These plants are standardized with fully automatic fail-safe controls, allowing unattended operation with remote start-up and monitoring 24/7. This enables quick response to even the slightest production interruption. Furthermore, the system provides virtually 100% uptime for H_2 product supply with the addition of a simple liquid backup tank.

2.2 Benefits

HYDROPRIME® plants offer a very viable alternative for relatively small (< 1 MMSCFD) hydrogen requirements. The plants are shop fabricated with high labor efficiency in a controlled environment, ensuring high quality control. The modular design of these plants reduces field construction, minimizes risk and provides faster project schedule. These plants also provide superior environmental and safety performance based on low emissions and reduced truck deliveries. Furthermore, the combination of low utility consumption and high reliability offers a very cost competitive solution for relatively small, industrial hydrogen consumers. By using this plant, Linde can provide 'over-the-fence' supply of H_2 molecules. In this arrangement, the end-user buys the molecule, under long term contract eliminating the need for any capital investment and responsibilities related to ownership such as installation, operation, maintenance, repair, insurances, manning and back-up of product in case of plant stoppage. Alternatively, these plants can also be owned and operated by the end-user.

3. Conclusion

HYDROPRIME®, a unique standardized steam methane reformer has been developed to reduce the cost and improve the reliability for relatively small industrial hydrogen consumers. HYDROPRIME® plants offer many advantages over traditional supply modes such as electrolytic plants, conventional steam methane plants, and truck-delivered bulk hydrogen. HYDROPRIME® plants are highly heat integrated, which translates into low operating cost. Furthermore, these plants offer high reliability, superior environmental and improved safety performance. HYDROPRIME® plants are fully automatic with fail-safe controls, allowing unattended operation with remote start-up and monitoring. In addition, these plants are modular with an open skid design, which provides easy site installation and excellent accessibility for maintenance. These plants provide a very competitive cost of ownership. HYDROPRIME® offers a flexible supply solution to the metals, glass, foods, chemicals, electronics and clean energy industries. The plants have demonstrated excellent results in actual commercial operation all over the world.

15 AIChE
Spring Meeting, Austin, TX

A Novel Two-Step Metal Oxide – Metal Sulfate Water Splitting Cycle for the Production of Solar Hydrogen

(Principal Author)
Rahul Bhosale
Department of Chemical Engineering
College of Engineering
Qatar University, Doha, Qatar
rahul.bhosale@qu.edu.qa

(Other Authors)
Anand Kumar
Department of Chemical Engineering
College of Engineering
Qatar University, Doha, Qatar
akumar@qu.edu.qa

Fares A. AlMomani
Department of Chemical Engineering
College of Engineering
Qatar University, Doha, Qatar
falmomani@qu.edu.qa

Shahd Gharbia
Department of Chemical Engineering
College of Engineering
Qatar University, Doha, Qatar
dd1202223@student.qu.edu.qa

Dareen Dardor
Department of Chemical Engineering
College of Engineering
Qatar University, Doha, Qatar
sg1000354@student.qu.edu.qa

Moustafa Hussein Ali
Department of Chemical Engineering

College of Engineering
Qatar University, Doha, Qatar
ma1203561@student.qu.edu.qa

Jamila Folady
Department of Chemical Engineering
College of Engineering
Qatar University, Doha, Qatar
jf1003686@student.qu.edu.qa

Shiva Yousefi
Department of Chemical Engineering
College of Engineering
Qatar University, Doha, Qatar
sy1106491@student.qu.edu.qa

Mehak Jilani
Department of Chemical Engineering
College of Engineering
Qatar University, Doha, Qatar
mj1104333@student.qu.edu.qa

Mashail Alfakih
Department of Chemical Engineering
College of Engineering
Qatar University, Doha, Qatar
ma090163@student.qu.edu.qa

[Insert here any copyright claims you wish to assert]

Prepared for Presentation at
American Institute of Chemical Engineers
2015 Spring Meeting
11th Global Congress on Process Safety
Austin, Texas
April 27-30, 2015

UNPUBLISHED

AIChE shall not be responsible for statements or opinions contained
in papers or printed in its publication

A Novel Two-Step Metal Oxide – Metal Sulfate Water Splitting Cycle for the Production of Solar Hydrogen

Rahul Bhosale
Department of Chemical Engineering
College of Engineering
Qatar University, Doha, Qatar
rahul.bhosale@qu.edu.qa

Anand Kumar, Fares A. AlMomani, Shahd Gharbia, Dareen Dardor, Moustafa Hussein Ali, Jamila Folady, Shiva Yousefi, Mehak Jilani, Mashail Alfakih

Department of Chemical Engineering
College of Engineering
Qatar University, Doha, Qatar

Keywords: Strontium Oxide – Strontium Sulfate Cycle, Water Splitting, Solar Hydrogen, Computational Thermodynamic Modeling

Abstract

This paper reports the thermodynamic analysis of solar H_2 production via two-step thermochemical Strontium oxide – Strontium Sulfate (SO-SS) water splitting cycle. The first step belongs to the exothermic oxidation of SrO via SO_2 and H_2O producing SrSO4 and H_2 and second step corresponds to the endothermic reduction of SrSO4 into FeO, SO_2, and O_2. The products, SrO and SO_2 can be recycled to step 1 and hence, reutilized for the production of H_2 via water splitting reaction. Thermodynamic equilibrium compositions of the thermal reduction and water splitting reactions were computed as a function of reaction temperatures. Second law thermodynamic analysis was performed to determine the cycle efficiency and solar to fuel energy conversion efficiency attainable without heat recuperation.

1. Introduction

As the direct dissociation of water is not practical due to the high operating temperature and gas separation problem, attempts are currently underway to achieve H_2 production via water-splitting reaction at lower operating temperatures and via bypassing the formation of H_2 and O_2 explosive mixture. Several thermochemical cycles such as iron oxide cycle [2-6], zinc/zinc oxide cycle [7-10], tin/tin oxide cycle [8, 11-13], mixed ferrite cycle [14-21], ceria cycle [22-24], sulfur-iodine cycle [25-27], and hybrid sulfur cycle [28-29] were investigated towards H_2 production via water-splitting reaction. Among these, the sulfur-iodine cycle and its variation the

Note: Do not add page numbers. Do not refer to page numbers when referencing different portions of the paper

hybrid sulfur cycle are more appealing as the required operating temperatures are lower as compared to other thermochemical cycles.

This paper proposes the utilization of the 'strontium oxide – strontium sulfate' (SO-SS) cycle for the production of solar H_2 via thermochemical water splitting reaction. In this study, the thermodynamic feasibility of this cycle was investigated and obtained results are presented in detail. At first, the thermodynamic equilibrium compositions during the solar thermal reduction of $SrSO_4$ under inert atmosphere and oxidation of SrO via water splitting reaction were determined. Furthermore, the maximum theoretical solar energy conversion efficiency of the SO-SS cycle is determined by performing the second law thermodynamic analysis over different solar reactor temperatures and with/without considering the heat recuperation.

2. Chemical Thermodynamic Equilibrium

2.1 Thermal Reduction of Strontium Sulfate

Figure 1 represents that solar thermal reduction of $SrSO_4$ into SrO, $SO_2(g)$ and $O_2(g)$ which is possible at 2300 K when 10 moles/sec of inert Ar flowrate is used. The thermal reduction reaction is represented as follows:

$$SrSO_4 \rightarrow SrO + SO_2(g) + {}^1/_2\, O_2(g) \qquad (1)$$

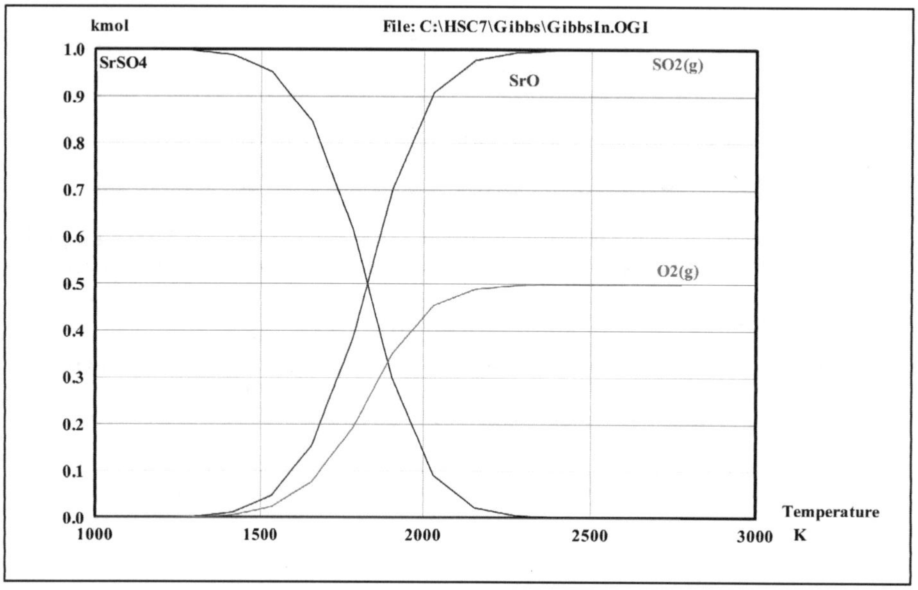

Figure 1. Solar thermal reduction of $SrSO_4$ (Ar = 10 moles/sec).

2.2 Water Splitting Reaction

Figure 2 represent variations in the equibrium composition for water splitting reaction (in absence of Ar). According to the results presented, the production of H_2 via oxidation of SrO by

SO_2 and H_2O is possible below 2000 K. It is further observed that at lower temperatures, $SrSO_4$ and H_2 are the major species, whereas the molar concentration of SrO, H_2O and SO_2 increases with the increase in water splitting temperature.

$$SrO + SO_2(g) + H_2O(g) \rightarrow SrSO_4 + H_2(g) \qquad (2)$$

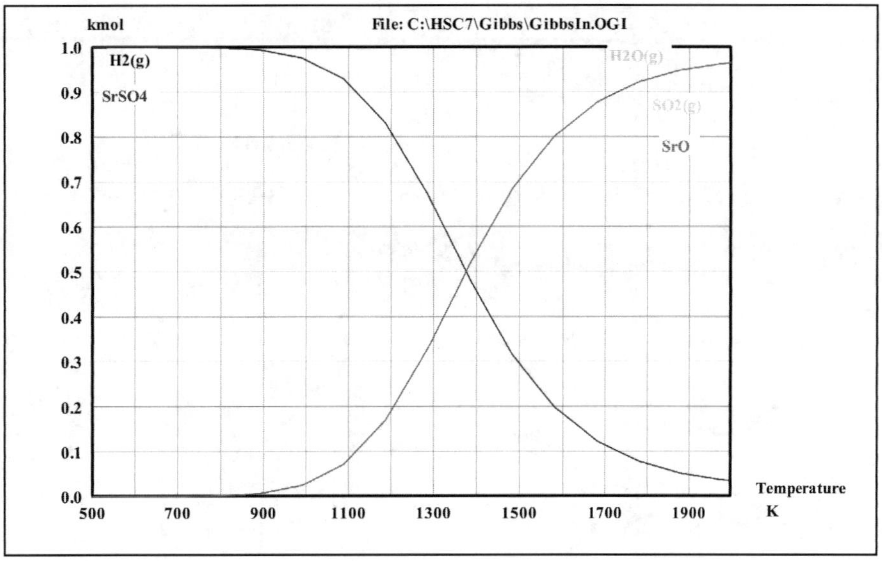

Figure 2. Water splitting reaction using SO-SS cycle.

3. Exergy Analysis

The second law thermodynamic analysis was used for the determination of the solar to fuel conversion efficiency of the SO-SS cycle. The process flow diagram for the SO-SS cycle is presented in Figure 3. The SO-SS cycle is comprised of a solar reactor, SrO oxidizer (H_2 generator), theoretical H_2/O_2 ideal fuel cell, a quench, and a gas separator. H_2 production is carried out at atmospheric pressure and at steady state conditions and also by neglecting any kind of viscous losses and variation sin the kinetic and potential energies. Energy required for the separation of gaseous products such as separation of SO_2 from O_2 and Ar are not accounted during the efficiency calculation. Also, the heat exchangers required for recovering the sensible latent heat are omitted from thermodynamic considerations.

In addition, several other assumptions are made such as follows:

- The solar reactor is assumed to be a perfectly insulated blackbody absorber
- No convective/conductive heat losses
- Effective emissivity and absorptivity equal to unity
- All the products separate naturally without expending any work
- All reactions reach 100% completion

The exergy analysis was carried out by following the methodology and governing equations derived and used previously for other MO based solar thermochemical cycles [7]. Thermodynamic properties were extracted from HSC software and databases and the analysis is normalized to the $SrSO_4$ molar flow rate (1 mol/sec) entering the solar reactor.

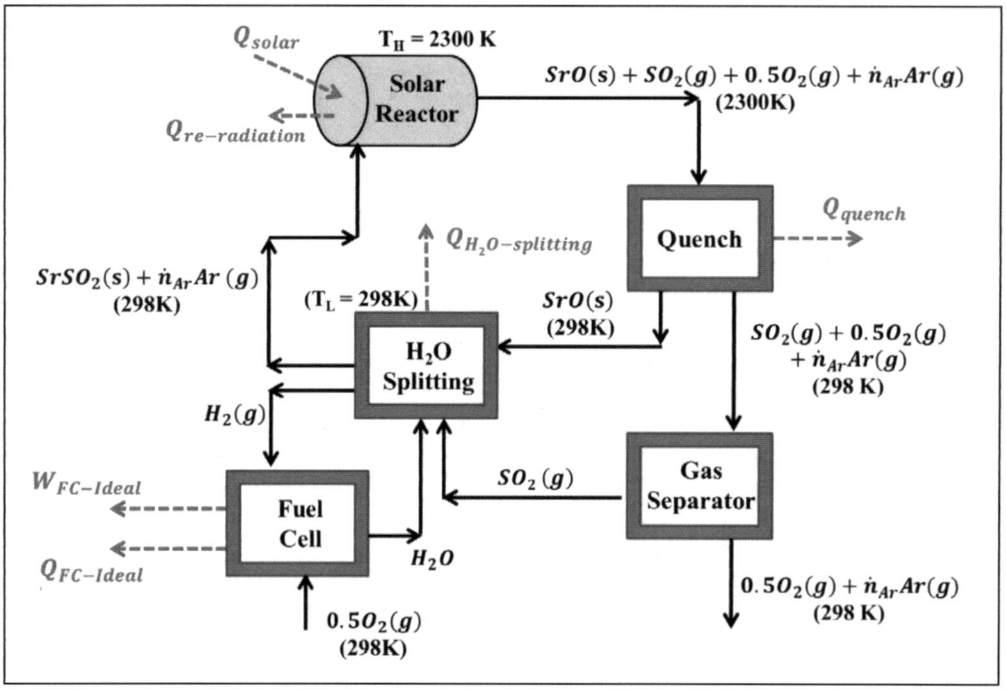

Figure 3. Process configuration for H_2 production via two-step solar thermochemical SO-SS water splitting cycle.

3.1 Solar energy absorption efficiency ($\eta_{absorption}$)

The solar energy absorption efficiency ($\eta_{absorption}$) is defined as the net rate at which energy is being absorbed by the solar reactor divided by the solar power input through the aperture of the solar reactor. The $\eta_{absorption}$ can be calculated as:

$$\eta_{absorption} = 1 - \left(\frac{\sigma T_R^4}{IC}\right) \qquad (3)$$

Where, σ = Stefan-Boltzmann constant, I = normal beam insolation $1(kW/m^2)$, C= solar flux concentration ratio of the solar concentrating system (5000 suns), $Q_{reactor-net}$= net energy absorbed in the solar reactor, Q_{solar} = solar energy input. For SO-SS water splitting cycle, at T_R= 2300 K, the $\eta_{absorption}$ is observed to be equal to 68.26%.

3.2 *Net solar energy absorbed for the operation of SO-SS cycle*

For the operation of solar thermochemical SO-SS water splitting cycle, following energy requirements associated with following steps needs to be fulfilled by solar energy input.

- Energy required for heating of Ar from ambient conditions upto T_R
- Energy required for thermal reduction of $SrSO_4$.

$$Q_{reactor-net} = Q_{SrSO_4-reduction} + Q_{Ar-heating} \qquad (4)$$

3.2.1 $Q_{Ar-heating}$

For complete thermal reduction of $SrSO_4$ at temperatures of 2300 K, 10 moles/sec of inert Ar flow rate is used as a carrier gas inside the solar reactor. Equation (5) gives the amount of energy needed to heat the inert Ar from ambient conditions upto thermal reduction temperature.

$$Q_{Ar-heating} = \dot{n}\Delta H|_{Ar(g)@298K \rightarrow Ar(g)@T_R} \qquad (5)$$

After solving this equation it was realized that 416 kW of heat energy is needed to heat the inert Ar from 298 K upto 2300 K.

3.2.2 $Q_{SrSO_4-reduction}$

The Energy required for the complete reduction of $SrSO_4$ (at 2300 K) into SrO, SO_2 (g), and O_2 (g) is given by following equation and observed to be equal to 854 kW.

$$Q_{SrSO_4-reduction} = \dot{n}\Delta H|_{SrSO_4@298K \rightarrow SrO + SO_2(g) + {}^1/_2 O_2(g)@T_R} \qquad (6)$$

At the end, the total net solar energy absorbed for the operation of SO-IS cycle is observed to be equal to 1270 kW (according to Eq. 4).

3.3 *Solar energy input to the solar reactor* (Q_{solar})

Based on the calculated $\eta_{absorption}$ and $Q_{reactor-net}$, the amount of solar energy input required to operate the solar reactor (Q_{solar}) dealing with the SO-SS water splitting thermochemical process can be determined by equation (7) and observed to be equal to 1860.68 kW.

$$Q_{solar} = \frac{Q_{reactor-net}}{\eta_{absorption}} \qquad (7)$$

3.4 *Radiation heat losses from the solar reactor*

Radiation heat losses and % re-radiation from the solar reactor conducting the thermal reduction of $FeSO_4$ can be calculated by equations (8) and (9):

$$Q_{re-radiation} = Q_{solar} - Q_{reactor-net} \quad (8)$$

$$\% \, re - radiation \, losses = \frac{Q_{re-radiation}}{Q_{solar}} \times 100 \quad (9)$$

Computational thermodynamic modeling indicate that radiation heat losses = 590.58 kW and % re-radiation = 31.74%.

3.5 Rate of heat rejected to the surrounding due to quenching

The product components coming out of the solar reactor includes solid SrO and gaseous SO_2, O_2, and Ar. To avoid the recombination, the products were quenched immediately to the ambient temperature (298K) by using a quench and the heat rejected by the quench is calculated by Eq. (10) which is equal to 821.93 kW.

$$Q_{quench} = -\dot{n}\Delta H|_{SrO(s)+\, SO_2(g)+0.5O_2(g)+Ar(g)@T_R \rightarrow SrO(s)\, +\, SO_2(g)+0.5O_2(g)+Ar(g)@298K} \quad (10)$$

3.6 Rate of heat rejected to the surrounding by SrO oxidizer

SrO generated via thermal reduction of $SrSO_4$ is transferred to the SrO oxidizer and reacted with the combined stream of SO_2 and H_2O at 298 K for the production of H_2. By assuming 100% oxidation of SrO producing $SrSO_4$ and H_2 via water splitting reaction, the rate of heat rejected to the surrounding by SrO oxidizer is estimated as 278.55 kW according to equation (11).

$$Q_{SrO-oxidizer} = -\dot{n}\Delta H|_{SrO+H_2O(g)+SO_2(g)@298K \rightarrow SrSO_4+H_2(g)@298K} \quad (11)$$

3.7 Rate of theoretical work performed and heat rejected by the ideal fuel cell

Solar thermochemical SO-SS water splitting cycle is completed with the addition of an ideal H_2/O_2 fuel cell with 100% work efficiency (assumption). This ideal fuel cell is added to the cycle to extract the maximum work from the net H_2 produced by this process. The rate of theoretical work performed by the ideal fuel cell is calculate by equation (12) and observed to be equal to 237.05 kW. Likewise, the rate of heat energy released by the ideal fuel cell is determined as equal to 48.56 kW [according to equation (13)].

$$W_{FC-Ideal} = -\dot{n}\,\Delta G|_{H_2(g)+0.5O_2(g)@298K \rightarrow H_2O(l)@298K} \quad (12)$$

$$Q_{FC-Ideal} = -(298) \times \dot{n}\,\Delta S|_{H_2(g)+0.5O_2(g)@298K \rightarrow H_2O(l)@298K} \quad (13)$$

3.8 Solar to fuel conversion efficiency

Cycle efficiency of the solar thermochemical SO-SS water splitting process is defined as the ratio of theoretical work performed by the ideal fuel cell to the solar energy input:

$$\eta_{cycle} = \frac{W_{FC-Ideal}}{Q_{solar}} \quad (14)$$

Furthermore, the solar-to-fuel energy conversion efficiency of the solar thermochemical SO-SS water splitting process is defined as the ratio of higher heating value (HHV) of the H_2 produced to the solar energy input:

$$\eta_{solar-to-fuel} = \frac{HHV_{H_2}}{Q_{solar}} \quad (15)$$

Where,

$$HHV_{H_2} = -\dot{n}\,\Delta H|_{H_2(g)+0.5O_2(g)@298K \rightarrow H_2O(l)@298K} \quad (16)$$

As per the HSC computation thermodynamic modeling, η_{cycle} and $\eta_{solar-to-fuel}$ equal to 12.74% and 15.36%, respectively.

4. Conclusion

Two-step solar thermochemical cycles for H_2 production via SO-SS based H_2O splitting reactions have been thermodynamically examined. Thermodynamic equilibrium calculations performed at 1 atm pressure and in absence of inert Ar indicate that the H_2 production via water splitting reaction is feasible below 2000 K. Similarly, the thermodynamic modeling also indicates that the thermal reduction of $SrSO_4$ into SrO, SO_2 and O_2 can be achieved at 2300 K when 10 mol/sec of inert Ar flow rate is used. The second law thermodynamic analysis indicate that indicate $_{\text{absorption}}$ = 68.26%, Q_{solar}= 1861 kW, $Q_{\text{reactor-net}}$ = 1270 kW, $Q_{\text{re-radiation}}$ = 591 kW, Q_{quench}= 822 kW, $Q_{\text{SrO-oxd}}$ = 279 kW, and η_{cycle}= 12.74%, and $\eta_{solar-to-fuel}$= 15.36%, respectively.

5. Acknowledgments

The authors gratefully acknowledge the financial support provided by the Qatar University Internal Grants QUUG-CENG-CHE-13/14-4 and QUUG-CENG-CHE-14\15-10.

References

[1] Agrafiotis C, Pagkoura C, Lorentzou S, Kostoglou M, Konstandopoulos AG. Hydrogen production in solar reactors. Catalysis Today 2007;127:265–77.

[2] Steinfeld A, Sanders S, Palumbo R. Design aspects of solar thermochemical engineering – A case study: two-step water-splitting cycle using the Fe3O4/FeO redox system. Solar Energy 1999;43:43–53.

[3] Kodama T, Nakamuro Y, Mizuno T. A two-step thermochemical water spitting by iron-oxide on stabilized zirconia. Journal of Solar Energy Engineering 2006;128:3–7.

[4] Gokon N, Murayama H, Umeda J, Hatamachi T, Kodama T. Monoclinic zirconia-supported Fe3O4 for the two-step water-splitting thermochemical cycle at high thermal reduction temperatures of 1400–1600°C. International Journal of Hydrogen Energy 2009;34:1208–17.

[5] Ishihara H, Kaneko H, Hasegawa N, Tamaura Y. Two-step water-splitting at 1273–1623K using yttria-stabilized zirconia-iron oxide solid solution via co-precipitation and solid-state reaction. Energy 2008;33:1788–93.

[6] Miller J, Allendorf M, Diver R, Evans L, Seigel N, Stuecker J. Metal oxide composites and structures for ultra-high temperature solar thermochemical cycles. Journal of Material Science 2008;43:4714–28.

[7] Steinfeld A. Solar hydrogen production via a two-step water-splitting thermochemical cycle based on Zn/ZnO redox reactions. International Journal of Hydrogen Energy 2002;37:611–19.

[8] Abanades S. Thermogravimetric analysis of CO2 and H2O reduction from solar nanosized Zn powder for thermochemical fuel production. Industrial & Engineering Chemistry Research 2012;51:741–50.

[9] Ernst F, Steinfeld A, Pratsinis S. Hydrolysis rate of submicron Zn particles for solar H2 synthesis. International Journal of Hydrogen Energy 2009;34:1166–75.

[10] Vishnevetsky I, Epstein M. Production of hydrogen from solar zinc in steam atmosphere. International Journal of Hydrogen Energy 2007;32:2791–2802.

[11] Abanades S, Charvin P, Lemont F, Flamant G. Novel two-step SnO2/SnO water-splitting cycle for solar thermochemical production of hydrogen. International Journal of Hydrogen Energy 2008;33:6021–30.

[12] Charvin P, Abanades S, Lemont F, Flamant G. Experimental study of SnO2/SnO/Sn thermochemical systems for solar production of hydrogen. AIChE Journal 2008;54:2759–67.

[13] Vishnevetsky I, Epstein M. Tin as a possible candidate for solar thermochemical redox process for hydrogen production. Journal of Solar Energy Engineering 2009;131:021007-1–021007-8.

[14] Scheffe JR, Li J, Weimer AW, A spinel ferrite/hercynite water-splitting redox cycle. International Journal of Hydrogen Energy 2010;35:3333–3340.

[15] Bhosale RR, Shende RV, Puszynski JA, H2 generation from thermochemical water-splitting using sol-gel derived Ni-ferrite. Journal of Energy and Power Engineering 2010;4:27–38.

[16] Neises M, Roeb M, Schmucker M, Sattler C, Pitz-Paal R. Kinetic investigations of the hydrogen production step of a thermochemical cycle using mixed iron oxides coated on ceramic substrates. International Journal of Energy Research 2010;34:651–61.

[17] Bhosale RR, Shende RV, Puszynski JA, H2 generation from thermochemical water-splitting using sol-gel synthesized Zn/Sn/Mn-doped Ni-ferrite. International Review of Chemical Engineering 2010;2:852–62.

[18] Lorentzou S, Zygogianni A, Tousimi K, Agrafiotis C, Konstandopoulos AG. Advanced synthesis of nanostructured material for environmental applications. Journal of Alloys and Compounds 2009;483:302–05.

[19] Bhosale RR, Khadka RP, Shende RV, Puszynski JA. H2 generation from two-step thermochemical water-splitting reaction using sol-gel derived SnxFeyOz. Journal of Renewable and Sustainable Energy 2011;3:063104-1–063104-16.

[20] Varsano F, Padella F, Alvani C, Bellusci M, Barbera A. Chemical aspects of the water-splitting thermochemical cycle based on sodium manganese ferrite. International Journal of Hydrogen Energy 2012;37:11595–601.

[21] Bhosale RR, Shende RV, Puszynski JA. Thermochemical water-splitting for H2 generation using sol-gel derived Mn-ferite in a packed bed reactor. International Journal of Hydrogen Energy 2012;37:2924–34.

[22] Scheffe JR, Steinfeld A. Thermodynamic analysis of cerium-based oxides for solar thermochemical fuel production. Energy & Fuels 2012;26:1928–36.

[23] Abanades S, Flamant G. Thermochemical hydrogen production from a two-step solar-driven water-splitting cycle based on cerium oxides. Solar Energy 2006;80:1611–23.

[24] Gal A, Abanades S. Dopant incorporation in ceria for enhanced water-splitting activity during solar thermochemical hydrogen generation. Journal of Physical Chemistry C 2012;116:13516–523.

[25] Lin X, Zhang Y, Wang Z, Wang R, Zhou J, Cen K. Hydrogen production by HI decomposition over nickel–ceria–zirconia catalysts via the sulfur–iodine thermochemical water-splitting cycle. Energy Conversion and Management 2014;84:664–70.

[26] Murphy IV J, O'Connell J. Process simulations of HI decomposition via reactive distillation in the sulphur-iodine cycle for hydrogen manufacture. International Journal of Hydrogen Energy 2012;37:4002–11.

[27] Dehghani S, Sayyaadi H. Energy and exergetic evaluations of Bunsen section of the sulfur–iodine thermochemical hydrogen production plant. International Journal of Hydrogen Energy 2013;38:9074–84.

[28] Lin S, Flaherty R. Design studies of the sulphur trioxide decomposition reactor for the sulphur cycle hydrogen production process. International Journal of Hydrogen Energy 1983;8:589–96.

[29] Hinkley J, O'Brein J, Fell C, Lindquist S. Prospects for solar only operation of the hybrid sulphur cycle for hydrogen production. International Journal of Hydrogen Energy 2011;8:589–96.

399843 Mitigating Ethylene Plant Flaring Risk during Equipment Switching Transitions By Dynamic Simulation

Abstract:

Equipment switching operations involving complicated operator procedures can increase the risk of flaring which can lead to environmental incidents and economic losses. Although flaring plays an important safety role in that it protects the operators and plant during process upsets, mitigating the risk of flaring is vitally important to improving process safety and sustainability. One example of equipment switching transition in ethylene plants is switching between solid desiccant dryers, commonly an operator procedure accomplished manually. The simulation was used to determine the conditions that would lead to a faulty dryer switching transition and identify the steps that when performed incorrectly increase the risk of flaring.

In this work, the dryer switching process was analyzed and simulated using the dynamic process simulation tool Honeywell UniSim Design. The transition of important process variables from the beginning of a dryer switch to the end were modeled by dynamic simulation. A general operating procedure was developed for the standard dryer switch procedure and various operator induced fault cases were developed to describe dryer switch transition faults based on potential procedural errors. The monitored variables were selected based on their sensitivity to faults during the dryers switching. Simulation results showed that some of the operator induced fault cases could result in natural gas header pressure spikes leading to flaring.

Copyright © American Institute of Chemical Engineers. All rights reserved.

397520 Sherlock Holmes, Why Trees, Bow Ties, and Investigating Process Incidents

Abstract:

When investigating crimes, Sherlock Holmes was an expert at observing important facts and deducing from them significant insights into what had happened. When investigating process incidents, PSM professionals similarly need to observe and develop important facts about what happened and deduce from them what might have caused the incident and how it can be prevented in the future. Are there things we can learn about incident investigation from Sherlock? This paper will review some key methods for investigating process incidents, applying appropriate insights from Sherlock. In particular, use of the why tree method for determining the root causes of incidents, and use of the bow tie method to better understand the contributions from failed or missing layers of protection, will be discussed.

Copyright © American Institute of Chemical Engineers. All rights reserved.

11TH GLOBAL CONGRESS ON PROCESS SAFETY

The Ten Commandments of Risk Based Process Safety

Robert Rosen
CCPS Staff Consultant
2088 Kingsview Road
Macungie, PA 18062
bobrosen@ptd.net

Prepared for Presentation at
American Institute of Chemical Engineers
2015 Spring Meeting
11th Global Congress on Process Safety
Austin, Texas
April 27-29, 2015

UNPUBLISHED

The Ten Commandments of Risk Based Process Safety

Robert Rosen
CCPS Staff Consultant
2088 Kingsview Road
Macungie, PA 18062
bobrosen@ptd.net

Keywords: Commandments, Risk Based Process Safety, Safety

Abstract

There are many critical points for inclusion in a Risk Based Process Safety program. This paper focuses on ten of them. While a person could come up with many more, these are the most important in the author's experience.

1. The Ten Commandments of Risk Based Process Safety - Intro

Instruction of many "Foundations of Risk Based Process Safety (RBPS)" courses has resulted in the realization that the volume of material presented in RBPS training could be overwhelming. The need for a simplified tool to ensure critical points are understood and actions follow upon the "Ten Commandments of Risk Based Process Safety" were developed to leave with the participants to highlight the concepts of the course. It is important to understand the breadth of each "commandment", for full integration and application. The ultimate intention is that each item will serve as a stimulus for taking action upon returning to your workplace.

2. The Ten Commandments of Risk Based Process Safety

Each commandment is listed below with an explanation of the relative concept.

I. Thou shalt always honor thy container.

Businesses that are diligent about keeping chemicals and energy in their respective containers in order to eliminate unplanned releases, significantly reduce the likelihood of

a process safety incident. In order to create a low likelihood of a process safety incident the following are required; proper operating procedures, a strong mechanical integrity program, appropriate training and routine observations.
It is also very important to include all containers in a process. Many significant incidents occur with finished materials storage tanks, raw material storage tanks, and others. Transportation containers need to also be included in your process safety management activities. A transportation incident can result in a process safety incident which may severely damage not only a company's reputation but can also affect the lives of others, especially those in the surrounding communities. Due diligence is necessary when ensuring that the appropriate containers are used and inspected, and that all transportation rules and guidelines are followed.

II. Thou shalt always maintain a sense of vulnerability.

Complacency equals incidents. The potential for a process safety incident is always present. This concept needs to be a critical part of the mindset of every employee and every contractor. In order to keep this active in any organization it requires consistency in practice form everyone in the organization. Safety awareness is not the job of just one or two people. It needs to be on the minds of every employee. The thought that it can't happen here, or that it has never happened before need to be dispelled.

III. Thou shalt eliminate normalization of deviation.

Normalization of deviation is accepting operations which are outside of set limits. One example is when the temperature limits on a reaction is set at 75 degrees C, plus or minus 2 degrees C. Over time the process is allowed to run consistently at 76 degrees C, then 77 degrees C, and then at 78 degrees C. This condition of constantly running at higher than the original set limit has reached a level of acceptability. Over time, it has the potential of further exceeding the original limit if no action is taken. At some point it may reach a temperature where a process safety incident is experienced. Designing a process without strict adherence to design standards is another area where normalization of deviation tends to exist. If a deviation is necessary in any circumstance, steps need to be taken such as management of change, supervisory approval and subsequent documentation to change the previously set limits. Advancements in technology as well as knowledge from experience (either on-site or from field experts) will dictate whether a deviation is acceptable or not. One individual should not determine whether a deviation is allowable. A deviation needs to be researched and based upon sound industry practice.

IV. Thou shalt know thy chemistry.

It is extremely important to know and understand the chemistry of everything under your control. This knowledge includes: side reactions, operating limits and why these limits are what they are, stability, reactivity, and other properties. This knowledge enables a person to make changes without additional risk. It must be determined if additional testing is required for a process change. This information also provides the "why" answers to those who operate the process and instills upon them commandment II, Thou shalt always maintain a sense of vulnerability.

V. Thou shalt educate, train and drill employees.

This covers everything from evacuation to regaining control of potential runaway reactions. As changes are made in a process, this thought becomes extremely important. Do not forget there is a big difference between driver education and driver training. The same concept applies to Risk Based Process Safety. After the education and training is complete, the drill is performed to ensure that the education and training was effective. Adjustments to training content or procedures, or retraining, may be required to ensure full demonstration of knowledge by all participants. The use of a pre test and post test method may be beneficial in determining the effectiveness of the training delivered and help in identifying areas of weakness. Annual retraining is another area that should be considered. This may go into more depth than the initial training.

VI. Thou shalt create and nurture a strong risk based process safety culture.

Without a strong safety culture all other efforts will eventually fade away. A common phrase used when teaching RBPS is "will you act or perform the same, even if no one is watching". The culture needs to be all encompassing, from the CEO and Board Members throughout all levels of the entire organization. Process safety is a way of life. It factors into every task and assignment. There must be a culture that allows for open and sometimes uncomfortable communication. A strong process safety culture may save a company from its ability to exist or not. Without a strong process safety culture, the whole program may deteriorate to a point of ineffectiveness and result in a process safety incident.

VII. Thou shalt recognize those who exemplify process and occupational safety.

In the past, and even now, occupational safety goals and objectives are commonplace. Some examples include recordable incidents, lost time incidents, and near miss reports. It is important that I use the term recognize rather than reward. In process safety it is important to recognize the attitude, leadership and performance of individuals. Recognition may take many forms such as a mention at a staff meeting, a verbal thank you for your idea when walking through the plant, or a written note to the individual with a copy in their personnel file. Process safety incidents usually occur much less frequently. An order of magnitude or two less than occupational safety incidents would not be uncommon. Rewards are usually financial in nature and reserved for items such as cost control, production records and improved quality.

VIII. Thou shalt not tolerate omissions in documentation.

This goes back to culture. Signatures, check marks or other written requirements must be completed. If not, it begs the question of what else is not completed as it should be? One example is the omissions on permits. There are many instances where permit omissions have resulted in significant losses. These omissions are also an indication of a lack of operating discipline. Keep in mind the adage that if it is not documented, it did not occur.

IX. Thou shalt not manage from behind thy desk.

Visibility of all levels of management is critical to ensure a strong safety culture and reduce risk. Managers need to communicate, demonstrate their commitment and maintain availability for questions. A safety audit once revealed an intolerable situation of free flowing solvent into an open vessel, tank 314. When the plant manager was approached and told of the situation, he replied that he did not believe the situation existed because he had issued an edict that free flowing solvents will not be allowed. When the auditor offered to show him the problem and he replied, "Where is tank 314?" This example is just one of many where management , at all levels, need to be visible on a daily basis as a minimum.

X. Thou shalt not violate rules.

There are only two choices that you have with a rule. Follow it or change it. With this mindset in place you have demonstrated a strong safety culture and reduced the likelihood of a process safety incident having a severe effect on your operation. It is a good idea to thoroughly review all rules on a regular basis to ensure their appropriateness. Discipline may have a role when rules are violated.

Conclusion

The major point to be made is one cannot rest when you are involved with process safety. Bad things can happen to good companies. What needs to be done is to ensure that your company does not experience a process safety incident. Look for the warning signs. Talk to everyone. Establish the culture so that process safety is always considered, whether it be a very trite item or a major revamp of a process.

Developing Credible Scenarios for a PHA

Nestor A. Paraliticci Jr.
Provenance Consulting, LLC
P.O. Box 3329
Borger, TX 79008
nestor.paraliticci@provenanceconsulting.com

Prepared for Presentation at
American Institute of Chemical Engineers
2015 Spring Meeting
11th Global Congress on Process Safety
Austin, Texas
April 27-29, 2015

UNPUBLISHED

AIChE shall not be responsible for statements or opinions contained
in papers or printed in its publications

Developing Credible Scenarios for a PHA

Nestor A. Paraliticci Jr.
Provenance Consulting, LLC
P.O. Box 3329
Borger, TX 79008
nestor.paraliticci@provenanceconsulting.com

Keywords: PHA, causes, consequences, scenarios

Abstract

OSHA's Compliance Guidelines and Recommendations for the PSM Standard states "A PHA is an organized and systematic effort to identify and analyze the significance of potential hazards associated with the processing or handling of highly hazardous chemicals."[1] It goes on to say "A PHA is directed toward analyzing potential causes and consequences of fires, explosions, releases of toxic or flammable chemicals and major spills of hazardous chemicals." [1] OSHA does not give any additional guidance; they just refer you to 18 additional "Sources of Further Information". I would like to present a process on how to develop scenarios (causes and consequences), tips on creating credible and realistic scenarios and the pitfalls to avoid during the process, commonly encountered during a PHA.

1. Introduction

What does it take to develop a credible scenario? A team of pessimistic employees? A wild imagination? No, all that is needed is a systematic approach and an experienced team to develop scenarios that are plausible and realistic. We will start with a brief review of the PHA process in order to develop that systematic approach. Then we will work on how to identify causes and some pitfalls to avoid during the process. We will then focus on how to determine credible and plausible consequences, pitfalls to avoid when determining consequences, and looking at some examples of what to expect from good and bad scenarios, to better understand the benefits of spending the time needed to develop credible scenarios.

2. Process Hazard Analysis (PHA)

Now OSHA describes a PHA as "...a thorough, orderly, systematic approach for identifying, evaluating, and controlling the hazards of processes involving highly hazardous chemicals."[1] And what does this all mean? Basically OSHA is letting us know what needs to be done but doesn't say exactly how to do it. So let's begin by looking at what a good PHA looks like. There are several items that should be covered when performing a PHA, hazards of the process, consequences of failure of engineering and administrative controls, and a qualitative evaluation of the possible safety and health effects of these failures on employees. Please keep in mind this is not an all-inclusive list of items that need to be covered but covering just these items will give you a good solid foundation to your PHA.

2.1 Hazards of the Process (aka Process Hazards)

Simply put a process hazard is the chemical or physical characteristic of a material, system, or process that has the potential for harm your employees or the environment. When reviewing the hazards of the process, we would focus on the process, what it contains, how it reacts and what the inherent dangers of the process are. For example, if we are analyzing a process that contains light hydrocarbons, butane or propane, we already know light hydrocarbons are extremely flammable, they can form vapor clouds that can go on until the find an ignition source and flash creating huge fires. Also depending on the process the light hydrocarbons are going through, they may contain toxics (H_2S, benzene, etc.) that can make the stream a health hazard if personnel are exposed to it. These are just a couple of examples of process hazards, we can go on and on for hours talking about things like reactivity, temperature swings, vibrations, etc. One other aspect of process hazards we would look at is evaluating these hazards for the different modes of operation. Do these hazards existing during normal operations or are they more prevalent during non-routine operations, for example during start-up, shutdown, turnaround operations or maybe there is batch operation that is not performed very often. We need to make sure we include these non-routine operations in our review because we have seen time and time again many incidents happen during these times.

2.2 Consequences of Failure

'What happens if...?" I cannot begin to tell you how often I have used this phrase, a hundred, a thousand, a million times? It's even gotten to a point that I catch myself saying it when I'm talking to my kid at home, but this is what every good PHA starts with. Identifying the true consequence of a failure of engineering or administrative controls is the bulk work of any PHA; stated another way "What can go wrong?" You have a flow controller on the bottom of a tower and it malfunctions open or it is inadvertently opened by an operator, would you have a gas blow-by and potentially over pressure downstream equipment or would you have a loss of level and cavitate a pump leading to a seal leak? The consequence of this scenario can lead to personnel injury, environmental release, equipment damage, and/or business interruption. So why it is important to know what happens? Mainly because if you do not know or do not identify the true consequence of a failure, how do you know if you are protected or have enough protection against that hazard? In the next couple of sections of this presentation, we will be talking about identifying causes and determining consequences in more detail.

2.3 A Qualitative Evaluation (aka Risk Ranking)

During the PHA, a qualitative evaluation of the possible safety and/or health effects of the failure of engineering or administrative controls on employees and/or the environment, in other words we are "risk ranking" the scenarios of the system under review. When you Risk Rank a scenario, you are assigning numerical value describing the potential harm to people or the environment taking into account the likelihood of the event happening. And by the end of the PHA, you have then identified multiple high risk scenarios that could result in injuries to people, environmental events, property loss, and/or business interruptions. And in doing so, the PHA provides information which you can use to make decisions on improving safety, reducing the risk of the processes, and at times help with budgeting, ensuring funds are properly allocated or even if doing a project PHA is the project is cost effective.

2.4 Snapshot of the Complete PHA Process

As you can see in Figure 1 below, there is a systematic way of going through a PHA. First you determine what the cause of your scenario is. All of the causes in you process should be identified and analyzed during the PHA. Once you have identified a cause, then the PHA Team should then talk about all credible and plausible consequences of concern. These consequences of concern do not solely have to be safety related, it could also include environmental impact, property loss, or business interruptions. We will talk in more detail about this part of the process in the following sections. After the consequence is determined then the safeguards are identified - these are any device, system, or action that would interrupt the chain of events following the cause that would prevent or mitigate the consequence. Once you have the safeguards identified then the discussion should turn to "are you sufficiently protected against the risk identified", if so then you have an acceptable risk. If not, then the Team can come up with recommendations to help prevent the consequence from happening or mitigate the "Risk" to an acceptable level.

As you can see, if you don't have a good cause then do you have a valid consequence? And if you don't have a valid consequence, are you spending your time and resources on protecting what needs to be protected?

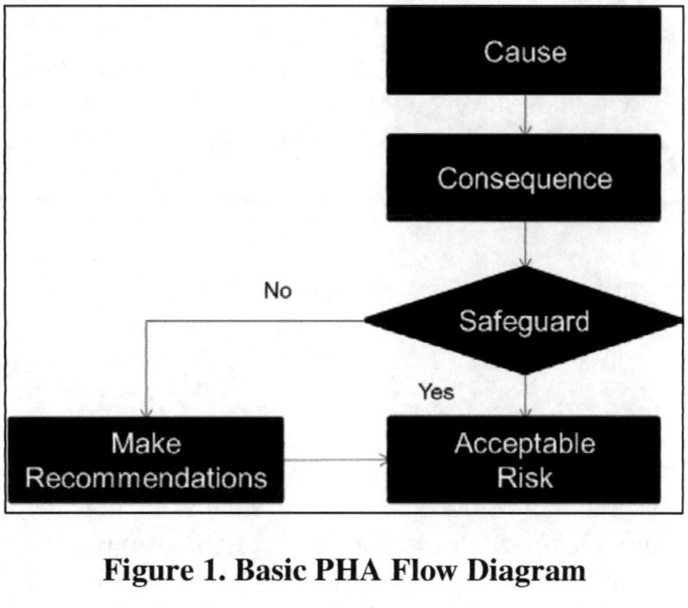

Figure 1. Basic PHA Flow Diagram

There are many different methodologies you can use to conduct a PHA, but for our discussions today I will be referring to the HAZOP Methodology which is a more extensive review as it analyzes each cause, valve by valve of the process; nowadays this methodology is the most widely used.

3. Identifying a Cause

"In war, events of importance are the result of trivial causes." - Julius Caesar[3]

3.1 Identifying Causes

Are we at war? No, but like war, in our industry, if we do not properly identify or understand the hazards we work with, we are faced with the possibility of fires, explosions and releases of toxic chemicals. So where do we start? By first determining what "causes" can end up leading to a hazardous consequence. Merriam-Webster Dictionary defines a "cause" as "something or someone that produces an effect, result, or condition: something or someone that makes something happen or exist"[2]. Earlier I stated that all causes should be identified, so where do we start? Start with the "something or someone" stated above, this represents the human and/or equipment failures that produces an effect, result, or condition but keep in mind at this point we don't know what the consequence is yet or even if it is a consequence of concern. The following are some tools you can use to help identify some causes for your process:

1. Process Descriptions – get familiar with the process and process conditions, start by identifying process hazards of your system under review and what can cause me to deviate from the established process conditions?
2. Process Flow Diagrams/Block Flow Diagrams – once familiar with the process, you can start identifying the major equipment that you will be working with, get a better overall understanding of the process and start the discussions on how the equipment fails.
3. Process & Instrumentation Diagrams (P&IDs) with Deviation Tables ("No Flow", "More Temp", etc.) – use the P&IDs for the process under review as a guide to help identify causes, another way to look at this is by thinking of the deviations as "Root Causes", and then identify on the P&IDs the specific equipment that would cause these deviations.

3.2 Tips on Determining Causes

I believe that since determining the causes of scenarios for the process under review is usually the easiest part of the process, it is easy to get complacent with the process at the beginning which could cause issues down the line. Here are some items to keep in mind when identifying causes:

1. Is it a credible cause? Sometimes the location of the equipment, the type of equipment, or assumptions made will determine if it is a valid cause of not. You do not want to waste time on items which are not realistic, for example you are looking at a 12" block valve in a middle of a pipe lane that is 20 feet in the air with no access points, is an operator going to inadvertently going to close/open that valve? PHA Team discussions are essential at this point.

2. Is it a cause or a consequence? It's easy to list a consequence as a cause (not a good practice for a HAZOP) which makes it difficult to determine what you are protecting against? For example, I have seen where the cause is listed as "Tank is overfilled." then the consequence is loss of containment, fire and personnel injury. Then when it comes to listing the safeguards what are you protecting against operator error, malfunctioning equipment, pump shutdown? Are you properly protected? PHA Team discussions are essential at this point.
3. Do not lump causes…initially. Sometimes it's easy to lump several valves on a line as being the same but they may have completely different consequences and even if they have the same consequences they may need different types of safeguards. For example you are on one P&ID and list a valve being misaligned open and evaluated that scenario but further down the process you see a control valve on another P&ID, it's very tempting to say, "Oh that will cause the same consequence we talked about several scenarios ago." and then lump it as another cause. Then not realizing that the alarm on the flow indication controlling that valve was used a safeguard for the scenario, now you have just lost a safeguard and have a scenario that is not properly protected. PHA Team discussions are essential at this point.
4. If you are pre-populating documentation in preparation for the PHA – don't assume you got it all! Talk with the Team members to see if there are any other causes in the system you are reviewing, the drawing may be incorrect or there may be another cause outside of the P&ID you are currently looking at, double check you are covered.

Did you notice what these tips all had in common? Yes, PHA Team Participation. As PHA Facilitators we may have a good understanding about processes and hazards but at times we have no idea what the system looks like out in the field, where the valves are, how long it takes to actually respond to an alarm, and the small quirks it takes to operate the system and these are just some of the questions that can be resolved by having good discussions between Team members. An engaged Team will help make the PHA process easier and more effective.

3.3 Documenting Causes

Now that we know how to identify causes, how do we capture these causes on paper? There is nothing more frustrating than opening an old PHA to review some scenarios because of an issue in the plant and looking at the cause and seeing "Block valve closed." Then realizing you have 20 drawing of the system and not knowing quite sure which of the 50 or so valves in the system the cause refers to. Sure you can spend the next half hour looking at the consequence and the safeguards of the scenario and making an educated guess as to which valve they were referring to but now you have wasted time researching something that should only have taken minutes to review. So here are some tips on how to document these causes:

1. Be specific about name/description of the cause – put down valve size, tag numbers, equipment name and number, etc.
2. Be specific of the failure of the cause – i.e. Valve misaligned closed, exchanger tube leak, Control valve malfunctions open, etc., this will help identifying the scenario at a later date easier

3. Document location and drawing number – put down the grid location and drawing number to make finding and identifying the cause at a later date easier, this helps a lot when you have to flip through a couple of hundred drawings looking for one cause.

This may seem time consuming at the time, but remember, people will be looking at this document for years and the more information you have the easier it will be to find what you are looking for.

4. Determining the Consequence

Michael Jordan once said "I never looked at the consequences of missing a big shot... when you think about the consequences you always think of a negative result." [3]

Unfortunately, when we talk about consequences in our industry we have to do the exact opposite of the philosophy which made Michael Jordan great. In order for use to succeed, we need to not only think about those negative results but also how to fix them and that is the main reason we need good consequences so we know what could happen and what is needed to properly mitigate the issue.

4.1 Developing a Consequence

OSHA states that the purpose of the PSM Standard is for "…preventing or minimizing the consequences of catastrophic releases…"[1] is this the best evaluation approach to take? Catastrophic by definition means the nature of a catastrophe, or disastrous event; calamitous. But what constitutes "catastrophic" – having 1 fatality, 2, or 5, hurting 10, 20, 30 people, contaminating 10, 20, 30 miles of coastline? Or simply losing a day's pay or your job because of an incident that happened at work? I would say that any of these would easily qualify as catastrophic especially if it directly affects you. But in our industry do we want to look at only "consequences of catastrophic releases"? It would be easy to overlook smaller scenarios that would cause smaller damage but if stopped might have prevented "The Big One". As you may know from incident investigations, there are typically several causes to an incident, so think of it this way, would that incident have happen if we mitigated just one of those causes? I believe a better approach is looking for consequences which are credible and plausible and then dealing with those consequences. And at times these credible and plausible consequences could be those worst case scenarios.

In developing a consequence there are several things you should take into consideration. First, all active safeguards (prevention and/or mitigation) associated with the scenario should be assumed to fail; we want to make sure we get an accurate depiction of what could happen. Safeguards will come into play later in determining the likelihood of events and when that time comes we will determined if they are effective to help mitigate the consequence we just developed. Secondly, consequences should be evaluated by looking for the realistic potential consequence, as discussed above. A realistic consequence is not normally the "ultimate worst case" consequence (but it could be) and must be plausible and credible. And lastly, once the consequence has been determined, consider the location of the event and adjust the consequence accordingly. The specific location of a release scenario can either reduce or increase the overall impact of the event, for example if you have a pipe gasket leak in the middle of a process area is that bigger hazard than having the same gasket leak in a pipe run in a tank field? And how do you resolve all of these items, PHA Team Participation! Again, these should be experienced people with real world knowledge of the process under review, if you don't get their participation you will not have a good PHA and as a result a potentially dangerous place to work.

4.2 Tips on Identifying Consequences

Unlike identifying causes, developing the consequences of you scenarios is not only the most important part of the PHA but also the hardest. Here are some tips on getting good results:

1. All active safeguards fail! No I am not repeating myself; it's been my experience this is the hardest part of the process to go through. Remember you have a PHA Team with years of experience and they are trained to look at the process and see how it is supposed to work and react if it doesn't look right. Now we are asking them to pretend you don't react, to pretend that alarm that they have always counted on doesn't work, or their fellow co-workers makes a mistake. It difficult to change that thought process so make sure you always remind them those safeguards aren't working.
2. How far do you go? We can basically run any event down through the entire process every time but is that worth it? Remember plausible and credible, concentrate on what would realistically happen and if you stop that event you will most likely stop the rest from happening.
3. If you are not sure, ask! Your typical PHA Team requirement is for someone with expertise in engineering and process operations along with a person knowledgeable on the PHA Methodology. That leaves out a lot of other disciplines, i.e. reliability, design, environmental, instrumentation, which the people in the room may not have the experience with. So if you come across a piece of equipment the Team doesn't feel comfortable talking about the failures or you do not have the proper engineering design parameters, stop and call someone. The Team will have more confidence in the PHA if they believe they are getting good information.
4. Know your consequences of interest. Make sure before you begin your PHA, that the Team is all on the same page as to what the PHA is going to be focusing on, is it safety, environmental, asset damage, business interruption, etc. You want to make sure the team focuses on what needs to be completed and not waste time on items that are not of interest.
5. PHA Team Participation! PHA is basically a group of people coming together and brainstorming about issues in their facility. If they are not talking because they are afraid or shy, make them feel comfortable and express to them that we are all here to create a safer work environment and their opinions and thoughts are valuable.

4.3 Documenting Consequences

Just like documenting causes, a well-documented consequence is very important. At the time of the PHA when the PHA Team members are in the zone, they know why and how the event happens and can explain in detail why they think its occurring, but wait for a month or a year and ask the same person why the consequence occurs and it may be difficult for them to explain the reasoning behind it. So good documentation helps preserve the thoughts and reasoning behind these consequences. Even during the PHA, if you have a very vague consequence, for example: loss of containment of a flammable and/or toxic material, could the severity differ from a flammable liquid release to a toxic material release? Wouldn't it be clearer if it stated, loss of containment of a flammable liquid (gasoline) into a containment area, no you know exactly what was spilled and where it was spilled so you can determine how bad of a scenario this is and what safeguards you have to mitigate this particular scenario. Here are some tips to remember:

1. Be specific about names/descriptions of the equipment– put down valve size, tag numbers, equipment name and number etc.
2. Be specific of the sequences of events – from initiating cause to the final element, knowing the sequence of events will help in the latter part of the PHA when identifying safeguards and potential mitigations, remember we are trying to break the link in the sequence of events. This will also help to understand the scenario a couple of years down the line.
3. Document location and drawing number – put down the grid location and drawing number of the final element to make finding and identifying the location of the event at a later date easier.

Again, this may seem time consuming at the time, but remember, people will be looking at this document for years and it will save a lot of heartache if it is easy to understand.

5. Conclusion

"Shallow men believe in luck or in circumstance. Strong men believe in cause and effect." - Ralph Waldo Emerson[3]

So what now? We have identified valid causes, documented credible and realistic consequences what are the benefits of all the time we just spent? Following are some benefits of developing these good scenarios, keep in mind these are not in any particular order of importance and it's not an all-inclusive list.

- Verified you have the proper safeguards to mitigate event – without knowing what the true consequence of a scenario is, you cannot determine if the safeguards you have in place are adequate to protect you against the outcome. This not only translates to peace of mind of having a safer workplace but also money savings from limiting/preventing injuries, asset damage, and business interruption.
- PHA Team has identified high risk areas in your process – what better way to have a way to focus your time, money and resources and tackling the problems you know can hurt you the most first.
- PHA Team has determined the proper mitigation for the specific consequences discuss during the PHA – again this is the peace of mind one has knowing that the next time you go to work you have that reassurance that it is a safer place to work at because of the PHA.

- Morale Booster – this is something that is hard to quantify, but if a PHA Team feels good about the PHA and see the positive changes that can occur because of the PHA, they can say they made a difference and share with their co-workers what they found and how they are going to correct it – the process does work.

We hear about industry incidents that hurt people or damage equipment and we ask ourselves "What happened?" Or "How could that occur?" Well if we develop the scenarios well to the point we have identified the hazards and what we have or need to protect ourselves against the hazards, then hopefully people won't have the chance to ask those questions about or sites.

6. References

Reference in order of first citation

Web resources:

[1] U.S. Department of Labor, Occupational Safety & Health, Available at https://www.osha.gov/, Accessed on November 13, 2014.

[2] Merriam-Webster Dictionary, Available at http://www.merriam-webster.com/, Accessed on November 13, 2014.

[3] Brainy Quote, Available at http://www.brainyquote.com/, Accessed on November 13, 2014.

400174 Reactive Production Scheduling for Ethylene Cracking Furnace System

Abstract:

Thermal cracking furnace system is the most critical section of an ethylene plant, where multiple furnaces in parallel are used to break various hydrocarbon feedstock (e.g. naphtha, ethane, and propane) into smaller-molecular hydrocarbon products. Due to its performance-decaying operational characteristic, the cracking furnace system has to experience periodic hot and cold maintenance, which consequently upset the furnace system operation and the downstream process operations. Also, in practice, because the feed supply conditions change constantly, it is necessary to implement a reactive scheduling strategy which can smartly reschedule the furnace operations with respect to any new feed delivery.

The previously published works are either focused on simulation and control aspects, or simply scheduling the furnace for maximum average daily profits without incorporating some realistic operational constraints. In the proposed scheduling model, most of the major scheduling issues concerning a cracking furnace system have been addressed, such as multiple feedstock, multiple furnaces, hot shutdowns for decoking operations, cold shutdowns for mechanical maintenance, decaying cracking performance, non-simultaneous shutdowns, split cracking and recycled C2/C3 cracking. Specially, to mitigate the flow rate upsets caused by furnace shutdowns to the downstream operations, the developed scheduling model considers different measures commonly applied in industry, such as additional furnace capacity makeup. Furthermore, the scheduling model is constructed in a reactive and iterative way to seamlessly make transitions from current schedule to reschedule under any delivery of new feeds and on a rolling forward scheduling horizon. The efficacy of the developed scheduling model is demonstrated by some case studies.

Copyright © American Institute of Chemical Engineers. All rights reserved.

400225 An Innovative Process Design for Ammonia Synthesis Based on Self-Heat Recuperation Technology

Abstract:

Ammonia is not only an important chemical raw material in industry, but also a potential energy carrier for on-board vehicular hydrogen storage. The modern ammonia synthesis process is mainly based on Haber-Bosch process, which involves the reaction of gaseous nitrogen with hydrogen on Fe-based catalysts at high temperature and high pressures. Due to the thermodynamically limited low conversion rate (usually 10%–20%) of hydrogen-nitrogen mixture to ammonia, ammonia synthesis plants are commonly implemented as recycle process, in which the unreacted reactants are separated and recycled.

The energy requirement of ammonia synthesis process strongly depends on the loop design of ammonia synthesis. As the conversion rate is just 10 to 20%, to produce 1 kg ammonia, 4- 6 kg nitrogen and hydrogen gas is usually recycled in the process. Thus, a large amount of energy is required for the ammonia synthesis compressor and refrigeration cycle compressor, even in the modern heat-integrated processes. Therefore, to reduce the energy consumption of ammonia synthesis is desirable.

In this study, we investigated the feasibility of applying self-heat recuperation technology to the ammonia synthesis process, and proposed a novel ammonia synthesis process from an energy saving viewpoint. Both thermal and cold energy were circulated in the process, largely reduced the exergy destruction. The conventional and proposed ammonia synthesis processes were simulated with a commercial software, PRO II 9.2. The results showed that the energy requirement for the proposed ammonia synthesis process was reduced by around 50% against the conventional process.

Copyright © American Institute of Chemical Engineers. All rights reserved.

398060 PHA Facilitation and Teamwork Dynamics: Or Why Not All Hazops Are Created Equal

Abstract:

Process Hazard Analysis (PHA) is a fundamental element of every process safety management program. The various techniques (what if, HAZOP, HAZID, LOPA) used to evaluate the risks associated with a process, project, or particular piece of equipment, are well established and in many companies they have been standardized and templates developed for their execution. Yet, not two PHAs are the same. The fact that a PHA is always a team effort where ideas are shared can lead to very different outcomes depending on the facilitator skills, and the team members and their attitudes. Based on the author's direct experience a PHA can be regarded as a very powerful safety tool or it can be one of the worst punishments for a gung ho engineer or operator. Since all qualitative and semi quantitative PHA techniques are structured around brainstorming it is important to think about how to generate, share, and parcel out these ideas. The risk of turning a PHA into a boring and endless exercise is very real, but the opportunity of using the analysis sessions as a stage where the PHA team members can showcase their vast knowledge can be a very rewarding one. This paper collects observations from the authors from great PHAs and from some of really bad ones, it presents ideas on how to think outside the HAZOP box, and how to empower and engage the review team members.

Copyright © American Institute of Chemical Engineers. All rights reserved.

396840 Assimilating Design Formulation and Design Review into a HAZOP - Lessons Learned and Additional Tips

Abstract:

There have been numerous guidebooks written for the application of Hazard and Operability (HAZOP) Studies to mature designs, e.g., operating facilities and capital projects where the design is complete. For these systems, key process safety information is typically available, but for fast-moving capital projects that is oftennot the case, and in addition there are often many contrasting issues:

- Design flexibility allows for more extensive changes at early stages of the project. Even the ability to cost-effectively make larger numbers of smaller changes can have a measureable improvement in safety and decreased operational risk.
- Less-detailed design information at early stages of the project can result in an incomplete safetyassessment.
- Design changes can have a more significant negative impact on project schedule later in the project cycle.
- Designers can benefit from feedback obtained from Plant Operations Personnel and from the thought process invoked by the HAZOP approach.

These issues represent fundamental contrasting priorities. Although applying a designed-to-be-detailed analysis tool like a HAZOP Study at early stages of the project is a fundamental challenge and presents many dilemmas to the HAZOP Study Facilitator, the application of a phased approach can help achieve the true objectives of the HAZOP Study, which is to make as many safety improvements as practical to minimize risk.

The "double-edge sword" is by incorporating HAZOP Study approaches early in the design process, important safety improvements can be done more cost-effectively, allowing limited project funds to have a greater impact in lowering net risk, but it may require some "re-visiting" of the HAZOP Study. The alternative (doing a final HAZOP Study, after all of the design work is complete) obviates the need for any re-assessment of safety issues, but it also eliminates the practical ability to implement any design changes, unless the risk is very high. Thus, if carefully managed, the introduction of HAZOP concepts early in the design cycle, evolving the level-of-detail with design maturity, and the performance of focused updates to the HAZOP Study, can be used to most cost-effectively minimize the risk associated with the project design. This approach can also be used to minimize last-minute design changes that could have had the potential to greatly increase project costs and result in project delays.

Our objective is to allow the audience to walk away from this with new concepts, practical examples, and charts and graphs that they can share with capital project management to allow a shift in paradigm that uses a HAZOP Study in a new way that provides for safer process facilities, while likely streamlining project schedules and budgets. In this way, the HAZOP is not perceived as a project impediment, but is seen as support of key capital project goals and objectives.

This paper will build on the insights presented at the Eighth GCPS, with some additional lessons-learned and application tips addressing the latest design challenges and increased application of Layer of Protection Analysis. Another key focus for this paper will be tips on absorbing key safety insights from the system designers and structuring the D-HAZOP for use by the Owner/Operator throughout the life of the facility.

Copyright © American Institute of Chemical Engineers. All rights reserved.

How to Survive an OSHA National Emphasis Program Audit

James E. Johnstone
Contek Solutions LLC
6221 Chapel Hill Blvd., Ste. 300 Plano, Texas 75093
JIMJ@CONTEKLLC.COM

Shannon Mitchum and David Zink
Contek Solutions LLC
6221 Chapel Hill Blvd., Ste. 300 Plano, Texas 75093
SMITCHUM@CONTEKLLC.COM and DZINK@CONTEKLLC.COM

Copyright 2015 by Contek Solutions LLC

Prepared for Presentation at
American Institute of Chemical Engineers
2015 Spring Meeting
11th Global Congress on Process Safety
Austin, Texas
April 27-29, 2015

UNPUBLISHED

AIChE shall not be responsible for statements or opinions contained
in papers or printed in its publications

How to Survive an OSHA National Emphasis Program Audit

James E. Johnstone
Contek Solutions LLC
6221 Chapel Hill Blvd., Ste. 300 Plano, Texas 75093
JIMJ@CONTEKLLC.COM

Shannon Mitchum and David Zink
Contek Solutions LLC

Keywords: Audits, Compliance, OSHA and Process Safety

Abstract

The Occupational Safety and Health Administration (OSHA) instituted a "National Emphasis Program" to conduct both targeted and random inspections of facilities subject to the Process Safety Management (PSM) regulations. OSHA responded to several high profile incidents by initiating a program to help assure that facilities handling highly hazardous chemicals are meeting the PSM regulations in an effort to reduce incidents. OSHA is currently inspecting or auditing PSM facilities and will certainly issue citations if the facility does not meet all of OSHA's requirements. This paper presents the results of the most commonly found deficiencies at the facilities and also provides information that can be used to make sure that a facility's PSM program will indeed pass an OSHA inspection under the National Emphasis program.

The Process Safety Management standard is "performance" based which means that companies must demonstrate compliance through programs and policies in place. Auditing a "performance based" system can be abstract as the auditing protocols may not be definitive. Both the operators of facilities and the regulators rely on recognized and generally accepted industry practices and past letters of interpretation from OSHA to help determine if a facility's PSM program is meeting the requirements of the regulations.

However, there are still definitive requirements in the PSM standard that plant operators must comply with. Given these facts, it is critical that a facility undergoing a PSM review make sure that programs and policies designed to meet PSM requirements are fully documented, communicated and implemented.

OSHA initiated the "PSM Covered Chemical Facilities National Emphasis Program" in 2011. Over the past three years many plants have undergone PSM compliance reviews. Industry has learned from these reviews where common vulnerabilities exist in their programs, more about what OSHA is looking for in terms of compliance and how to improve PSM efforts.

This paper presents information learned by both industry and OSHA about how PSM compliance efforts can be improved. The techniques presented in the paper, to improve PSM programs, enables the facility operator to make sure that their PSM compliance efforts are in line with industry norms and OSHA's expectations for compliance. The result of implementing the findings and recommendations found in this report will help industry to improve process safety performance and regulatory compliance resulting in fewer PSM related incidents.

1. Overview of the NEP Program

In July of 2009, OSHA [1] implemented a pilot version of the National Emphasis Program (NEP) for PSM covered chemical facilities. The pilot program was used by OSHA for programmed inspections in Regions I, VII and X only. OSHA decided to institute the NEP pilot program as a "new" approach for inspecting PSM covered facilities. Previously, OSHA relied on the CPL-02-02-045, ("Process Safety Management of Highly Hazardous Chemicals – Compliance Guidelines and Enforcement Procedures") and realized that only a few Program Quality Verifications ("PQV") inspections could realistically be done each year. The new NEP approach was implemented to allow for a greater number of inspections by OSHA through better allocation of resources.

Two years later, in November of 2011, OSHA sent out a notice [2] instituting a nationwide program replacing the pilot program. The new program modified the original 2009 NEP program and was launched OSHA-wide.

2. Will my facility be inspected?

OSHA targeted certain types of facilities to be inspected. The sources of determining which facilities to "target" for the inspections included:

1. Level 3 facilities designated by the EPA's Risk Management Program (RMP)

2. Explosive Manufactures

3. OSHA's Integrated Management Information System (IMIS) which is an online database containing OSHA's investigation summaries

4. Knowledge of OSHA local office representatives of local facilities. Often, facilities are made known to the local OSHA office representative through press reports of incidents, announcements of new facilities or if the facility is in a very active region for PSM facilities.

Based on the four sources used to select facilities, it is surmised that OSHA can literally select any facility to conduct a NEP program audit on which it so chooses. No PSM covered chemical facility can be immune from the NEP inspections.

OSHA has also set a goal of conducting 3-5 "programmed" inspections per year for each area office. OSHA currently has roughly 90 area offices and there are states with their own State OSHA programs are obligated to participate in this program also. Consequently, OSHA's goal is to inspect 270 to 450 sites per year. There are over 7,000 PSM covered facilities in the United States [3] which roughly means that OSHA is looking at inspecting between 4 to 6% of the nation's PSM facilities each year.

Your facility may also be selected for a NEP inspection if it is selected as part of the "Site-Specific Targeting Plan" (SST) or is caught up in the "un-programmed" inspection methodology. "Un-programmed" inspections result from either a complaint for in response to an incident or to a potentially catastrophic event.

3. Who will OSHA send to inspect my facility?

Inspections by OSHA will be carried out by either a single OSHA employee or a team of OSHA inspectors. OSHA has set up qualification guidelines for their inspection personnel which include a combination of training and experience with PSM facilities. The qualification guidelines try to match up the inspector's expertise with particular portions of the PSM standard. For example an inspector with little experience ("Level 3") may only evaluate compliance with the employee participation, training, contractors, hot work permitting, incident investigations and the emergency planning and response portions of the PSM requirements.

4. What is OSHA's "Dynamic List?"

A key part of the NEP was the development of "dynamic list" of questions to be used for the inspections. The lists are provided to the inspector via OSHA's Directorate of Enforcement Programs intranet website. The questions are not provided to the public, and are subject to change at any time.

The questions are structured so that the OSHA inspector will answer: "yes," "no" or "NA." Any "no" answer means that the facility did not meet the requirements of the question and may indicate non-compliance and result in a citation being issued.

5. Can OSHA Issue Citations Outside of PSM during a NEP Inspection?

Yes, the inspectors can certainly and have often cited facilities for non- PSM standards during the inspections regardless of whether they were specifically addressed in the NEP. For this reason, it is important to make sure that facility makes every effort possible to be sure that all OSHA standards are being followed.

6. Common Citations

Data from the Occupational Safety and Health Administrations inspection data website [4] was used to determine the results of twenty three different OSHA "planned" inspections on chemical facilities resulting in 100 different citations. The citations issued were between January 1, 2013 and December 31, 2014.

The results indicate that there was essentially an even split between PSM (1910.119) related citations and citations issued for other sections of the Code. Below is a graph showing that 49% of the citations were PSM related and 51% were related to other topics.

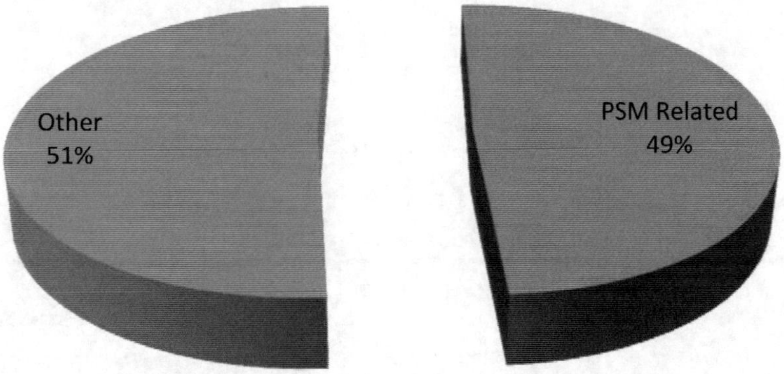

Figure 1: Breakdown of 100 Randomly Selected Citations

Of the citations issued for PSM, it was found that 70% could be accounted for by just four different categories: 1) Process Hazard Analysis, 2) Process safety Information, 3) Mechanical Integrity and 4) Pre-Startup Safety Review. Below is a pie-chart showing the breakdown of PSM citations.

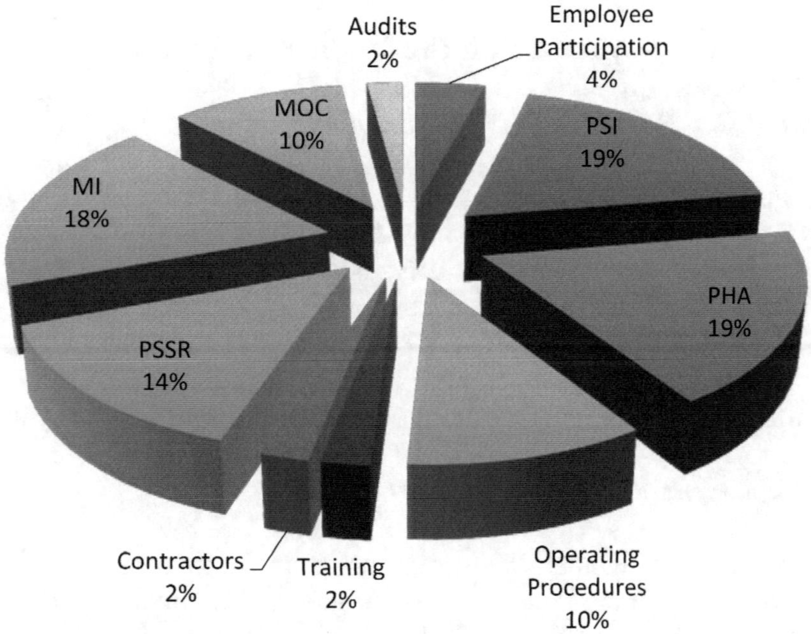

Figure 2: Breakdown of PSM related citations

Non-PSM related citations were much more varied in terms of topics referenced. The top four categories, comprising just 39% of the total were: electrical, PPE, confined space entry, and

HAZCOM. Below in Figure 3 is a detailed breakdown of non-PSM regulatory citations issued during the inspections.

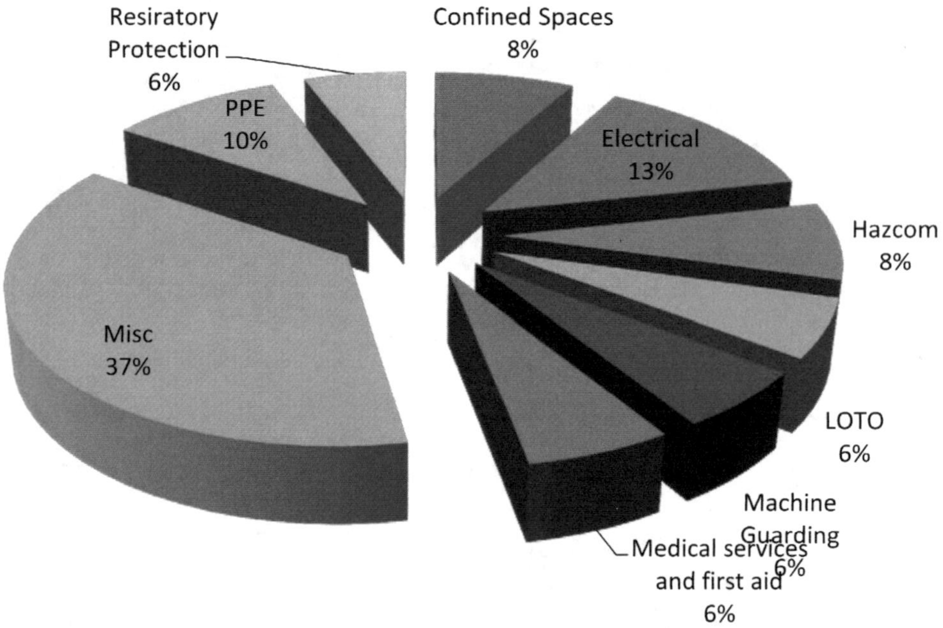

Figure 3: Breakdown of non-PSM related citations

7. How to Assure Compliance when the Inspector Arrives

7.1 Get Management "On Board" with NEP Compliance efforts

Perhaps the first and foremost step is to get the facility's management "on board" with the NEP program. Facility management needs to know that OSHA has this program in place and is actively using it to conduct inspections at all types of PSM facilities. Management needs to understand that facility programs need to be put in place to make sure that the facility is in compliance with all PSM requirements and how to react when the OSHA inspectors arrive to do a programmed or un-programmed inspection. The fact also that OSHA may issue citations for other regulatory infractions while walking around the facility needs to taken into account.

7.2 Develop and Implement a HSE Management System

It is clear that many citations against PSM programs stem from specific management processes not being completed. Management systems enable organization to better comply with regulations and other requirements that apply to their business. Management systems have at their foundation the assignment of roles and responsibilities. Assigned functional responsibilities increase the likelihood that regulatory requirements and their related business processes are completed on time, to a consistent and repeatable standard. In addition to roles and responsibilities companies can incorporate other key elements such as risk assessment, strategic planning and advocacy to assist in their efforts to understand the requirements of PSM, assign responsibilities and implement processes to better management them.

7.3 Inform employees and contractors about the NEP Program

Information about the OSHA NEP program also needs to be communicated to all employees and contractors. Employees need to know that their work can be subject to an OSHA inspection at any time. Work must always be conducted at a high standard and in a manner that is in conformance with PSM and other rules and regulations. Under the OSHA NEP "Instruction," the inspectors will solicit input from workers and their representatives including contractors. Employees that are well informed and can communicate the aspects of PSM that impact them, will be able to demonstrate to auditors and regulators alike that participation in the PSM program is strong and effective.

Contractors are not immune from the NEP inspection process. The NEP "Instruction" also requires the inspector to use the questions from the "Dynamic" list when evaluating contract employer compliance. The OSHA inspector will even choose an additional PSM-covered process at the site with contractors, if the "Selected Unit" first chosen does not have contractors.

Employees and contractors must not only be knowledgeable about PSM but also accountable for PSM so that they can accurately respond to the inspector's questions.

7.4 Institute a Self Audit Program

One of the best ways to get prepared for an OSHA NEP inspection is to institute a self auditing program. Typically, most PSM facilities utilize a third party every three years to make sure that the facility is in compliance with the PSM requirements.

To be prepared for an OSHA NEP inspection, it is often worthwhile to institute a program to conduct in-house audits on a very frequent basis. OSHA is looking to determine if the PSM program and various elements are reviewed on a regular basis and is still accurate. In-house auditing with documentation of such, establishes this effort. This could be simple as an in-house, quarterly audit of the facility's Hot Work permits or LOTO records.

In addition to the PSM protocols, the facility should be looking at all OSHA requirements as the OSHA NEP inspector may indeed write up a facility for violations outside of OSHA's PSM requirements. The more incidents an inspector finds not-compliant with PSM: the more the inspector with start looking for non-compliance in other areas.

There are many resources on the web that can help with protocols for both OSHA's General Industry requirements [5] and the PSM requirements and from the New Jersey Work Environmental Council [6].

7.5 Conduct a thorough Review of Your PSM Documentation

First and foremost, all PSM documentation should have an owner with specific requirements for the care and upkeep of the document.

The OSHA inspector will request to have access to all of the documents which are typically compiled by the employer. Please note that these documents may be requested prior to actually doing the site inspection and before the inspector opens the "Dynamic List" of questions. In other words, the inspector may request to see these documents even if they are not covered by the specific questions from the "Dynamic" list.

It is very important that facility management make sure that all of the documents are available, up-to-date, and meet current expectations for today's world. Below is Table 1, shows a listing of these documents that the OSHA inspector may ask to see either prior or during the inspection:

Table 1: Requested Documentation

1. Three years of OSHA logs for both the operator and contractors	7. P&IDs	13. Organizational chart showing PSM responsibility for each element
2. All contract employee injury and illness logs	8. Plot Plans	14. Initial and most recent PHAs including documentation showing the status of the action items
3. A list of all PSM- covered processes in the unit	9. Electrical area classification drawings	15. Safe upper and lower limits
4. Maximum intended inventories	10. Process narrative descriptions	16. PSM Incident reports and investigation reports for the operator and contractors
5. Description of the PSM Program	11. Safety system description (i.e. Cause and Effect Chart)	
6. Process flow diagrams	12. Design codes and standards used	

Most of the documents are fairly straightforward and readily available at most all facilities. However there are a few which can be problematic. These include:

Number 3 - A list of all PSM- covered processes in the unit: Facility operators need to make sure that their PSM documents include a list of processes that are covered. Some processes may be exempt under the "retail facilities" or the "normally unoccupied remote facilities" exemptions. Others may be exempt because the threshold quantity of chemical is not achieved, if there are less than 10,000 lbs. of a flammable fluids, fuels, or atmospheric tanks containing flammable liquids.

The location of any separate vessels, that are believed to be exempt from PSM, must be reviewed to make sure that release could not affect nearby non-interconnected PSM vessels.

Bottom line is that a document needs to be on file showing which parts of the facility are covered by PSM and which are not.

Number 4 - Maximum Intended Inventories: A simple spreadsheet will suffice showing the maximum working capacity of each vessel and tank located at the facility. It is also good to include the location, contents and quantity of materials stored in drums and small day tanks on the spreadsheet.

Number 12 - Design Codes and Standards Used: It seems that a single list of design codes and standards used often is not available. Reasons may be that the Company does not have rigid design codes and standards or that the various process units were built and installed by several different manufactures over the years. Usually, a single bulleted page showing the industry standard followed for the design and construction of the facilities will suffice in meeting this requirement.

Number 14 - Initial and most recent PHAs including documentation showing status of the action items: Most facilities will have the PHAs available, but documentation showing the status of the action items is often missing. Make sure that your list of actions recommended to resolve PHA findings is up to date.

Number 15 - Safe upper and lower limits: This information can usually be found in the cause and effect charts, in the operating procedures or in the information provided by the process unit engineering firm. It is best to compile this information on a single spreadsheet for the OSHA inspector to review.

7.6 Have a plan for when the OSHA inspector arrives

Each facility should have a plan in place for dealing with an OSHA NEP inspection. Having a plan should greatly reduce the stress on personnel assigned to assist the Compliance Safety and Health Officer (CSHO) assigned to review the facility. Having a plan in place, which plant personnel have been trained on, and even drilled on, can make the inspections go much easier for everyone.

The CSHO assigned to the facility will recognize that the facility "has its act together" if plant personnel are able to provide the documents, lead a tour and always treat the CSHO in a professional manner. Keep in mind that many of the CSHO's worked previously for industry before joining OSHA. Having a trained person assigned to help the inspector, able to answer questions and provide the needed documents can go a long ways in demonstrating how well the facility is managed and operated.

The NEP inspection plan should contain a number of key elements. These include:

- Identification of the lead person who will be responsible for assisting the CSHO during all aspects of the inspection.
- Identification of management personnel for the CSHO to ask about the PSM program at the facility. These personnel should be on-site the day of the inspection.
- Identification of operating personnel and contractors who the CSHO should interview about the PSM program at the facility.
- A records management system whereby all of the required documents can be provided quickly to the CSHO. Often this can be as simple as a three ring binder with a tab for

each different type of records or an easy to access online system that is straightforward and readily accessible. When given a document request by OSHA pre-inspection, be sure to print out hard copies (or samples of large documents) to have ready for the inspector to review. Also, the inspection is enhanced by having all the requested documentation already assembled in the work area for the CSHO.

- A work area should be prepared for the CSHO. Often this may be a conference room where the CSHO can examine materials. The facility may also offer internet access to the CSHO as a professional curtsey.
- Plan on how the CSHO should tour or inspect the plant. It is a good idea to have a tour route planned out ahead of time so that facility personnel can show the CSHO portions of the plant which are important to PSM compliance efforts. The CSHO may deviate from the planned tour area or ask to see specific equipment not in the tour area. Be professional and honor the request.

After the plan is developed, it is important that plant personnel be trained in the use of the plan. Additionally, it would be very desirable to conduct some mock either "table top" or actual drills on the plan so that all facility personnel know exactly what their role should be in the event that the facility undergoes an NEP inspection.

7.7 Make it part of your ongoing efforts

One of my bosses once told a group of us that he never had to worry about getting a speeding ticket. We were all in amazement and wondered if he had bought a new radar detector or had found a miracle application for his cell phone. He simply went on to state that, "I always drive the speed limit or less."

This is a great lesson for all of us working at PSM facilities and wondering what will happen if an OSHA NEP inspection were to occur. The best way to make sure of passing an inspection is to make following the PSM requirements just a part of your culture or "how you do things."

8. Conclusions

Many facilities have passed OSHA NEP inspections without a single citation. The first step is to make sure that the plant's management is "on board" with a robust PSM compliance program, including employees assigned the responsibility for pieces of the program, working under a well defined management system

Being in compliance not only prevents those malevolent citations from being issued, but makes for a much safer environment for all to work in. There are many tools and programs available to make sure that the facility is in compliance, but the best is just getting everyone, including contractors at the facility to take an active role in PSM. There are many checklists and self audit forms available to gauge how the facility is doing.

Another key factor is having all of the required plans, programs and records, complete and current; and readily available. This includes the dates of reports, training, drills, tests, etc., signature signoffs offs and when completed. Be sure that your document management system allows for PSM related documents to be easily found. Check all of your exiting files to make sure that nothing is missing and everything is there.

Having a plan in place for when the NEP inspector shows up is critical. The plan should address all of the key aspects of the inspector's visit. Plant personnel should know what their role is during the inspection through training and drills.

Preparation is the key to dealing with a NEP inspection. It is much better to assume that your facility will be inspected in the near future and be prepared accordingly. It is much worse to wonder what you should have done when the inspector is knocking at your door.

9. References

[1] 1 OSHA. (2009). *PSM Coverd Chemical Facilities National Emphasis Program.* Washington DC: U.S. Department of Labor.

[2] OSHA. (2011). *PSM Covered Chemcial Facilities National Empahsis Program.* Washington D.C. : U.S. Department of Labor.

[3] Kleindorfer, P. R., Feldman, H., & Lowe, R. L. (2000). *Accident Epidemiology and the U.S. Chemical Industry: Preliminary Results from RMP*Info**.* The Wharton School.

[4] United States Department of Labor. (n.d.). *Data and Statistics*. Retrieved January 29, 2015, from Occupational Safety & Health Administration: https://www.osha.gov/oshstats/

[5] U.S. Dept. of Labor. (n.d.). *Small Business Handbook*. Retrieved from OSHA: https://www.osha.gov/Publications/smallbusiness/small-business.html

[6] NJ Work Environmental Council. (2009). *Preventing Chemical Accidents*. Retrieved from OSHA: https://www.osha.gov/dte/grant_materials/fy08/sh-17813-08/compliance_checklist.pdf

400380 Analysis of the Capabilities of Cybersecurity Defenses

Abstract:

The threats to our critical infrastructure continue to evolve. As new cyberattacks are created and discovered, it is incumbent upon us to evaluate the capabilities of our defensive strategies and technologies against these new offensives.

This session will investigate and report on how the modern, targeted online and removable media threats to industrial cybersecurity fare against the most popular protections deployed to protect critical infrastructure, including: whitelisting, device control, next-gen firewalls, unidirectional gateways, the FLIP, and application data controls.

Copyright © American Institute of Chemical Engineers. All rights reserved.

FUNDAMENTALS OF AMINE SYSTEM DESIGN & OPERATION

Mahdi Nouri, CH2M HILL Oil, Gas & Chemicals
Michael Sheilan, Amine Experts

15 AIChE
Spring Meeting, Austin, TX

CH2MHILL.
amine eXperts

Agenda

- Introductory of Amine System
- Amine System Design Considerations
- Amine System Operation Considerations
- Amine System Material Selection & Corrosion Monitoring Points

15 AIChE
Spring Meeting, Austin, TX

Typical Block Flow Diagram of Gas Processing Plant

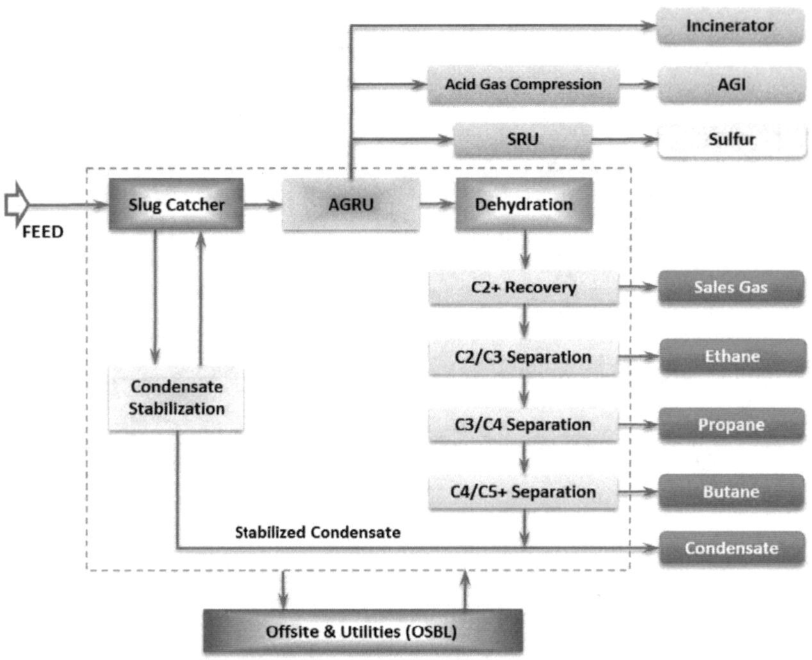

15 AIChE
Spring Meeting, Austin, TX

Why Sweeten?

- Safety and Environmental Consideration
- Treated Gas Specification
- Corrosion Rate Minimization
- Operation Performance and Flexibility

15 AIChE
Spring Meeting, Austin, TX

Treating System Processes

- Absorption Processes
  - Chemical Solvents
  - Physical Solvents
  - Hybrid Solvents
- Adsorption Processes
  - Molecular Sieve
  - Scavengers
- Membrane Treating

15AIChE
Spring Meeting, Austin, TX

Amine Chemistry

Alkanolamines have three functional group:

- Nitrogen atom
  - The site where acid gases react
- Alcohol (hydroxyl) group
  - Improves water miscibility, modifies the base strength and lowers vapor pressure (minimize losses)
- Alkane (hydrocarbon) arm
  - Separates the amine and hydroxyl group to provide chemical stability

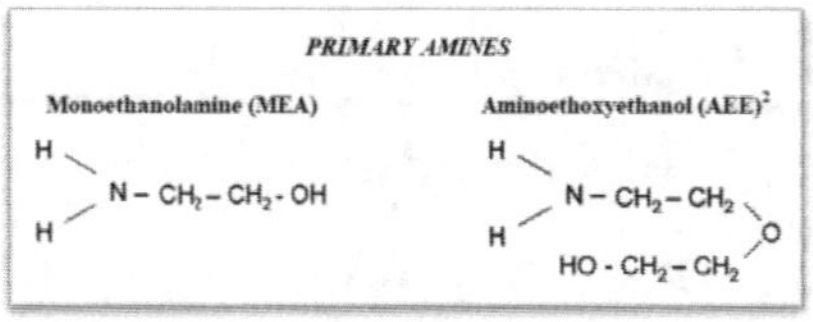

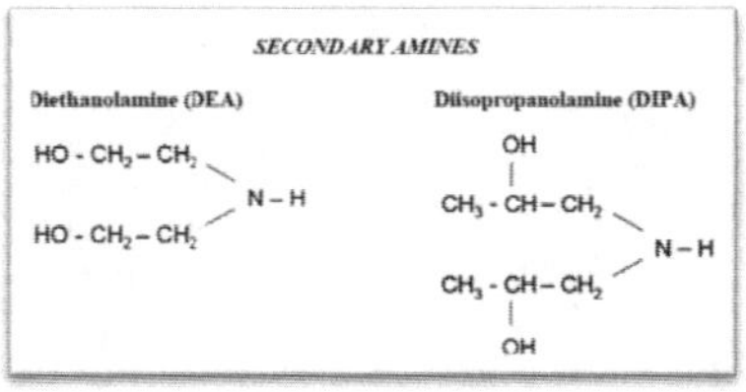

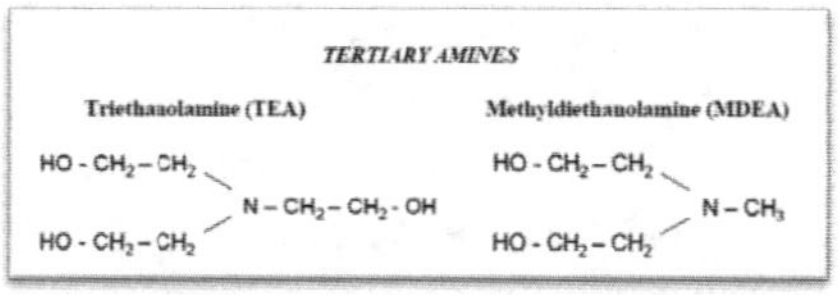

15AIChE
Spring Meeting, Austin, TX

Typical Guidelines for Generic Amine Solvents

	Amine Type	MW	Amine Strength	Rich AG Loading	Acid Gas Capacity
		lb/lbmole	Wt.%	mol/mol	lbmole/t
MEA	Primary	61	15 - 20	0.35	2.53
AEE (DGA)	Primary	105	30 - 60	0.35	3.68
DEA	Secondary	105	25 - 35	0.40 - 0.50	2.52
DIPA	Secondary	133	20 - 40	0.45 - 0.50	3.35
MDEA	Tertiary	119	35 - 50	0.45 - 0.50	4.16

15 AIChE
Spring Meeting, Austin, TX

Key Items for Selecting the Appropriate Solvent

- Grassroots vs. Existing Facilities
- Solvent Costs
- Sour Compound Removal Selectivity
- Sour Compound Removal Capability
- Amine Circulation Rate
- Energy Consumption
- Operating Stability e.g foaming
- Solvent Stability e.g. degradation
- Corrosion Rate

15 AIChE
Spring Meeting, Austin, TX

Amine System Design Objectives

- Meet Product Specifications
- Minimize the Amine Circulation Rate
- Minimize the Reboiler Duty
- Minimize the Corrosion Rate
- Optimize CAPEX / OPEX

15 AIChE
Spring Meeting, Austin, TX

AMINE SYSTEM DESIGN CONSIDERATIONS

15 AIChE
Spring Meeting, Austin, TX

CH2MHILL.

amine eXperts

Typical Amine System Process Flow Diagram

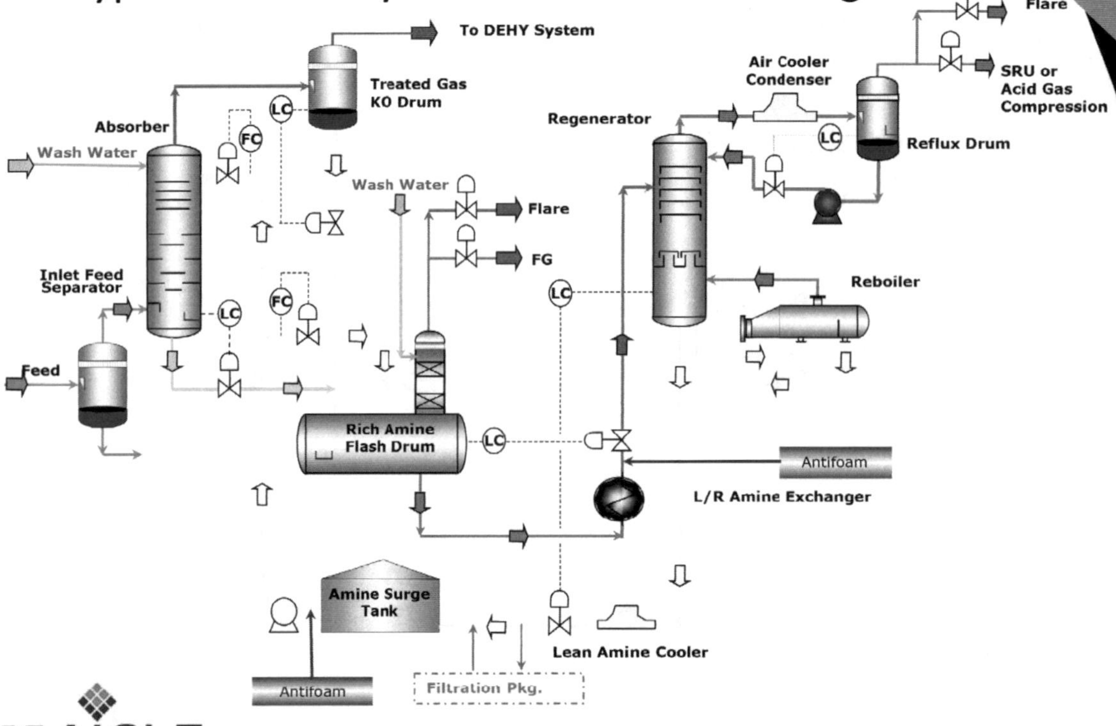

15 AIChE Spring Meeting, Austin, TX

Checklist: Design Considerations

- Meet the Treated Gas Specifications
- Acid Gas Quality in case of there is a SRU at downstream
- Thermodynamic Model
- Amine Solution Selection
- Amine Solution Concentration
- Acid Gas Loading
- Inlet Gas Pressure
- Regenerator Pressure
- Operating Temperature
  - Inlet Gas Temperature,
  - Lean Amine Temperature,
  - Rich Amine Temperature to Regenerator,
  - Bottom of Regenerator Column,
  - Reflux Temperature
- Reflux Ratio

15 AIChE Spring Meeting, Austin, TX

Inlet Feed Separator

- **Objective**
  - To collect liquid entrainments from upstream facilities, and to remove liquid hydrocarbons and other contaminants due to foaming tendency reduction through absorber column
- **Characteristics**
  - Vertical separator equipped with demister pad to prevent liquid droplets carry-over
- **Alternatives**
  - Vertical Cartridge Coalescer

15 AIChE
Spring Meeting, Austin, TX

Inlet Feed Separator and Cartridge Coalescer

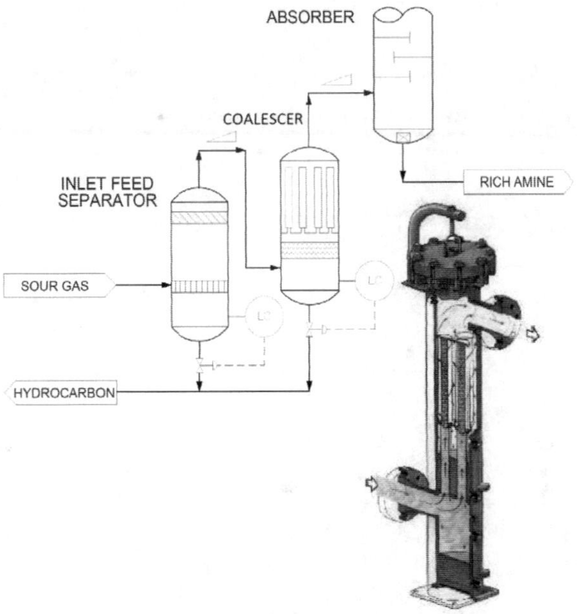

15 AIChE
Spring Meeting, Austin, TX

Absorber

- **Objective**
  - Acid gas removal from feed
- **Characteristics**
  - Fractionation column equipped with trays (generally around 20) or packing beds (generally 2 or 3 beds).
  - Lean amine should warmer than feed gas temperature to avoid hydrocarbon condensation. The recommended approach temperature is + 5 to 8 °C (10 to 15 °F).
  - Consider pressure differential instrument across the column
    - Sharp DP increase foaming
    - Slow DP indicates tray or bed fouling
  - Hydrocarbon and foaming layer can be detected by level glasses.

15 AIChE
Spring Meeting, Austin, TX

Treated Gas KO Drum

- **Objective**
  - To remove entrained amine from treated gas
- **Characteristics**
  - Vertical drum equipped with demister pad,
- **Alternative**
  - Cartridge Coalescer, Optional

15 AIChE
Spring Meeting, Austin, TX

Rich Amine Flash Drum

- **Objective**
  - To remove the absorbed liquid hydrocarbons and flash dissolved gas
- **Characteristics**
  - Horizontal three-phase drum due to hydrocarbon removal
  - Liquid retention of 15 to 20 minutes is recommended
  - Liquid hydrocarbons skimming system is required
  - Pressure let down by **Control Valve** or **Turbo-Expander**
  - Operating pressure is usually Fuel Gas Pressure. In oil processing facilities, separator should be operated at very low pressure for a better degassing of the solution

15 AIChE
Spring Meeting, Austin, TX

Typical Rich Amine Flash Drum Configurations

Calming & Internal Baffles

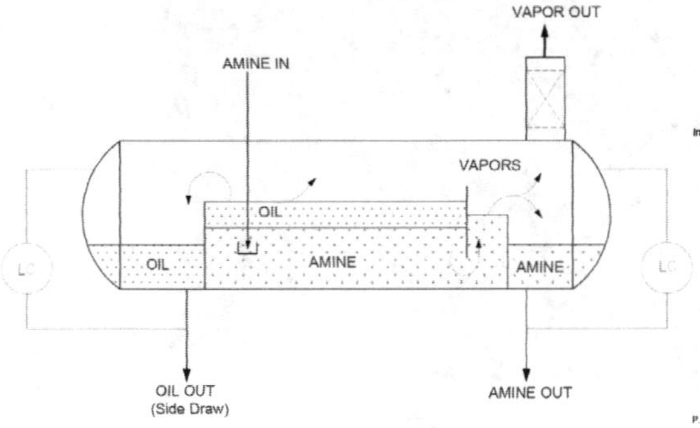

Internal Baffle Box

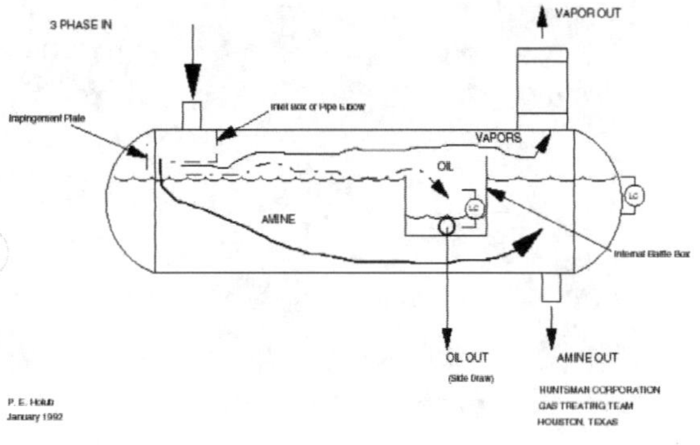

15 AIChE
Spring Meeting, Austin, TX

Rich/Lean Amine Heat Exchanger

- **Objective**
  - Heat Recovery (to preheat the rich stream to the regenerator)
- **Characteristics**
  - Shell and tube exchanger with rich solution tube side / lean solution shell side
  - Minimum vaporization in the tube side is recommended due to prevent corrosion
  - Lean solution inlet temperature between 110 to 130 deg.C
  - Velocity should be limited to around 1 m/s. Higher velocity can increase erosion and damage scales & lower velocity can foul equipment.
  - Couple of exchangers in series can be required.
- **Alternative**
  - Plate frame exchanger

15 AIChE
Spring Meeting, Austin, TX

What Would You Like to Pick?

	S&T Exchanger	Plate Frame
Fouling Issues / Maintenance	+	-
Upstream Filter Req.	+	-
Pressure Drop	-	+
Cost	-	+
Plot Space	- [1]	+
Heat Transfer	-	+

[1] couple of shells in series

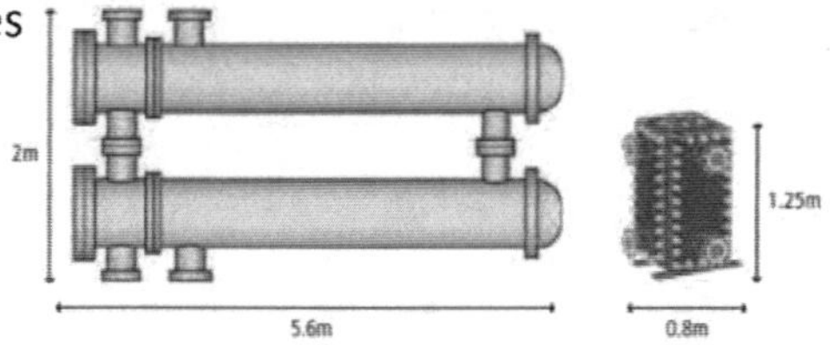

15 AIChE
Spring Meeting, Austin, TX

405575 Fundamentals of Amine Systems Operations

Abstract:

The Fundamentals of Acid Gas Removal provides an overview and a basic guide to design, operation, corrosion and troubleshooting for amine treating units.

Copyright © American Institute of Chemical Engineers. All rights reserved.

401757 Reduced Energy Requirements in Gas Treating Using Formulated Solvents

Abstract:

Monoethanolamine (MEA) is a broadly used alkanolamine solvent for acid gas treating applications, especially for refinery gas streams. Its use entails high operating and capital costs, due to relatively higher regeneration energy requirements and higher thermal degradation and corrosion propensities. With specially formulated solvents, these problems can be alleviated, causing significant cost savings to producers. Using an in-house plant simulator, we demonstrate reduced energy requirements with a formulated solvent for a typical gas treating requirement, compared to MEA. Additionally, solvent fill requirements are reduced, arising from lower solvent losses, owing to a lower volatility and higher thermal degradation resistance of the formulated solvent. For the case studied here, cost savings from reduced regeneration energy requirements alone is close to 50%, in comparison to MEA.

Copyright © American Institute of Chemical Engineers. All rights reserved.

401917 Ethical Considerations for Chemical Engineers

Abstract:

A summary of recent additions to Texas Engineering Rules, including the requirement for a one-time fingerprinting and criminal background check for all engineers.

Discuss the criteria for ethical decision-making as well as the specific ethical considerations mentioned in the Engineering Rules - to protect the public's health, welfare and property.

Address other aspects of employer/ employee principles or values that are important in the engineering profession, such as

- conflict of interest,
- confidential and /or proprietary information,
- gifts and entertainment,
- bribes and kickbacks and
- relationship with clients.

The presentation will include case studies and illustrative examples for discussion and an opportunity for questions.

KEY WORDS: Ethics, licensing, values

Copyright © American Institute of Chemical Engineers. All rights reserved.

401857 Writing a Successful Proposal

Abstract:

For over 40 years, Joe Reynolds lead a project team at Tracor, an Austin-based applied technology firm. Early on, he developed a strategy for leading his team in writing proposals that could beat competition from the top-tier defense contractors. His team's specialty was in designing leading-edge applied technology for the Defense Department.

After a good idea, a successful proposal continues with an evaluation of your company's strengths and weaknesses. Next, you size-up the competition's strong and weak points. As you complete the proposal, you want to describe any fatal flaws that your competition's solutions may have and show how your design avoids them.

KEY WORDS: strategy, proposal writing, winning

Copyright © American Institute of Chemical Engineers. All rights reserved.

VALUE CAPTURE FROM SHALE: INNOVATIVE PROCESSING IN A CONDENSATE REFINERY

Donald A. Eizenga, David J. Shecterle, Xin X. (Frank) Zhu
UOP LLC, a Honeywell Company
Des Plaines, Illinois, USA

Executive Summary

Recent increases in hydraulic fracturing operations have produced a significant amount of condensate liquids, particularly in the United States. Conversion of these light liquids into finished products for sale at maximum value is being considered in remote locations and via expansion of existing facilities. The quality of many condensates is such that minimal hydrotreating is required, but significant upgrading of the paraffinic naphtha is required to meet gasoline specifications. A condensate refinery is therefore expected to include reforming and isomerization of naphtha along with naphtha and diesel hydrotreating. In this presentation, novel synergized unit flowscheme solutions will be discussed. Three key ideas were studied as noted below.

Idea 1: The goal of the Alternate Condensate Fractionation Unit (CFU) Design was to reduce equipment count and save capex or to reduce energy consumption for opex savings especially when upstream of Combined Naphtha and Diesel Hydrotreating (CHT). Typically Naphtha and Diesel are separated in the CFU, but they would need to be combined and then re-separated for a CHT unit approach resulting in redundant fractionation. Innovative designs were considered to overcome the penalty of redundant fractionation. Elimination of columns was not found technically or economically viable. Energy benefits were found by optimization of the columns; however the benefits were still not sufficient to overcome the cost of redundant fractionation with a Combined Hydrotreating approach. The overall optimum economics were achieved when the hydrotreating units are individually optimized and a 3 column CFU design is employed.

Idea 2: The goal of the Combined Hydrotreating study was to minimize equipment count and thereby optimize the refinery economics. In the final analysis, the combination of naphtha and diesel in a single hydrotreating unit was found to have limited economic justification. Cost penalties were associated with reactor section pressure, catalyst volumes, and redundant fractionation of naphtha. Only limited efficient combinations of shared auxiliary systems were proposed so that each unit could be individually optimized.

Idea 3: UOP explored novel solutions for using isomerization with the UOP Platforming™ Process to determine an economic optimum configuration for gasoline production in the proposed unit. A number of options and synergies were evaluated to reduce capital costs and operating costs. Significant capital cost savings were achieved using innovative naphtha complex schemes relative to the base case separated unit configuration. A higher yield case was found to have highest NPV, while a lower CAPEX approach was evaluated which may make sense when a customer is severely capital constrained.

The market seeks technology with minimal capital investment as installation in remote locations could be a significant factor. UOP proposes to handle these issues with construction of modular units, which have been optimized for CAPEX and OPEX in consideration of the feed properties and product quality targets.

© 2015 UOP LLC. All rights reserved.

Introduction

While the various technologies used for hydraulic fracturing of shale to produce natural gas have developed over the last 40 years, the rapidly escalating energy prices of the last 10 years led to an ongoing boom in the use of the technology particularly in the United States. US shale deposits are distributed over a wide range of the country in a number of formations as illustrated in Figure 1.

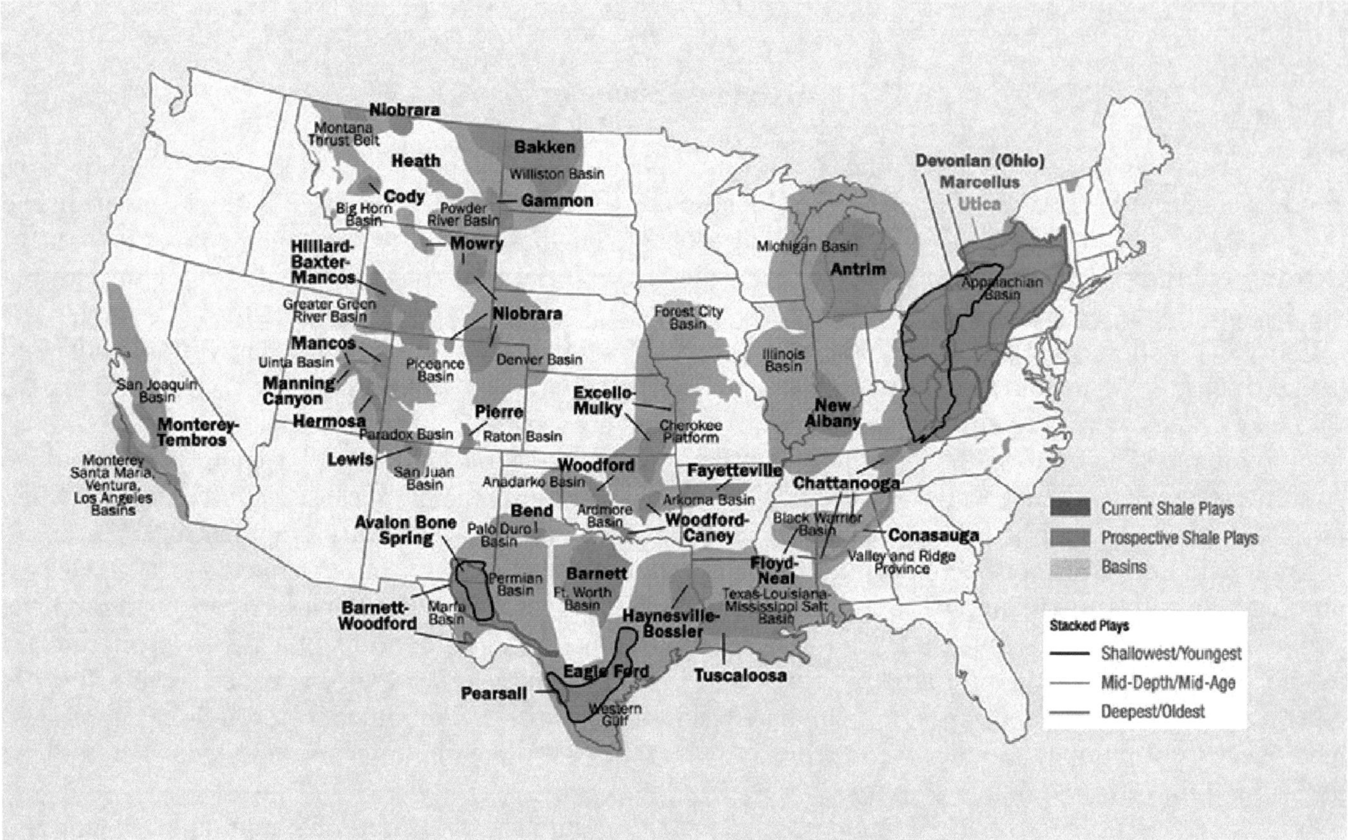

Figure 1: U.S.-Based Oil and Gas Shale Plays[1]

In the process of extracting gas from the wells, condensates and natural gas liquids (NGLs) are also extracted. These liquids include a full range of hydrocarbon constituents from propane to heavy residues that boil at and above 1000°F. The liquids are therefore similar to other petroleum crudes, but are lighter and generally sweet. For example, crudes are often differentiated by API gravity, with Heavy crudes averaging ~15° API, Medium at ~25° API, Light at ~35° API, and Extra Light at ~45° API. Condensates average 55° API, but there is a wide range of compositions depending on the source and the composition may change as the field matures. Figure 2 illustrates the stages of Drilling the well, Fracking, and then Production of gas and primary separation of liquids. Figure 3 shows a simplified condensate separation scheme.

© 2015 UOP LLC. All rights reserved.

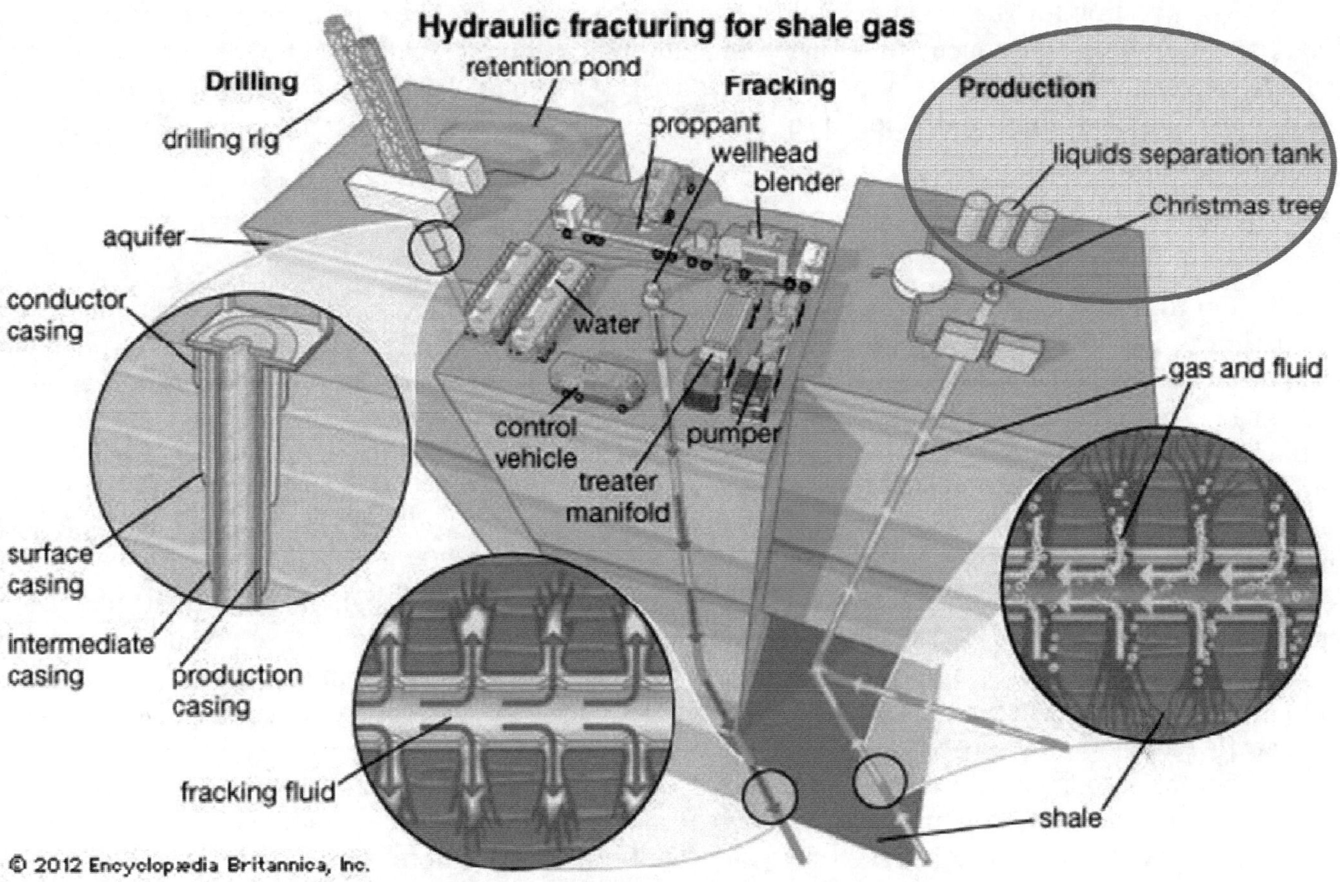

Figure 2: Fracking: Hydraulic Fracturing for Shale Gas[2]

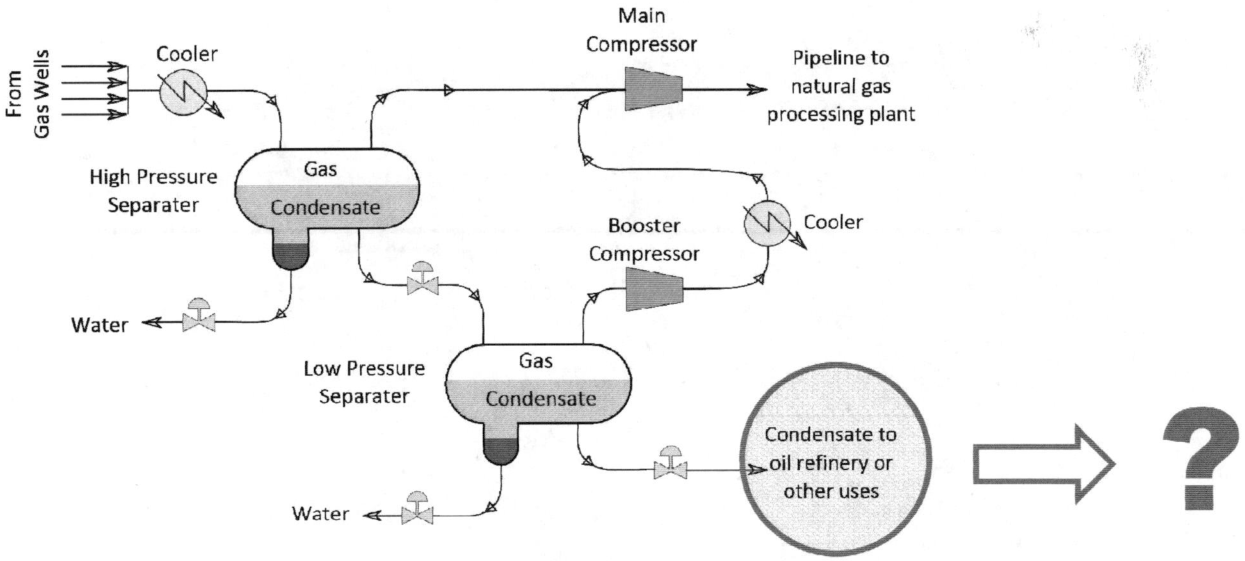

Figure 3: Simplified Condensate Separation[3]

© 2015 UOP LLC. All rights reserved.

The question for gas producers, midstream companies, and refiners is: *How best to process these liquids?* Certainly adjustment to refinery operation is required, and possibly revamp of existing equipment and/or addition of new process units (e.g. via a modular expansion). Alternatively, new processing facilities could be developed for local markets, which may be placed near the feed sources in remote locations.

Condensate Properties

In order to understand how best to process the condensate liquids, a good understanding of the properties is required. One key attractive property in the current market environment is that they are relatively low cost and sold at a discount to crudes, in part owing to the major recent increases in fracking. The other key attractive property is that they are light liquids, which make them ideal for refining into higher margin gasoline and diesel products. Recent estimates have predicted dramatic growth and sustained high levels of production of condensate; however this may be impacted by the high volatility in crude oil market pricing. Legislation of the sale of "crude oil" from the US may also have an impact.

Condensates may contain up to 10% LPG, and are typically high in both light and heavy naphtha. In fact, a survey of 116 worldwide condensates showed that LPG + Light Naphtha + Heavy Naphtha ranged from 30% up to over 90%, with a median near 60%. The remaining material is mostly in the distillate range, suitable for jet fuel and/or diesel fuel with appropriate processing. Condensates are mostly quite low in heavy residue range (e.g. boiling above 685°F).

For the recent study, a "Light Case" and "Heavy Case" feed were considered for a 25,000 BPSD Condensate Refinery as show in Table 1. The API gravity was 59.4 for the light case and 52.0 for the heavy case. Sulfur was low at 20-30 ppm vs. levels of up to 2% in more contaminated condensates and conventional crudes. Target products for the study included off gas which could be used as fuel gas for the complex, LPG, gasoline (or blending component) with an octane target of 87, Ultra Low Sulfur Diesel, and a low sulfur residue (e.g. for refinery FCC or Hydrocracking feedstock or fuel oil).

Table 1: Study Basis – Processing Objectives

Feed	**"Light Case"**	**"Heavy Case"**
Condensate (BPSD)	25,000	25,000
API / SG	59.4 / 0.741	52.0 / 0.771
Sulfur, ppm	22	31
Products (BPSD)		
Off Gas (MMBTU/hr)	171	151
Total LPG	2,800	2,000
Gasoline (R+M)/2 = 87	**12,600**	**10,300**
ULSD	**7,200**	**9,600**
Low Sulfur Residue	1,000	2,000

© 2015 UOP LLC. All rights reserved.

Ideas for Condensate Processing

For the base case of the study, a Mini-Refinery was proposed as shown in Figure 4, and the objective of the study was to analyze opportunities that might optimize the overall complex. The original premise of the study was that reduction of equipment count and minimization of CAPEX would be the most important objective. Three parts of the study will be summarized:

1. Condensate Fractionation Unit (CFU)
2. Separated vs. Combined Hydrotreating
3. Innovative Naphtha Complex Schemes

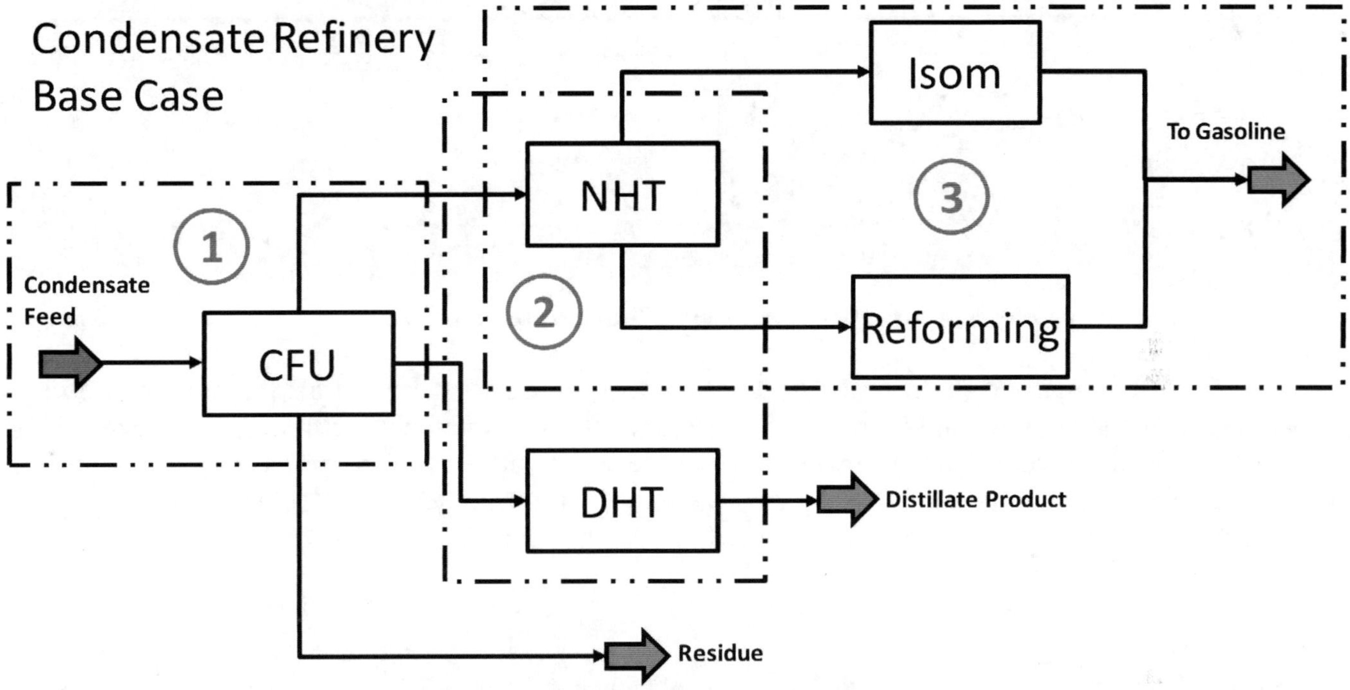

Figure 4: Mini-Refinery – A Big Picture

1. Variations in Fractionation

The first question posed when reviewing the Base Case Mini-Refinery was: "can we use just one column like a typical Crude unit?" Various options were then explored with a target of optimizing the design for expected use of downstream Combined Hydrotreating (CHT discussed below) and to minimize CAPEX and/or OPEX in general. The questions then became: 1. What is the best Condensate Fractionation Unit (CFU) configuration? 2. Can CAPEX and OPEX be reduced if Combined Hydrotreating (CHT) is used?

Upon further examination of the Crude column approach shown in Figure 5, issues quickly became apparent. While the Crude unit is typically a single column, the overhead gas is typically compressed and then processed in a Gas Concentration Unit. This requires a high CAPEX offgas compressor. Higher pressure can be used to keep the overhead as totally condensing, but there are constraints on the column whereby fouling is expected at the bottom of the column when high pressure leads to higher bottoms temperatures. The study did show that use of steam stripping and adjustments to the column pressure could handle a higher LPG and lower Residue feedstock. Surprisingly, however, the energy input requirements remained high and CAPEX savings were not substantial.

© 2015 UOP LLC. All rights reserved.

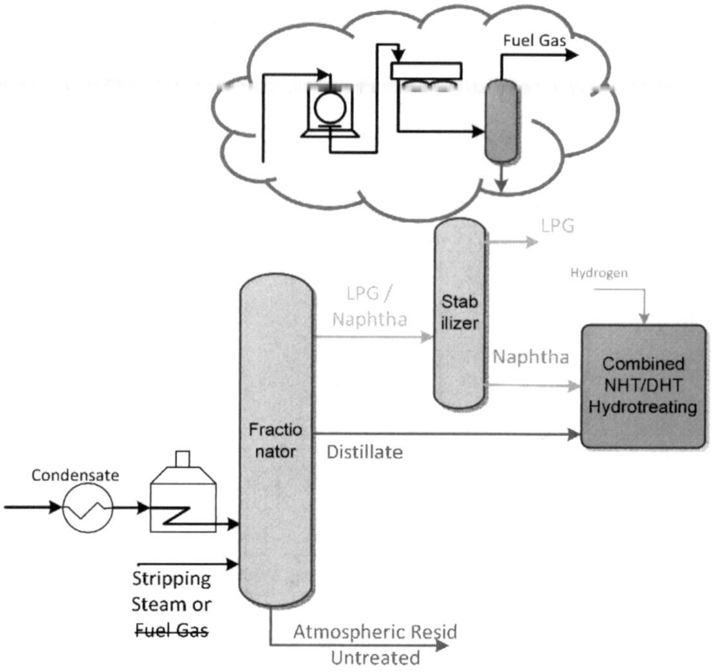

Figure 5: Crude Column Approach

A three column design approach to meet on spec product cuts could also avoid the use of an overhead compressor (shown in Figure 6). A separate Distillate Fractionator column allowed for higher pressure in the first column and lower pressure in the heavy oil column to effectively avoid fouling, however capital cost and operating costs were high and the rigorous separation achieved did not appear to be optimized if naphtha and diesel were going to be re-combined for a downstream CHT.

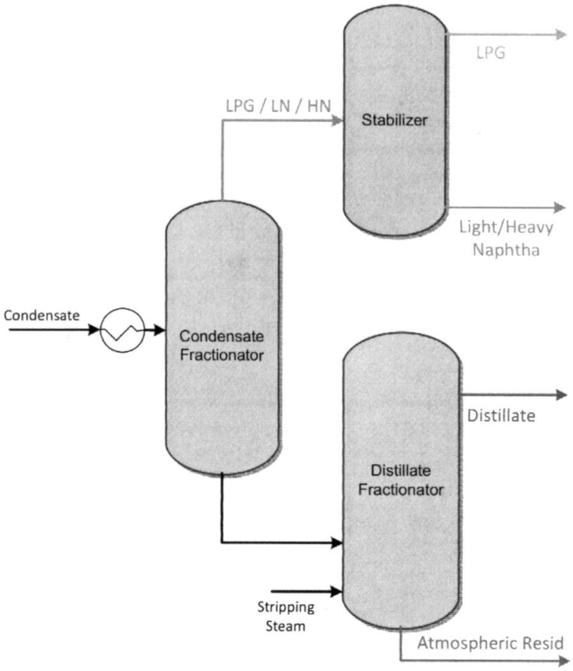

Figure 6: Three Column Approach

© 2015 UOP LLC. All rights reserved.

Relaxation of design targets for the product cuts were then studied, which included adjustment of the product draws, reflux, and heat exchanger network. Optimization of the configuration did allow for minimization of energy input and some CAPEX reduction. Two problems remained, however, which is that the naphtha and diesel products from the CFU were now lower in temperature and would require more energy in the CHT to re-heat, and the question of whether the Energy Savings in the CFU could compensate for the additional energy required by the CHT (relative to separate hydrotreating units) was not resolved.

OPEX savings were achievable in CFU, but the options had quite similar CAPEX. It was determined that the single Crude column approach was not in fact optimized and it did not provide a heat advantage. There was no clear winner without considering the entire complex. Once the overall analysis of CFU and CHT including redundant fractionation was considered, it was found that the energy savings in the CFU were offset by the higher energy costs of the downstream CHT.

2. *Combined Hydrotreating or Not?*

For the relatively easy hydrotreating requirements within the small condensate refinery study, it was initially proposed that the objective should be to minimize equipment by combining the feeds into a Combined Hydrotreating (CHT) unit as illustrated in Figure 7.

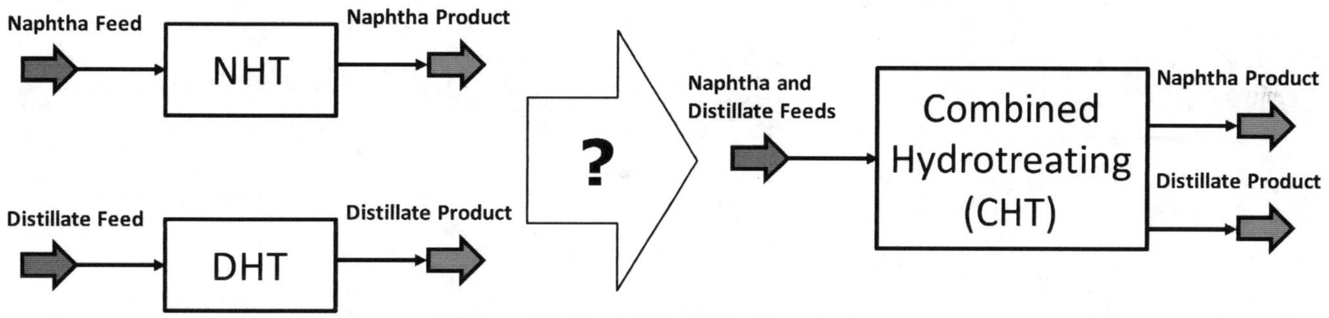

Figure 7: Combined Hydrotreating

The high level concerns about this approach were whether the naphtha was being over-processed, and also whether the cost of redundant fractionation of the naphtha from the distillate feeds could be overcome. Additionally, it was questioned what the best approach might be within the Diesel Hydrotreating (DHT) to optimize the separated configuration.

In the case of separate hydrotreating, each unit has optimized catalyst selection, pressure conditions, and fractionation to meet the product objectives. While Combined Hydrotreating can significantly reduce equipment count, full fractionation was required to cleanly separate the diesel product from the naphtha which required further processing. Estimated capital costs for the approaches were found to be similar. The main penalty with the CHT approach is the additional heater duty associated with the required Naphtha / Diesel fractionation. Even in combination with an optimized CFU approach as described above, the CHT approach did not show a Net Present Value advantage.

This was a surprising result, explained by the fact that savings in reducing a number of pieces of equipment were offset by higher costs in some of the more expensive equipment such as reactor section exchangers, reactors, and the fired heater reboiler. In this case, each individual unit could be better optimized and with shared auxiliary systems and limited integration.

© 2015 UOP LLC. All rights reserved.

3. *Innovative Naphtha Complex Schemes*

Due to the paraffinic nature of the study feed, both catalytic reforming and isomerization processes are required to meet the gasoline octane target (87 R+M / 2). A variety of isomerization unit technologies and configurations are available depending on the octane level required and any other limitations such as gasoline pool aromatic content. Isomerization processes can be used with the UOP Platforming™ Process unit, for which novel synergized solutions may be used to optimize the refinery economics as illustrated in Figure 8. Both product value and equipment capital costs need to be considered.

UOP capabilities in kinetic modeling of catalyst performance and process simulation of a variety of flowscheme options were used to optimize the unit. An economic analysis considering CAPEX, OPEX, yields, and cycle length impact was carried out for several solutions. Significant capital cost savings were achieved using innovative naphtha complex schemes relative to the base case separated unit configuration. A higher yield case was found to have highest NPV, while a lower CAPEX approach was evaluated which may make sense when a customer is severely capital constrained.

Even for a small complex where equipment minimization appeared to be a sound approach from a high level, catalyst and process conditions have a significant impact on the project economics.

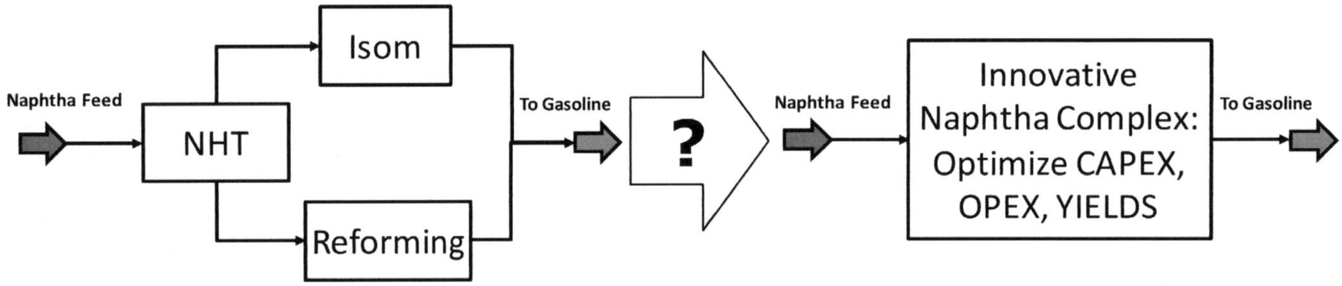

Figure 8: Innovative Naphtha Complex

Concluding Thoughts

In the construction of new plants, two choices are generally available: Stick built construction and Modular construction. There are a number of reasons why Modular may be preferred for the study case including:

- 25,000 BPSD capacity is ideal for modular scale
- Higher levels of quality control can be achieved in the design
- Shop fabrication and inspection lead to higher quality and improved schedule
- Modular is more suitable for remote locations where skilled labor is limited
- Significant schedule reductions and logistics savings are realized
- Projects can be executed with a Fixed Lump Sum Price

UOP and UOP Russell have significant experience in the design, construction, and delivery of Modular units. This includes natural gas liquids Separation units, isomerization Units, Merox™ Units, and catalytic process units such as NHT and reforming.

The recent expansion of shale fracking for gas and generation of condensate feedstocks has been a great opportunity and a challenge to produce high value refined products. In order to capture the value, a good understanding of the feed and processing technologies is required. As illustrated, optimizing is not the same as minimizing, and the effort should consider the whole effected complex.

© 2015 UOP LLC. All rights reserved.

The use of innovative naphtha processing approaches and potential for condensate refineries represent value added opportunities. UOP has the expertise to optimize complexes for a variety of feedstocks, and innovative solutions to monetize condensate. Use of a modular approach can bring significant advantages in project execution.

Acknowledgements

The authors would like to express sincere thanks to the following Honeywell UOP personnel who participated in the work described above - Joshua Kaye, Ron Long, Charlie Van Zile, Dick Hoehn, Dave Bachmann, Paul Steacy, Mark Benson, and Rich Russeff.

References

1. Photo: American Petroleum Institute; Patrick C. Miller, "API's Energy from Shale campaign features Bakken", thebakken.com, January 21, 2015
2. 2012 Encyclopedia Britannica, Inc. http://www.britannica.com/EBchecked/topic/1655635/fracking
3. http://en.wikipedia.org/wiki/Natural-gas_condensate

© 2015 UOP LLC. All rights reserved.

401828 Challenges in Processing of Shale Oil in FCC Unit

Abstract:

The introduction of Shale gas and oil in the recent years has had a strong impact on the Petroleum market in North America and around the globe. Due to attractive prices of Shale oil, refiners have taken advantage of this stream to improve their profit margin. The use of shale oil in FCC units comes with its own challenges as Shale oils are more paraffinic than conventional crudes and contain a different combination of contaminant metals. Shale oils also generally make lower coke, reducing the FCC regenerator temperature. While this is beneficial with respect to catalyst deactivation, it can become an operational limit, especially where the FCC is being operated at reduced severity in diesel mode operations. The high paraffinic nature of Shale oils leads to higher gasoline volume but lower octane and lower LPG olefinicity, resulting in poorer quality alkylation unit feed. Lowering the Rare Earth of the catalyst can help improve olefinicity, but a more flexible and effective approach is to add a high activity ZSM-5 additive, such as Super Z Excel. ZSM-5 additives boost gasoline octane and increase the olefinicity of C3 and C4 hydrocarbons. The usage of ZSM-5 in North American FCCUs has been increasing as a result of Shale oil processing.

As the metal contaminants of Shale oils, such as Fe, Ca, Na, K, are usually higher than in conventional crudes, their accumulation can lead to more accelerated deactivation of FCC catalysts. Fe present in the FCC feed usually builds up an impervious or glassy layer on the catalyst surface, usually manifested as Fe nodules on particles. This inhibits access to inner particle, and hence results in loss of conversion, deterioration of bottoms selectivity, and potentially catalyst circulation problem. In severe Fe poisoning cases, the emission of SOx in the flue gas also increases as the Fe carries over some sulfur from the reactor to the regenerator. These negative impacts are usually referred to as Fe poisoning. Ca, K and Na make Fe poisoning worse as they acts as fluxing agents along with the other contaminants, restricting access to the active components in the particle interior. To alleviate the impact of Fe, flushing equilibrium catalyst (Ecat) is typically used to dilute metal contaminants. Until recently, this flushing strategy was the only practical solution, however, effective metal trap additives, such as Cat-Aid, have been shown to mitigate the effects of Fe poisoning significantly. Electron microscopy has shown that Cat-Aid reduces Fe nodules and smoothens the catalyst surface, allowing hydrocarbons into and out of the catalyst interior. Thereby conversion recovers, improving gasoline and bottoms selectivity and lowering dry gas and H2 yields. In some operations, the need for addition of flushing Ecat has been completely eliminated and the fresh catalyst addition rate has also been lowered.

Copyright © American Institute of Chemical Engineers. All rights reserved.

401918 Bridging the Cultural Gap in International Business – At Home and Abroad

Abstract:

Working with management and employees requires more than simply 'speaking the language". Dr. Kelm will describe several cultural issues that often come up when doing business in Latin America or China. Similar problems also arise in European business settings. Cultural issues may arise in the context of language, environment, social organization, context, authority, non-verbal communication and time perception.Several case studies, based on interviews with Latin American and Spanish professionals, will be raised for discussion and problem solving.Whether you are planning to work for an international company, relocate to another country for work, or build a successful and effective international team based in the US, you will learn practical strategies for bridging the cultural gap.

Copyright © American Institute of Chemical Engineers. All rights reserved.

Shale Oils – Designing for Improved Crude Preheat Train Reliability and Performance

Dominic Varraveto, PE
Chief Process Engineer
& Director of Technology

Mark W. Lockhart, PE
Process Technology Manager,
Refining & Chemicals

Abyar Aejaz, EIT
Senior Process Engineer,
Process Group

15 AIChE Spring Meeting, Austin, TX

Paper - ID# 401061

BURNS & McDONNELL

Introduction / Overview

- Shale plays – locations and current production
- Tight oil and condensate – composition
- Design approach for crude preheat train design
  - Exchanger fouling and design
  - Desalter design and operation
  - Flash drum placement and design
  - Crude charge heater fouling and design
- Conclusions

15 AIChE Spring Meeting, Austin, TX

BURNS & McDONNELL

2

Shale Plays Location & Current Production

- **Shale plays - lower U.S. 48 map:**
  - Niobrara
  - Bakken
  - Monterrey
  - Haynesville-Bossier
  - Utica
  - Wolf Camp & Bone Springs
  - Eagle Ford

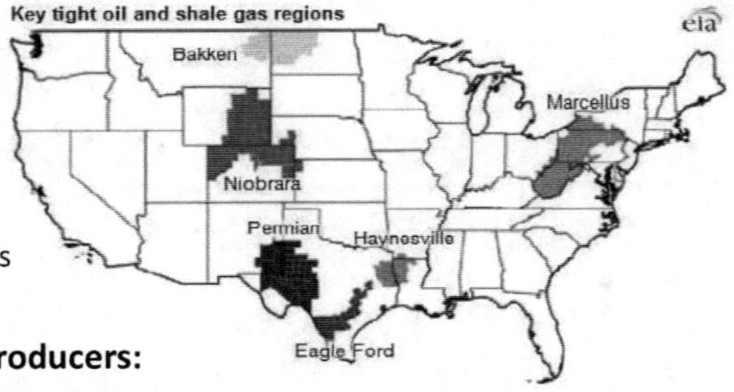

- **Largest liquid "Tight Oil" producers:**
  - Bakken in Montana / North Dakota
  - Eagle Ford in South Central Texas
- **Production:**
  - Bakken 2014E: **0.94 MMBPD**
  - Eagle Ford 2014E: **1.21 MMBPD**

Tight Oil & Condensate Composition

- **Tight oil and condensate are:**
  - Light and sweet
  - Sulfur: < 0.25 wt%
  - High naphtha content
  - Paraffinic
  - Heavy metals are low
  - High filterable solids

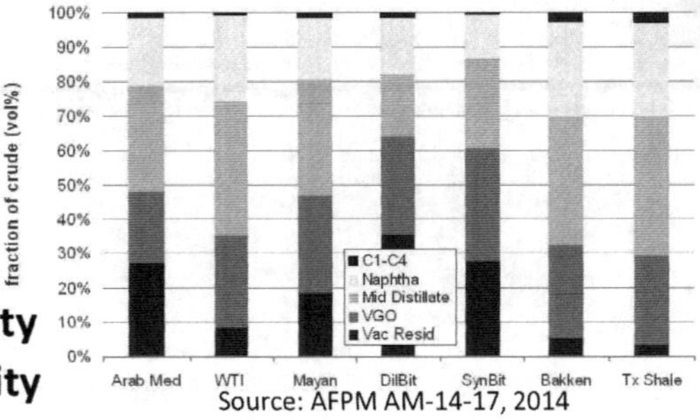

Source: AFPM AM-14-17, 2014

- **Tight Oil: ~40 – 55 API gravity**
- **Condensate: ~55 – 65 API gravity**
- **Processing Challenges from:**
  - Paraffinic nature – **wax deposition, asphaltene precipitation when blend crudes, exchanger and charge heater fouling**
  - Filterable solids – **exchanger and fired heater fouling**
  - Refiners lack of ability to clean equipment on-line

Crude Preheat Train - Design Approach for Reliability & Performance

- Exchanger Fouling and Design
- Desalter Design and Operation
- Flash Drum Placement and Design
- Charge Heater Fouling and Design

15 AIChE
Spring Meeting, Austin, TX

BURNS McDONNELL

5

Crude Preheat Train Overview

Design Considerations:

- **Exchanger Fouling and Design**

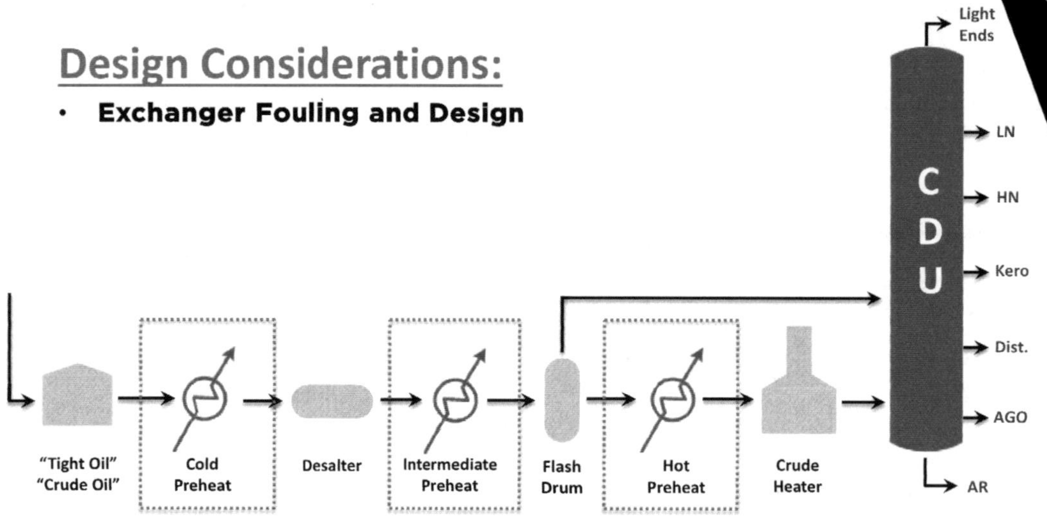

15 AIChE
Spring Meeting, Austin, TX

BURNS McDONNELL

6

Exchanger Fouling & Design

- **Tight oil and condensate considerations:**
  - High paraffin content - wax deposition
  - High filterable solids – solids deposition
  - Asphaltene precipitation potential - crude blending
  - Desalter design and operation – upsets and carryover
- Exchanger Types:
  - Shell and tube
  - Helical baffle
  - Refer to figures on next two slides

15 AIChE
Spring Meeting, Austin, TX

BURNS & McDONNELL

7

Exchanger Types – Shell and Tube (S&T)

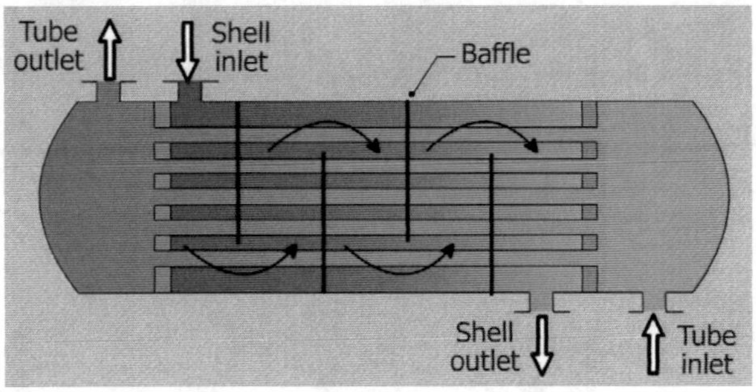

- Tube side – high velocity
- **Shell side:**
  - Lower velocity vs helical baffle exchanger (HBE)
  - Higher pressure drop (same velocity) vs HBE
  - Segmental baffles - potential for deposition

15 AIChE
Spring Meeting, Austin, TX

BURNS & McDONNELL

8

Exchanger Types – Helical Baffle (HBE)

- Tube side – high velocity
- **Shell side:**
  - Can achieve higher velocity vs S&T
  - Can achieve lower pressure drop with high velocity (without vibration)
  - Reduced dead zones – extended run times

15 AIChE Spring Meeting, Austin, TX

BURNS & McDONNELL

9

Exchanger Selection Methodology

- **Exchanger Selection – to start:**
  - Optimization / process simulation is complete
  - Configuration / number of heat exchangers defined
  - Hot and cold streams, flowrates, and exchanger duties defined
- Methodology
  - Assess causes of fouling in the cold, intermediate and hot train
  - Based on causes, assess: can hot and cold fluids be placed on the shell and/or the tube side for S&T and Helical baffle exchangers (HBE)?
  - Use exchanger modeling tool (i.e. HTRI, EDR, etc.) with prescribed design requirements and conduct comparison

15 AIChE Spring Meeting, Austin, TX

BURNS & McDONNELL

10

Exchanger Selection Methodology

- **Methodology (continued)**
  - Apply design margins, fouling factors, exchanger cleaning requirements
  - Tabulate results for S&T and HBE's while placing the hot and cold fluids on each of the applicable sides of the exchanger
  - Compare area, pressure drop, on-line cleaning considerations and installed capital and/or lifecycle costs
  - Based upon results - select exchanger type and fluid placement
- **Other considerations – bigger picture**
  - Exchanger pressure drop
  - Pumping horsepower
  - Equipment design pressure
  - Metallurgy
  - Seek existing exchanger fouling data

15 AIChE Spring Meeting, Austin, TX

BURNS & McDONNELL

11

Exchanger Fouling & Design – Cold PHT

EXAMPLE

- Causes of Fouling:
  - **Fouling is more likely to occur on cold crude side**
  - **Cause: Wax build-up, high filterable solids deposition**
- Design Considerations:
  - Maintain high fluid velocity (>5 fps)
  - Minimize dead spots on crude side (i.e. NOT place on shell side for S&T option)
  - Recognize potential to reduce ΔP for crude if placed on the shell side.
  - Other (include design margins, fouling factors, exchanger cleaning requirements)
- **Exchanger Selections:**
  - **Tabulate and compare results**
  - **Continue through all cold PHT exchanger services and make selections**
  - **Apply "Other considerations" for final selections**

15 AIChE Spring Meeting, Austin, TX

BURNS & McDONNELL

12

Exchanger Fouling & Design – PHT Exchangers

- Causes of Fouling:
  - **Cold Train:**
    - Fouling is more likely to occur on cold crude side
    - Cause: Wax build-up, high filterable solids deposition
  - **Intermediate Train**
    - Fouling potential for the cold crude and hot stream sides
    - Cold (crude) side - desalter operation/upsets with carryover of emulsions or solids, possible asphaltene precipitation
    - Hot side – potentially higher fouling streams (AGO, AR, VR, etc.)
  - **Hot Train:**
    - Highest temp in train - fouling potential for the cold crude and hot stream sides
    - Cold (crude) side - desalter operation/upsets with carryover of emulsions or solids, higher temps result in higher fouling with asphaltene precipitation potential
    - Hot side – higher fouling streams at high temps, in particular AGO, VGO, Atm and Vac resid

15 AIChE Spring Meeting, Austin, TX

BURNS McDONNELL

13

Exchanger Fouling & Design - Summary

To achieve the most cost-effective, safe, and reliable PHT design - Evaluate and determine a balance between:

- Exchanger fouling
- Fluid velocity criteria
- Exchanger pressure drop
- Placement the cold and hot fluids
- Exchanger metallurgy
- On-line cleaning
- Pumping head
- Equipment design pressures

15 AIChE Spring Meeting, Austin, TX

BURNS McDONNELL

14

Crude Preheat Train Overview

Design Considerations:

- Exchanger Fouling and Design
- **Desalter Design and Operation**

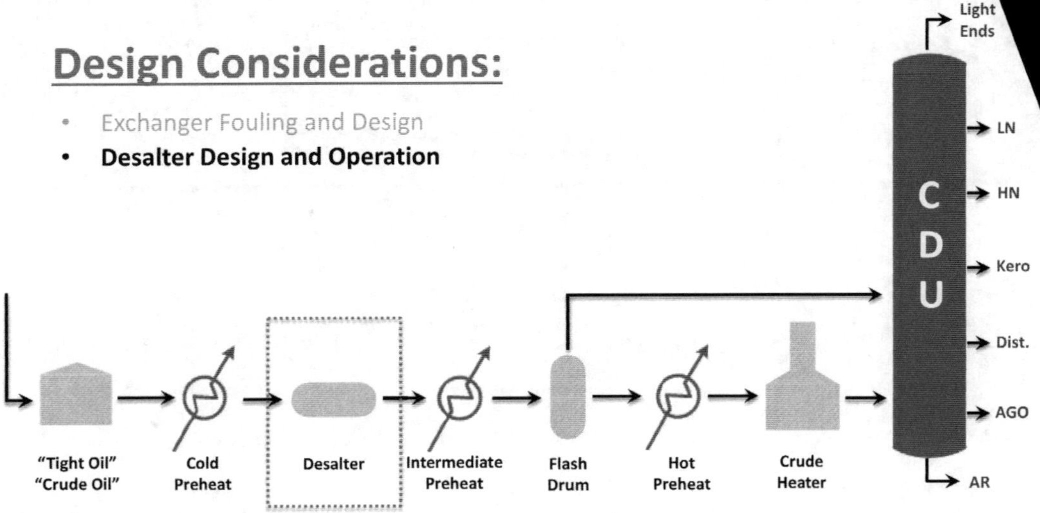

15 AIChE Spring Meeting, Austin, TX

BURNS & McDONNELL

15

Desalter Design & Operation

- **Purpose & Function of Desalter:**
  - Reduce salt content and remove solids
  - Wash Water mixed with Oil. Separates by differential gravity enhanced by electric field
  - Resulting Oil is clean and dry. Brine contains salts , sediment
- **Design Considerations:**
  - Feed characteristic including paraffinic-asphaltene compatibility
  - Temperature to reduce viscosity, limited by water solubility in oil & mechanical equipment
  - Corresponding Pressure needed to maintain liquid phase and prevent vaporization
  - Tendency to form emulsions and need for addition of chemical dispersants and demulsifiers
  - Solids handling capacity internal and external oily brine processing
- **Tight Oil Desalter Features:**
  - Two-Stage has higher removal efficiency than Single Stage. Reduces chance of carryover, Can be serviced w/o total shutdown
  - Robust Mudwash system for high solids and stabilized emulsions.
  - Wash Water pH control to assist in removal of amine added during transportation (to lower H_2S content in tight oil)

15 AIChE Spring Meeting, Austin, TX

BURNS & McDONNELL

16

Desalter Design & Operation

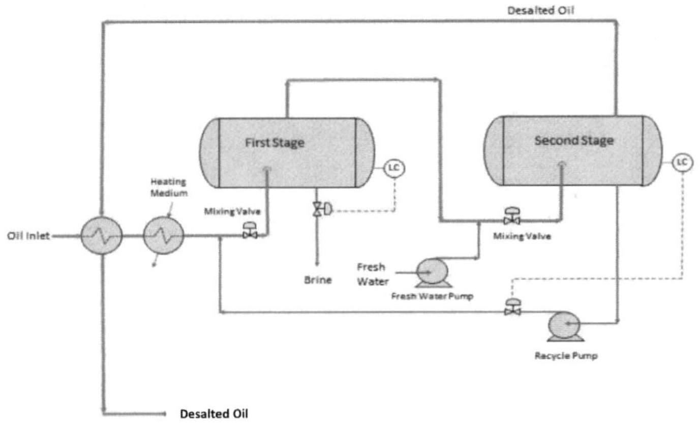

Source: Piping Engineering

Crude Preheat Train Overview

Design Considerations:

- Exchanger Fouling and Design
- Desalter Design and Operation
- **Flash Drum Placement and Design**

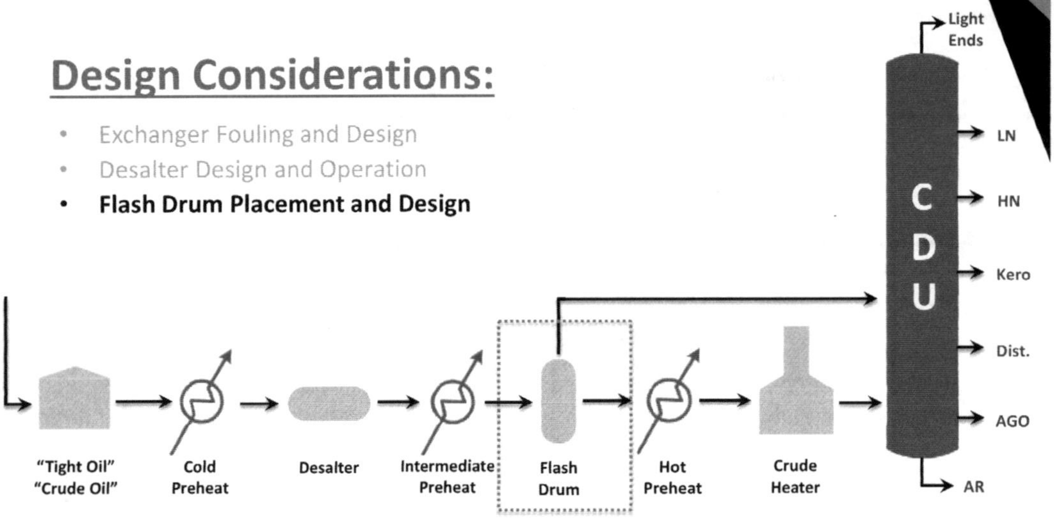

Flash Drum Placement & Design

- Location of vapor feed to column
  - Option 1: Flash zone where fired heater transfer line connects to the crude tower
    - Results in a larger crude tower bottom section diameter.
    - Vapor stream in flash zone acts as a quench requiring a higher heater outlet temp. to maintain the desired flash zone temp.
  - Option 2: Taken to an appropriately higher section in the tower where temp. and composition of the flash drum vapor is similar.
    - For light oil flashed at 400 °F, vapor return location would be above the distillate section and below the naphtha section.
    - Required heater duty and crude tower diameter could be reduced in this scenario.

Flash Drum in CPHT

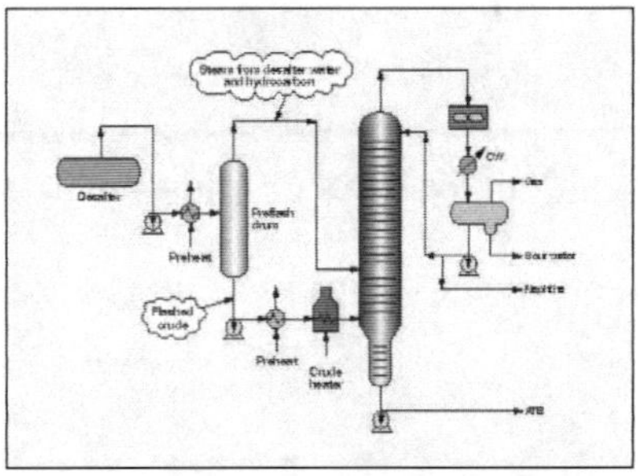

Source: Process Consulting Services

15 AIChE Spring Meeting, Austin, TX

BURNS & McDONNELL

Flash Drum Placement & Design

- Option 3: Flash tower with trays and overhead system
  - Does NOT return any flashed vapor to the crude tower.
  - More complex and costly than a flash drum
  - Typical for a revamp due to:
    - Limitations on existing crude tower
    - Requirement of additional throughput and improved separation

Flash Tower in CPHT

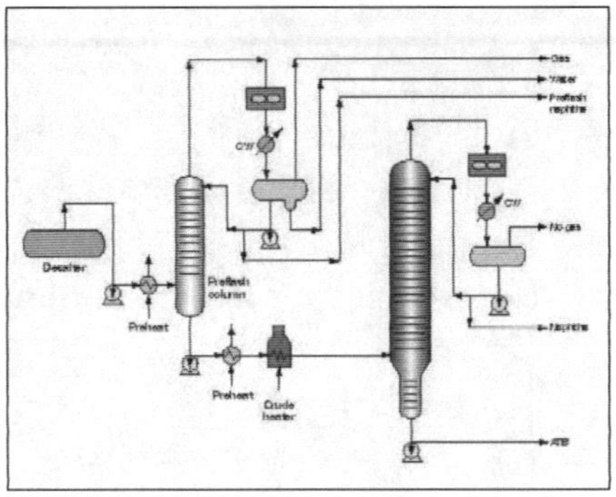

Source: Process Consulting Services

15 AIChE Spring Meeting, Austin, TX

BURNS & McDONNELL

Crude Preheat Train Overview

- **Design Considerations:**
  - Exchanger Fouling and Design
  - Desalter Design and Operation
  - Flash Drum Placement and Design
  - **Charge Heater Fouling and Design**

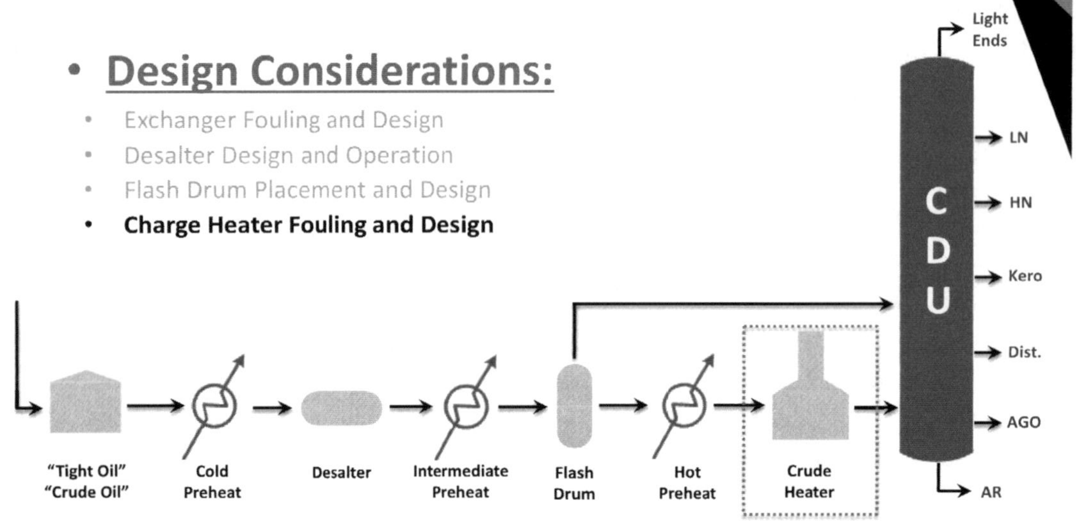

15 AIChE Spring Meeting, Austin, TX

BURNS & McDONNELL

21

Charge Heater Fouling & Design

- Causes of fouling:
  - Asphaltene deposition from blending crudes
  - Desalter carryover of emulsions and solids can also contribute
  - High tube metal temperatures (TMT)
  - High residence times
  - High percent vaporization (tube running dry)
  - → Cracking reactions with coke laydown
- Design Considerations:
  - Even radiant heat flux – avoid localized hot spots
    - See figures to right
  - High mass velocity – lower residence time
    - Consider velocity steam injection at multiple locations

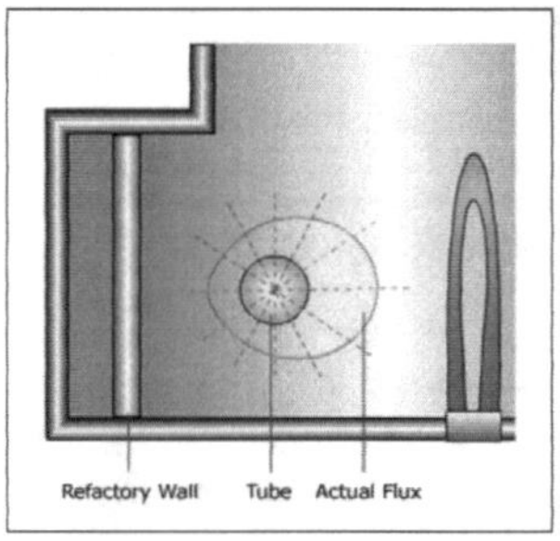

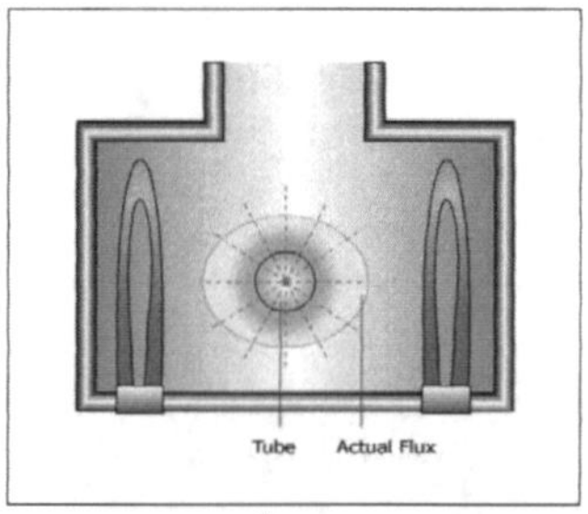

15 AIChE Spring Meeting, Austin, TX

BURNS & McDONNELL

22

Charge Heater Fouling & Design

- Design Considerations (cont'd):
  - Max percent vaporization (50 – 60%)
    - Lights content of tight oil exacerbates
    - Can recirculate atm resid
  - Include on-line decoking or pigging of tubes
- Charge Heater type:
  - Double fired – even heat flux
    - All floor or wall and floor burners
    - Floor burners alone can accomplish even heat flux and is lower cost
  - Multiple cells – for on-line cleaning

15 AIChE Spring Meeting, Austin, TX

BURNS & McDONNELL

23

Charge Heater Fouling & Design

- Charge Heater Type:
  - Figure to right features:
    - Two-cell, box cabin with floor burners
    - Ability to isolate each cell for on-line decoking and/or pigging (smart pigs)
    - Design may or may not include:
      - ID, FD fans, with SCR and common stack

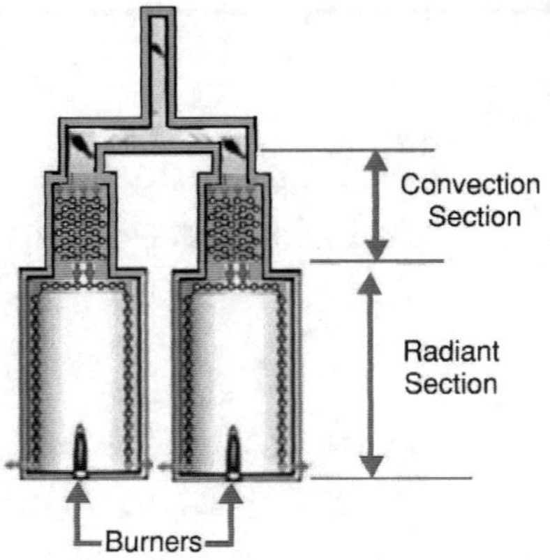

15 AIChE Spring Meeting, Austin, TX

BURNS & McDONNELL

24

Charge Heater Fouling & Design

- Other Design Considerations:
  - Carefully consider charge heater design margin
    - Allows operation at higher capacity when cleaning one cell
  - Fouling in fired heater tubes is less desirable than fouling in heat exchanger
    - Strive for max heater inlet temperatures
  - Vaporization at heater outlet affected by
    - Flash zone temperature and pressure
    - Consider recirculation of atm resid

15 AIChE
Spring Meeting, Austin, TX

BURNS McDONNELL

25

Conclusion

- **Tight oils and condensates are presenting unique challenges for design of new CPHT's and in operating a existing CPHT's**
- **Existing crude units are experiencing high fouling in heat exchangers and charge heater tubes leading to unplanned outages and loss of production**
- **Crude unit reliability can be increased in new unit design, in crude unit retrofits and/or in adjusting operating parameters in existing units**

15 AIChE
Spring Meeting, Austin, TX

BURNS McDONNELL

26

Conclusion

- This paper has provided specific design approaches for the CPHT to address the unique challenges when processing tight oil and condensates, including:
  - Exchanger fouling and design
  - Desalter design and operation
  - Flash drum placement and design
  - Fired heater fouling and design

Questions and Answers

- Burns & McDonnell – 100% Employee Owned!

Dominic Varraveto
dvarraveto@burnsmcd.com
816.822.4282

Mark W. Lockhart
mlockhart@burnsmcd.com
832.214.1975

Abyar Aejaz
aaejaz@burnsmcd.com
832.214.1947

BURNS McDONNELL

15 AIChE
Spring Meeting, Austin, TX

Supplemental Slides

- Transportation and Storage
- Crude Tower Corrosion
- Exchanger Bypassing
- Desalter Operation
- Charge Heater Fouling

15 AIChE
Spring Meeting, Austin, TX

BURNS McDONNELL

Crude Preheat Train Overview

Design Consideration:

- **Transportation and Storage Considerations**

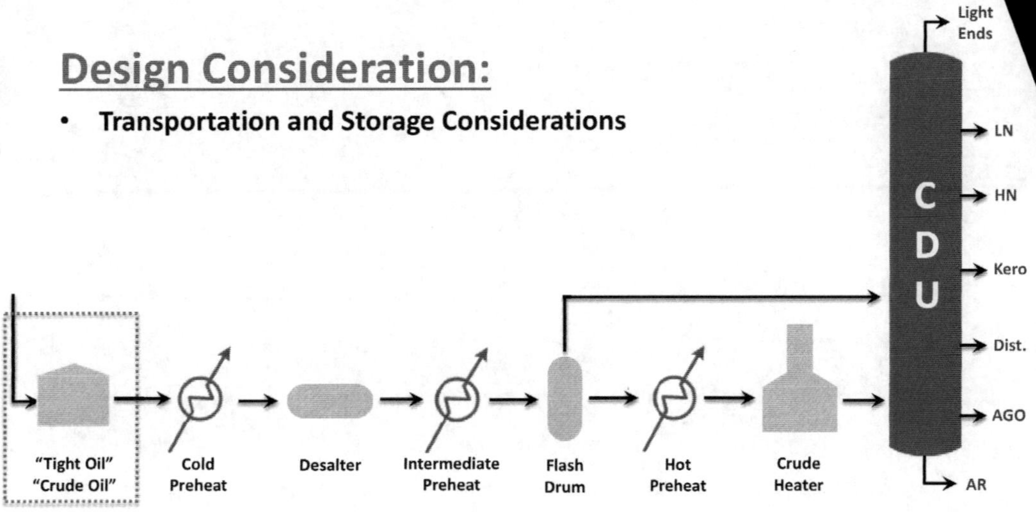

15 AIChE Spring Meeting, Austin, TX

BURNS & McDONNELL

Transportation and Storage

- **Plugging in piping and exchangers can occur caused by wax deposits from paraffinic crudes**
- **Potential for more cleaning required**
  - Pigging pipelines, chemical additions and dispersants
- Crude Storage Tanks
  - Sludge build-up
  - Wax build-up on walls and inside transfer lines
  - Can increase pressure drop in lines
  - Considerations: install heating coils, internal mixers, or recirculation
- Caution when blending crudes
  - Destabilization of asphaltenes when mixing tight oils with high asphaltene crudes
  - Could result in stable emulsions and sludge formation

15 AIChE Spring Meeting, Austin, TX

BURNS & McDONNELL

Crude Preheat Train Overview

Design Consideration:

- **Crude Tower Corrosion**

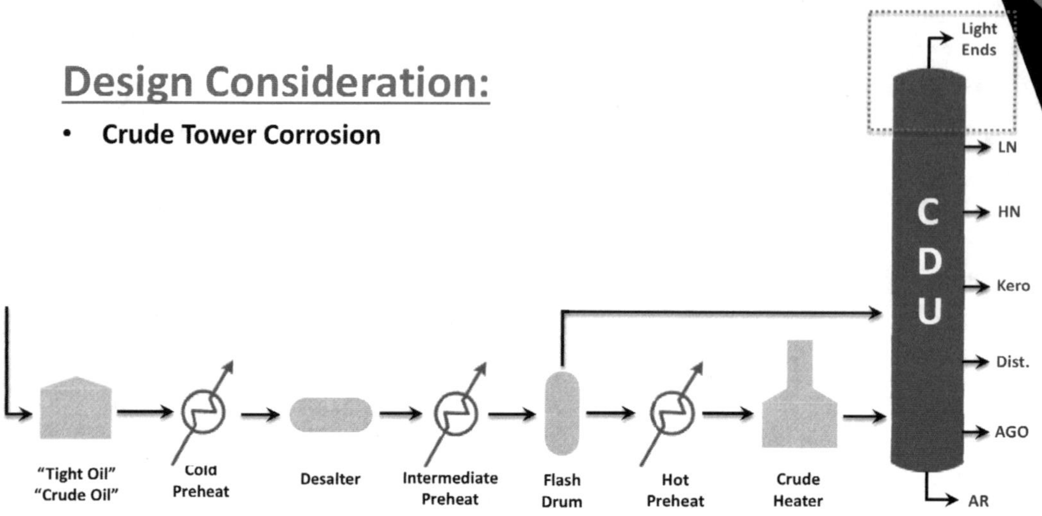

15 AIChE Spring Meeting, Austin, TX

BURNS & McDONNELL

Crude Tower Corrosion

- **Crude tower overhead corrosion - causes:**
  - Light metal salts in crudes (i.e. $CaCl_2$, $MgCl_2$, etc.)
    - Hydrolyze in charge heater – form metal and chloride ions
    - Recombine and form HCl with H_2O condensation in crude tower overhead
    - Tight oil has lower salt content - corrosion and fouling risk lower
  - Amine chloride salts can form – under deposit corrosion potential
  - For tight oil, tramp amines sometimes added during transportation to scavenge H2S

15 AIChE Spring Meeting, Austin, TX

BURNS & McDONNELL

Crude Tower Corrosion

- **Crude tower and overhead system design considerations:**
  - Operate at a temperature above the water dew point
  - Water circulation and washing to inlet of overhead condenser – wash salts into solution
  - Tramp amines – operate desalter at lower pH
  - Add corrosion inhibitor into overhead system
  - Metallurgy selection for overhead system – consider alloy for:
    - Upper tower shell and overhead line
    - Upper trays in upper section
    - Overhead condenser

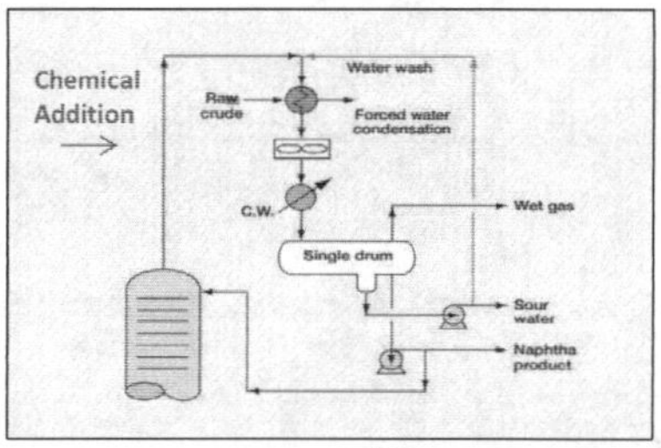

15 AIChE
Spring Meeting, Austin, TX

BURNS McDONNELL

Crude Tower Corrosion

- **Crude tower shell and equipment:**
  - **Tight oils and condensates contain lower sulfur**
    - Less passivation on equipment surface
    - Increased risk of naphthenic acid corrosion in high temp zones
  - **Tight oil and condensates naphthenic acid low**
  - **Use care in metallurgy selection**
    - For current and <u>future feed slates</u> consider level of sulfur and naphthenic acid
    - Consider use of cladding in mid and upper portions of the column shell

15 AIChE
Spring Meeting, Austin, TX

Exchanger Bypassing

- **Crude tower shell and equipment:**
  - Tight oils and condensates contain lower sulfur
    - Less passivation on equipment surface
    - Increased risk of naphthenic acid corrosion in high temp zones
  - Tight oil and condensates naphthenic acid low
  - Use care in metallurgy selection
    - For current and future feed slates consider level of sulfur and naphthenic acid
    - Consider use of cladding in mid and upper portions of the column shell

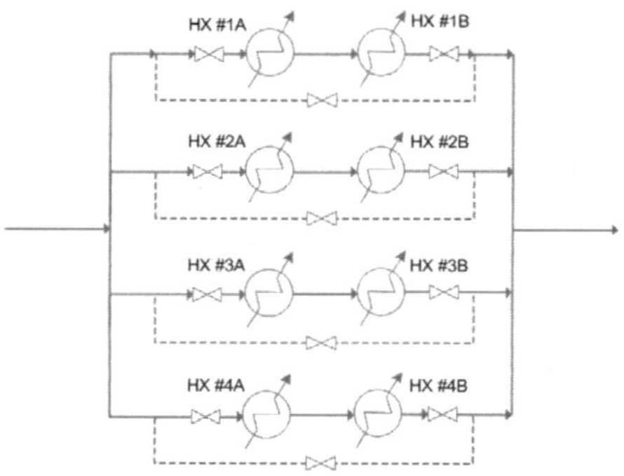

15 AIChE Spring Meeting, Austin, TX

BURNS & McDONNELL

Exchanger Bypassing

- For exchangers in series, one exchanger or shell can be bypassed if necessary valves and bypasses are included in the design
  - Crude unit rates affected minimally
    - As long as valves and bypass are adequately sized
    - Downstream equipment may need evaluation when hot streams bypassed.

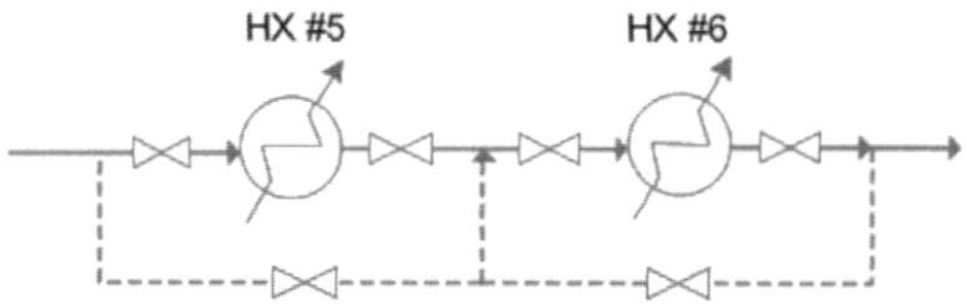

- Desalter inlet temperature controlled by the temp. controller reducing bypass flow when upstream exchanger being cleaned.

15 AIChE Spring Meeting, Austin, TX

Desalter Operation – Wash Water

Having a sufficient quantity of suitable quality wash water is crucial to achieving successful desalter performance.

- The wash water rate should be high enough so that the salt specification of the dry crude can be met while considering the residual basic sediment and water (BS&W) in the oil.
- Higher wash water rates are required for high salt crudes as well as heavy viscous crudes that are difficult to dehydrate. Typical wash water rates are in the 1 to 3 volume % range with rates of 5% or more needed for hard to treat crude oils.
- Wash water of sufficient quality to avoid stable emulsion formation should not contain coke particles, suspended solids, iron sulfide and/or emulsified oil. The wash water pH should be between 7 and 9 and should not contain surfactant chemicals or excess caustic soda, which could stabilize emulsions. If the wash water pH is above 9, provisions for acid injection for pH control should be included.
- Other wash water contaminant limitation guidelines are less than 200 ppm of ammonia to avoid overhead system fouling, less than 0.02 ppm oxygen and less than 1 ppm fluoride.
- Wash water should also be low in sodium salts, suspended solids and hardness.
- Treated water like steam condensate or boiler feed water that meet the quality requirements noted above can be used as fresh desalter wash water make up. It is also possible to reuse refinery water streams like sour water stripper bottoms and vacuum tower ejector condensate as desalter wash water.

15 AIChE
Spring Meeting, Austin, TX

Desalter Operation

Other desalter design considerations include:

- Ensuring sufficient mixing energy is applied to the combined oil and water feed.
- Chemical addition of dispersants and demulsifiers to wet or improve particulate affinity for the water phase and to neutralize the tendency of precipitated asphaltenes from agglomerating and trapping water molecules that can create a stabilized emulsion
- Neutralizing amines used during transportation to lower the hydrogen sulfide content in tight oil that can cause corrosion in the crude tower overhead system
- Handling solids including mudwashing
- Processing emulsions for oil recovery and solids disposal
- The impact of brine on waste water treatment

15 AIChE
Spring Meeting, Austin, TX

BURNS McDONNELL

Desalter Operation

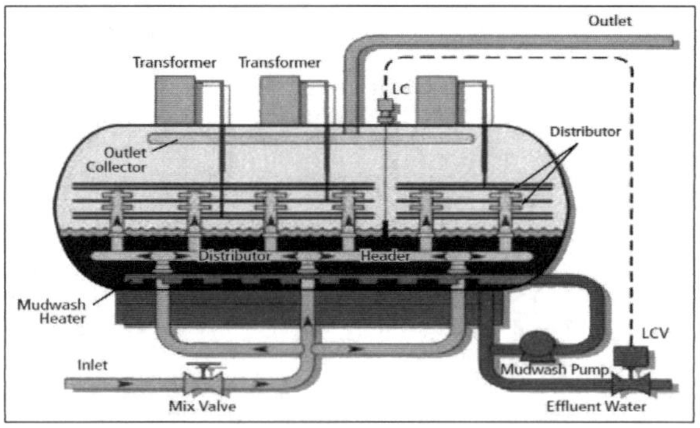

The desalter should be monitored and controlled closely during normal operation. Chemical addition rates, wash water rates, mud-washing frequency and other parameters should be monitored and adjusted regularly. For adequate salt removal, the operating temperature range should also be closely monitored. As noted previously higher operating temperature decreases viscosity resulting in quicker settling of coalesced water droplets.

15 AIChE
Spring Meeting, Austin, TX

BURNS & McDONNELL

Charge Heater Fouling – Existing Unit

- **Even radiant heat flux:**
  - Adjust burners for even flame patterns
  - Minimize tube impingement
- **Tube temperature monitoring:**
  - TI's and pyrometer readings to measure tube wall temps
  - Can use with burner adjustments to validate results
  - Indicator of when to reduce rates and/or clean tubes
- **Heater outlet temperature:**
  - Can this be reduced – how effect tower performance?
  - Utilize previous operating data – balance of outlet temp and fouling achievable?
- **Charge heater tube velocity:**
  - Can heater tube residence time be reduced?
  - Possibilities include - Increase operating rate, allow more lights from the flash drum, add velocity steam, recirculate atm resid

15 AIChE
Spring Meeting, Austin, TX

BURNS & McDONNELL

Charge Heater Fouling – Existing Unit

- **Percent vaporization in tubes:**
  - Determine percent vaporization (<60-65%?)
  - Can lead to dry out of tubes and increased cracking reactions
  - Possibilities to reduce percent vaporization include – lower heater outlet temp, lower amount of lights from flash drum, increase crude tower op pressure, recirculate atm resid to heater inlet
- **Lower firing rate:**
  - Tube temps increase with fouling
  - May need to lower operating rate
  - Can heater inlet temp be increased?
  - Evaluate preheat train – how improve performance, absorb more heat from hot streams?
  - Exchanger cleaning justified?
  - Maximizing heater inlet temp can move fouling out of heater and into exchangers which can be cleaned more easily

15 AIChE Spring Meeting, Austin, TX

BURNS & McDONNELL

Charge Heater Fouling – Existing Unit

- **Review desalter performance and operation:**
  - Tight oil – high and variable amounts of filterable solids
  - Crude blending can cause asphaltene precipitation
  - Review operation closely – water circulation rates, good mixing, frequency and adequacy of mud washing.
  - Review chemicals injection program with chemicals supplier.
  - Chemicals utilize may include – emulsion breaker, wetting agent, acid, crude stabilizer, metal deactivator, dispersant, naphthenic acid inhibitor, adjunct dispersant, neutralizer, filmer and corrosion inhibitor
  - Chemical supply vendors:
    - May have new and better chemicals for your situation
    - May provide bench testing work for refiners on blends, chemicals injection rates, pH control, etc.
- **Crude and tight oil blending**
  - May need to blend to sufficiently fill out downstream units and meet economic objectives
  - More data exists and chemical suppliers are accumulating more data relative to crude blending, asphaltene precipitation and equipment fouling
  - May be prudent to adjust blends and find an optimum for your plant's situation
  - Evaluation may be tied into the chemicals program evaluation above

15 AIChE Spring Meeting, Austin, TX

BURNS & McDONNELL

Charge Heater Fouling – Existing Unit

- **Learn from plant data and equipment inspections**
  - Plant operating data is invaluable when charging tight oil
  - Heat exchangers should be monitored – where is fouling occurring and why? Can adjustments to flow be made to increase velocity (i.e. parallel branches)
  - Perform equipment inspections when taken out of service. Take samples of fouling material for lab analysis to further understand the nature of the fouling
- **Capital investment options**
  - Options are found throughout this presentation that can be evaluated as ways to invest at different levels to improve reliability and operation in an existing CPHT
  - An evaluation can be made as to how to get the maximum benefit from a reasonable investment.
  - Examples include:
    - Adding exchanger bypasses to allow cleaning on-line
    - Replacing or modifying key exchanger services that are experiencing the most fouling (i.e. increase velocity, change exchanger type, etc.)
    - Install new desalter internals or add a second desalter
    - Add a pre-flash drum
    - Add on-line smart pigging capability to fired heaters
    - Modify crude tower internals

15 AIChE
Spring Meeting, Austin, TX

BURNS McDONNELL

401936 Effect of Nanosize on Adsorptive and Catalytic Properties of NiO Nanoparticles Towards Heavy Hydrocarbons: Kinetics, Thermodynamics and Mechanistic Insights

Abstract:

The application of nanoparticles in the oil and gas industries has increased in recent years because of their unique physical and chemical properties. Notwithstanding the great deal of interest in the use of nanoparticles as adsorbents and catalysts for enhancing heavy oil upgrading and recovery, still the effect of nanoparticle size on their adsorption and catalytic behaviour towards heavy hydrocarbons remain unclear. This study focused on the effect of NiO nanoparticle size on adsorption and post-adsorption catalytic decomposition of heavy polar hydrocarbons, exemplified by Quinolin-65 (Q-65) and Violanthrone-79 (V-79). In this study, different nanosizes of NiO nanoparticles, ranging from 5-80 nm, were prepared by thermal dehydroxylation of $Ni(OH)_2$ following the hydrothermal technique. Using batch adsorption equilibrium measurements of Q-65 and V-79 dissolved in toluene coupled with thermogravimetric analysis, we studied the effect of nanoparticle size on the adsorptive and catalytic behavior of the in-house prepared NiO nanoparticles. Adsorption isotherms were described by the Langmuir model and the Solid Liquid Equilibrium model (SLE), and the catalytic behavior of the nanoparticles were compared on the basis of the effective activation energies trends using the isoconversional method of Ozawa−Flynn−Wall (OFW). In addition, computational modeling was conducted to understand the chemistry behind the differences on adsorption capacity and catalytic activity of the different sizes of NiO nanoparticles. In case of V-79, on the surface-area-normalized basis, results obtained in this study revealed that the adsorptive capacity increased with increasing nanosize of NiO particle, while the oxidation activation energy decreased with decreasing of size of NiO particle. On the other hand, an optimum nanosize of NiO nanoparticle was found for Q-65 for which a maximum adsorption capacity and minimum activation energy were obtained. For the case of binary solutions (V-79 and Q-65), for all sizes of NiO nanoparticles, NiO showed higher selectivity towards V-79 than to Q-65 indicating a selective adsorption; which was backed up by the computational modeling The experimental adsorption capacity of V-79 was enhanced in the presence of Q-65, presumably due to new interactions of V-79 in the presence of Q-65 (entangling of aliphatic chains between Q-65 and V-79 seems to reduce the adsoption energy according to the computational study) .

This study confirms that nanosize plays vital roles in selective adsorption and subsequent catalytic decomposition of heavy hydrocarbons. The study opens up new possibilities for manipulating adsorptive and catalytic performance of metal oxide nanoparticles, and widens our general understanding of their surface functionality and chemical activity.

Copyright © American Institute of Chemical Engineers. All rights reserved.

Crude and Condensate Export – Distillation Requirements Defined and Explained

Mark W. Lockhart, PE
Process Technology Manager,
Refining & Chemicals

15 AIChE
Spring Meeting, Austin, TX

BURNS & McDONNELL

Introduction / Overview

- **PART 1 – Condensate Distillation Requirements Defined**
  - Bureau of Industry and Security (BIS) - Rules for:
    - Exporting crude oil
    - Distillation requirements for exporting condensate
  - Shale Play Production Locations
  - Tight Oil and Condensate Composition
  - Crude and Condensate Stabilization
  - Distillation Spectrum
  - Distillation Requirements Met – Light Naphtha Production?
  - Various Possibilities Exist Based Upon:
    - Upstream, Midstream and Downstream Perspective
    - Logistics, Location and Economics
- **Refer to Part 2**

15 AIChE
Spring Meeting, Austin, TX

BURNS & McDONNELL

Introduction / Overview

- PART 2 –Location, Logistics and Economics →Upstream, Midstream and Downstream Perspective
  - Upstream
    - At the Wellhead
  - Midstream
    - Gas Plant Facility
    - Condensate Splitter Facility
  - Refinery
    - Add a Condensate Splitter
    - Add a Crude Unit
- Conclusions
  - Economics and Logistics – Scale and Infrastructure Requirements

15 AIChE Spring Meeting, Austin, TX

BURNS McDONNELL

PART 1

Distillation Requirements Defined

15 AIChE Spring Meeting, Austin, TX

BURNS McDONNELL

Bureau of Industry and Security (BIS) FAQ Page

FAQs – Crude Oil and Petroleum Products December 30, 2014

- 1. What are the restrictions on the export of crude oil?
- 2. How do I determine if my commodity is crude oil?
- 3. Is lease condensate considered crude oil?
- 4. What is required in order for liquid hydrocarbons to have been "processed through a crude oil distillation tower"?
- 5. Is an export license required to export refined petroleum products?
- 6. According to the Regulations, "foreign-origin" crude oil can be exported (with a valid license) as long as it is not "co-mingled" with domestic crude oil. What is the acceptable level of domestic crude that can be mixed with foreign crude and still be eligible for export?

http://www.bis.doc.gov/index.php/policy-guidance/faqs

15 AIChE Spring Meeting, Austin, TX

BURNS & McDONNELL

What is Required in Order for Liquid Hydrocarbons to Have Been "Processed Through a Crude Distillation Tower"?

The factors BIS will consider in reviewing commodity classifications to determine whether the product has been "processed through a crude oil distillation tower" include, among others:

(1) Whether the distillation process materially transforms the crude oil, by using heat to induce evaporation and condensation, into liquid streams that are chemically distinct from the crude oil input; (DHECLC)
(2) The change in API gravity between the input of the process and the output of the process; (ΔAIO)
(3) The change in percentage of different types of hydrocarbons between the input and output of the process; (Δ%HIO)
(4) Whether the streams resulting from distillation have purposes other than allowing the product to be classified as exportable petroleum products, such as use as petrochemical feedstock, diluent, and gasoline blendstock; (SDP EPP)
(5) Whether the distillation process utilizes temperature gradients and has significant internal structures, such as trays or packing, and differentiated output streams; (TG&IS DS)
(6) Whether the distillation uses towers with more mechanical complexity and heat, higher residence time, internal structures that promote condensation and better separation, and a consistent quality liquid streams (also called cuts or fractions) than equipment used to separate vapors and liquids for transportation needs. (QLS>T)

15 AIChE Spring Meeting, Austin, TX

BURNS & McDONNELL

Shale Plays – Locations

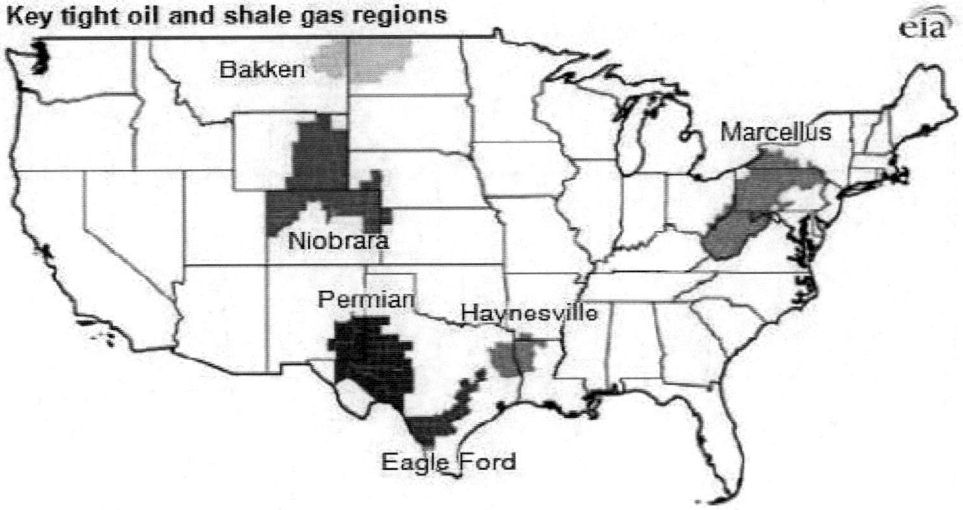

15 AIChE
Spring Meeting, Austin, TX

BURNS & McDONNELL

Tight Oil & Condensate Composition

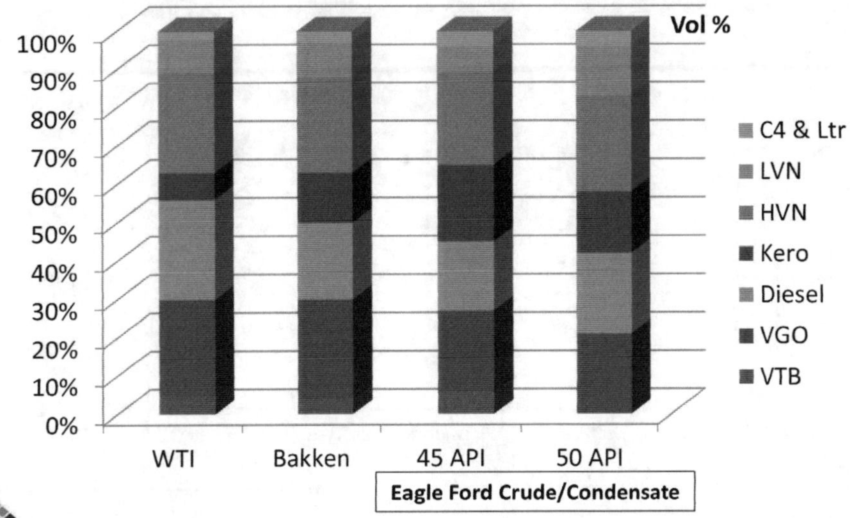

15 AIChE
Spring Meeting, Austin, TX

BURNS & McDONNELL

Crude/Condensate Stabilization (Wellhead)

Figure 1: Condensate Stabilizer Flow Diagram

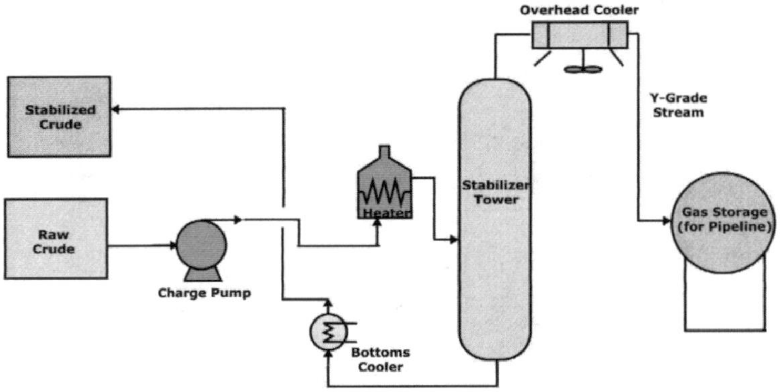

Source: Turner, Mason & Company, Blog 9-8-2014

15 AIChE Spring Meeting, Austin, TX

BURNS & McDONNELL

Distillation Spectrum

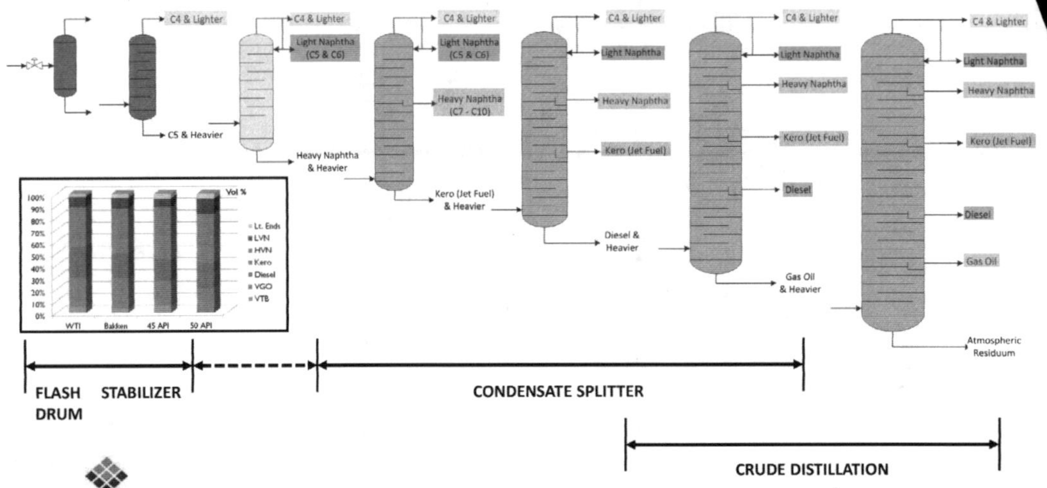

15 AIChE Spring Meeting, Austin, TX

BURNS & McDONNELL

Distillation Requirements Met?
Light Naphtha Production

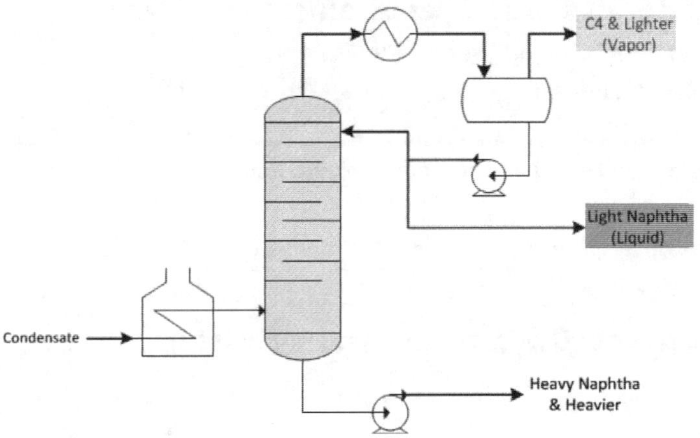

Complies w/ BIS

(1) ☑ (DHECLC)
(2) ☑ (ΔAIO)
(3) ☑ (Δ%IHO)
(4) ☑ (SDP EPP)
(5) ☑ (TG&IS DS)
(6) ☑ (QLS>T)

15 AIChE Spring Meeting, Austin, TX

BURNS McDONNELL

Distillation Requirements – BIS Disclaimer

In reference to the six (6) requirements on the FAQ page, BIS also lists the following disclaimer:

"These factors are not intended to be categorical or exhaustive. In reviewing commodity classification applications, BIS will look at the particular circumstances of each application to determine whether the output of a process can be considered a petroleum product under the current regulatory definition."

15 AIChE Spring Meeting, Austin, TX

BURNS McDONNELL

Conclusions

- **PART 1 – Condensate Distillation Requirements Defined**
  - Bureau of Industry and Security (BIS) sets the Rules
  - Crude and Condensate Stabilization
  - Distillation Spectrum
  - Distillation Requirements Met – Light Naphtha Production?
  - Distillation – Location Possibilities
  - Condensate Distillation Requirements Defined – Example!

- **Refer to Part 2 for Upstream, Midstream and Downstream Perspective**

15 AIChE Spring Meeting, Austin, TX

BURNS & McDONNELL

PART 2

Location, Logistics and Economics

Upstream, Midstream and Downstream Perspective

15 AIChE Spring Meeting, Austin, TX

BURNS & McDONNELL

Distillation Spectrum

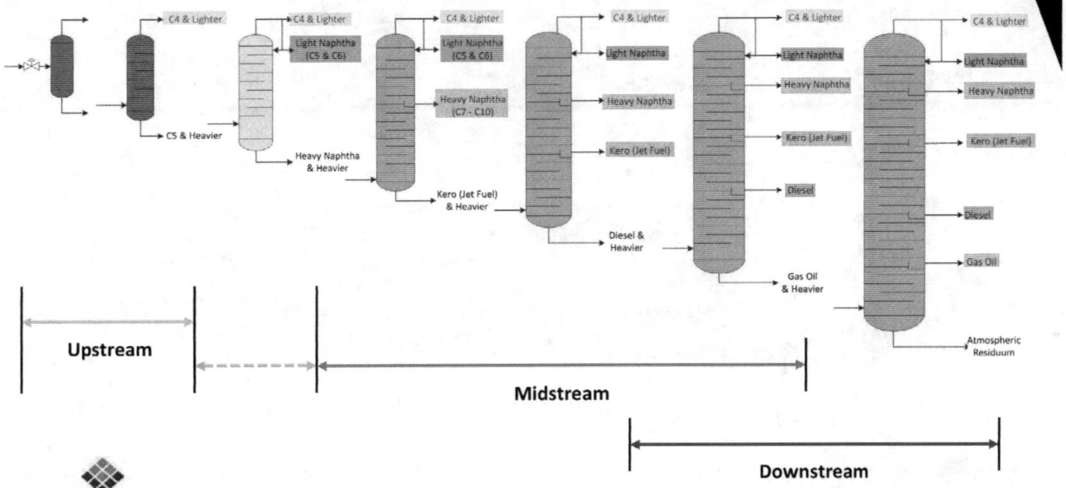

15 AIChE Spring Meeting, Austin, TX

BURNS McDONNELL

Upstream

15 AIChE Spring Meeting, Austin, TX

BURNS McDONNELL

Natural Gas - Transportation System

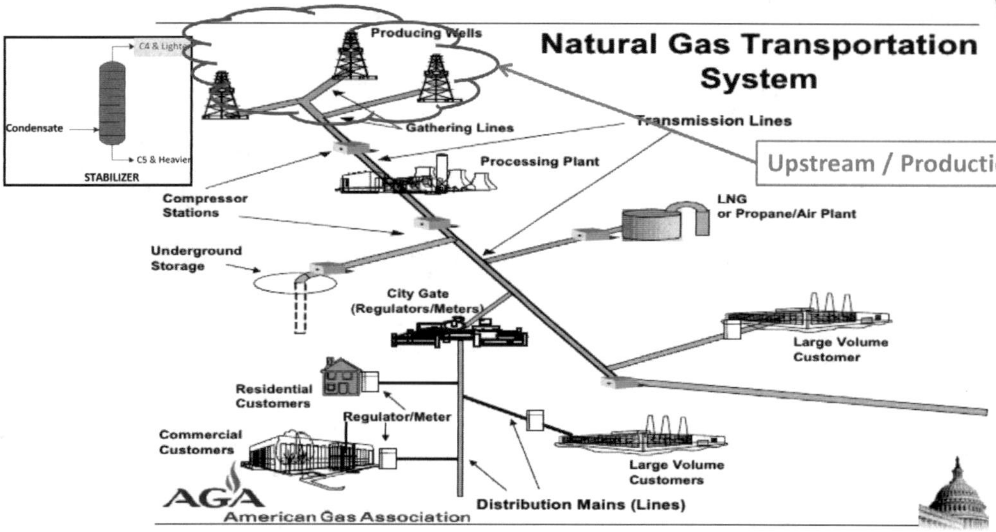

15 AIChE Spring Meeting, Austin, TX

BURNS & McDONNELL

Upstream - At the Wellhead

Wishful Thinking: If stabilizer were acceptable for "distillation", export of condensate is fairly straight-forward and ready to go

Logistics:

- Existing infrastructure, tankage and support systems
- Existing pipelines
- Still need export infrastructure/facility:
  - Tankage
  - Docks for ship loading
  - Ability to use existing export facility

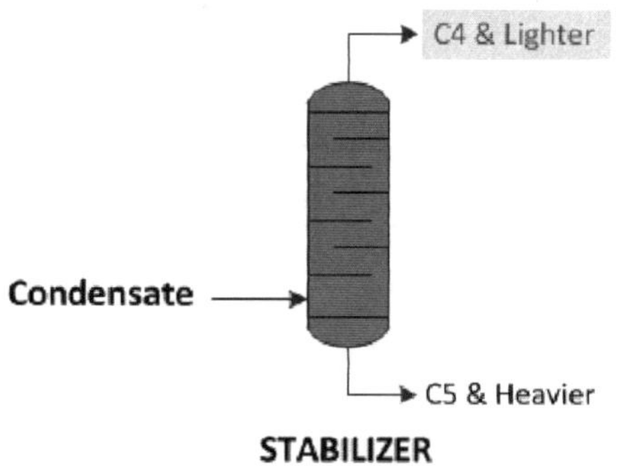

15 AIChE Spring Meeting, Austin, TX

BURNS & McDONNELL

Midstream

15 AIChE
Spring Meeting, Austin, TX

BURNS McDONNELL

Natural Gas - Transportation System

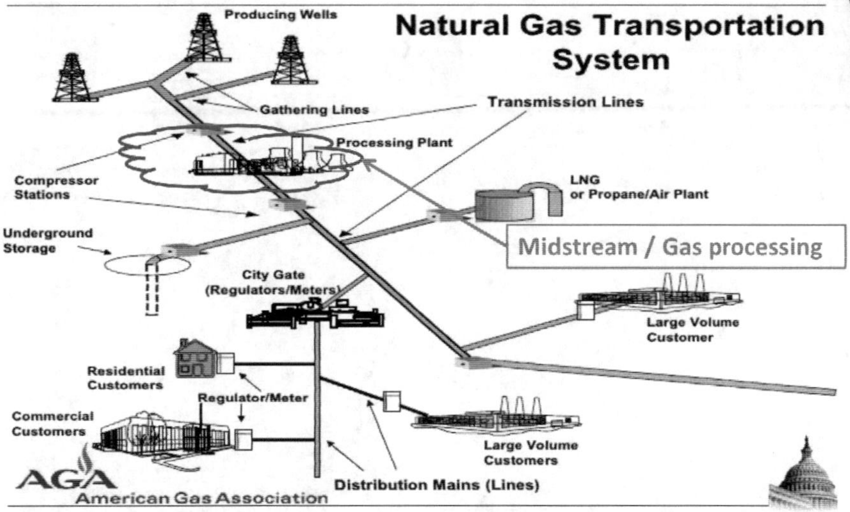

15 AIChE
Spring Meeting, Austin, TX

BURNS McDONNELL

Gas Plant Facility

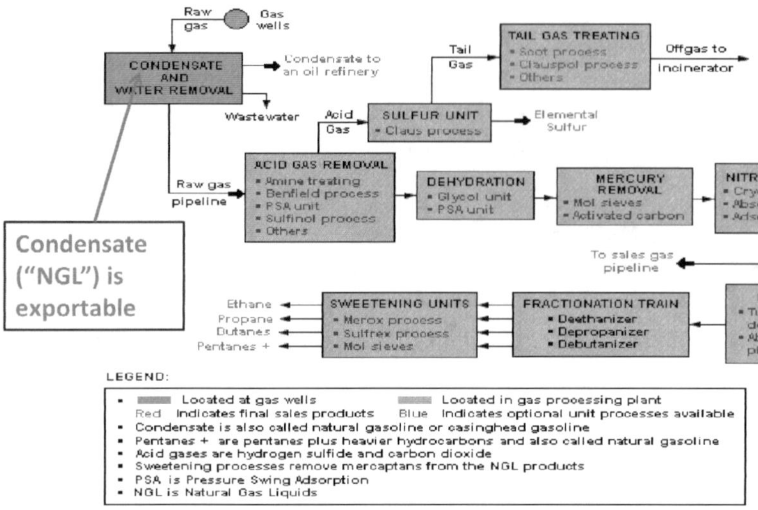

15 AIChE
Spring Meeting, Austin, TX

BURNS & McDONNELL

Condensate Splitter Facility

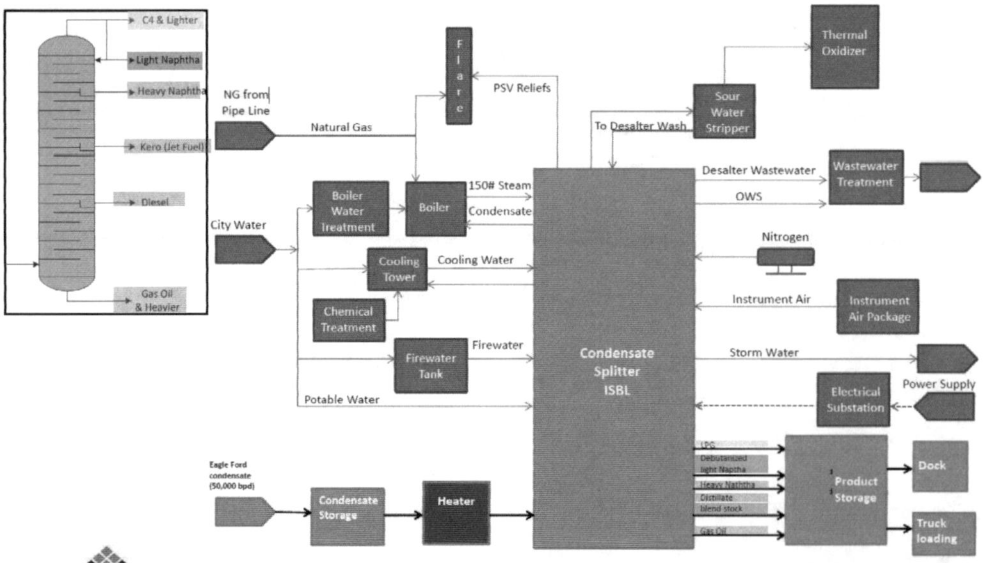

15 AIChE
Spring Meeting, Austin, TX

BURNS & McDONNELL

Kinder Morgan – Condensate Splitter Facility

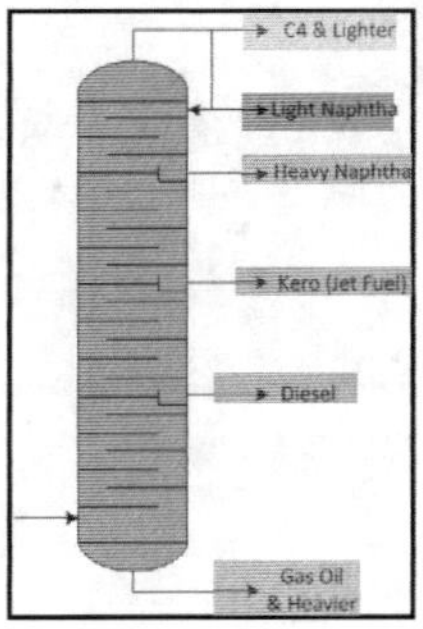

Kinder Morgan Condensate Processing Facility

Kinder Morgan is building a petroleum condensate processing facility near its Galena Park terminal on the Houston Ship Channel. The approximately $360 million project which is supported by a long-term, fee-based agreement with BP North America, involves Kinder Morgan constructing two units to split condensate into its various components, such as light and heavy naphtha, kerosene, diesel and gas oil. The facility's total capacity is 100,000 barrel per day (bpd) and can be further expanded pending additional market interest. Kinder Morgan's investment also includes the company building almost 2 million barrels of storage capacity at its Galena Park Terminal for product that will be transported from the condensate facility through new piping, manifolds and pumps.

The splitter facility combined with the Kinder Morgan Crude and Condensate Pipeline system which transports product from the Eagle Ford to Houston and associated Houston Ship Channel storage facilities offer customers unparalleled connectivity to the full range of Gulf Coast markets including refineries, chemical companies, gasoline blenders, outbound pipelines and marine facilities.

The company expects the new storage tanks and the condensate facility's first unit to be in service in March 2015, with the second unit coming online in the summer of 2015.

Source: Kinder Morgan Website

15 AIChE Spring Meeting, Austin, TX

BURNS McDONNELL

Downstream

15 AIChE Spring Meeting, Austin, TX

BURNS McDONNELL

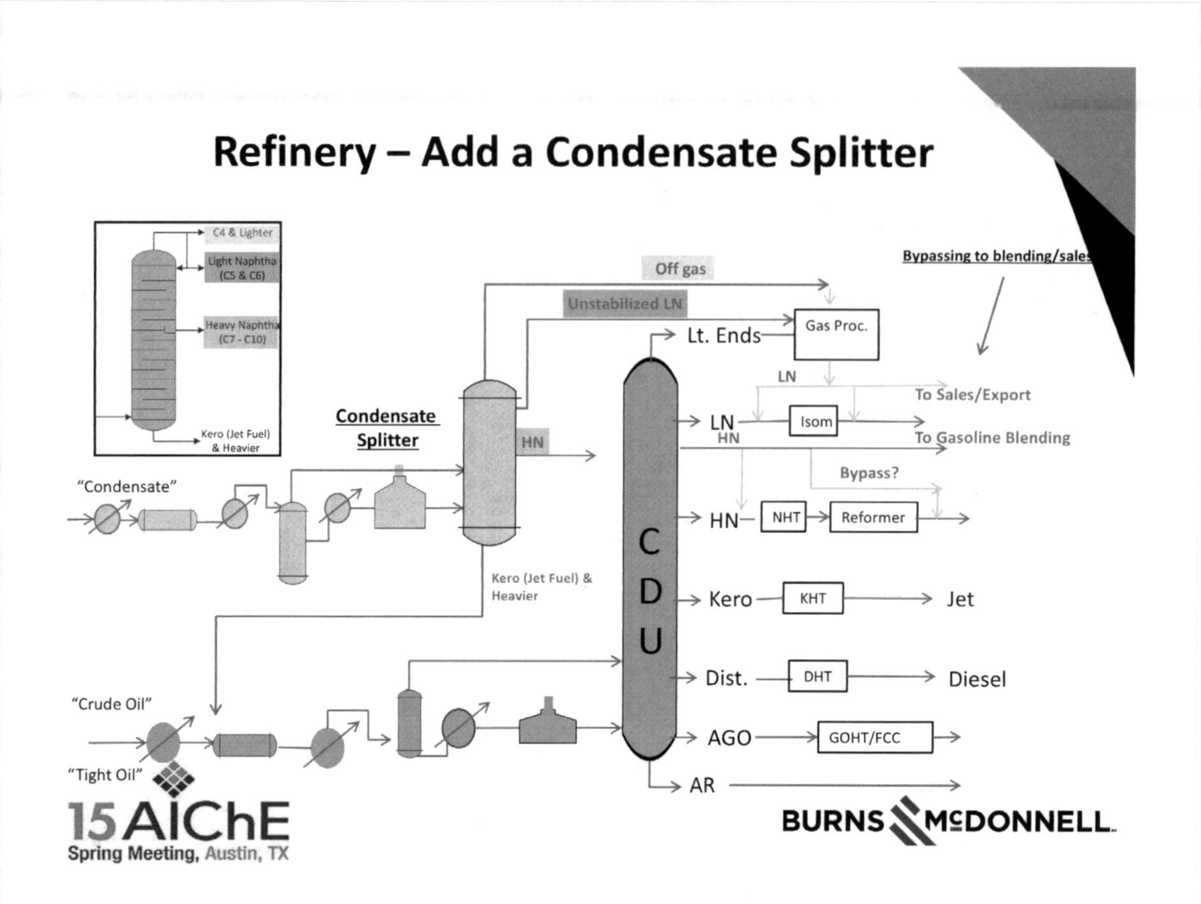
Refinery – Add a Condensate Splitter
C4 & Lighter
Light Naphtha (C5 & C6)
Heavy Naphtha (C7 - C10)
Kero (Jet Fuel) & Heavier
Condensate Splitter
"Condensate"
Off gas
Unstabilized LN
HN
Kero (Jet Fuel) & Heavier
Bypassing to blending/sales
Gas Proc.
Lt. Ends
LN
Isom
To Sales/Export
To Gasoline Blending
HN
Bypass?
NHT
Reformer
C
D
U
Kero
KHT
Jet
Dist.
DHT
Diesel
AGO
GOHT/FCC
AR
"Crude Oil"
"Tight Oil"
15 AIChE
Spring Meeting, Austin, TX
BURNS McDONNELL

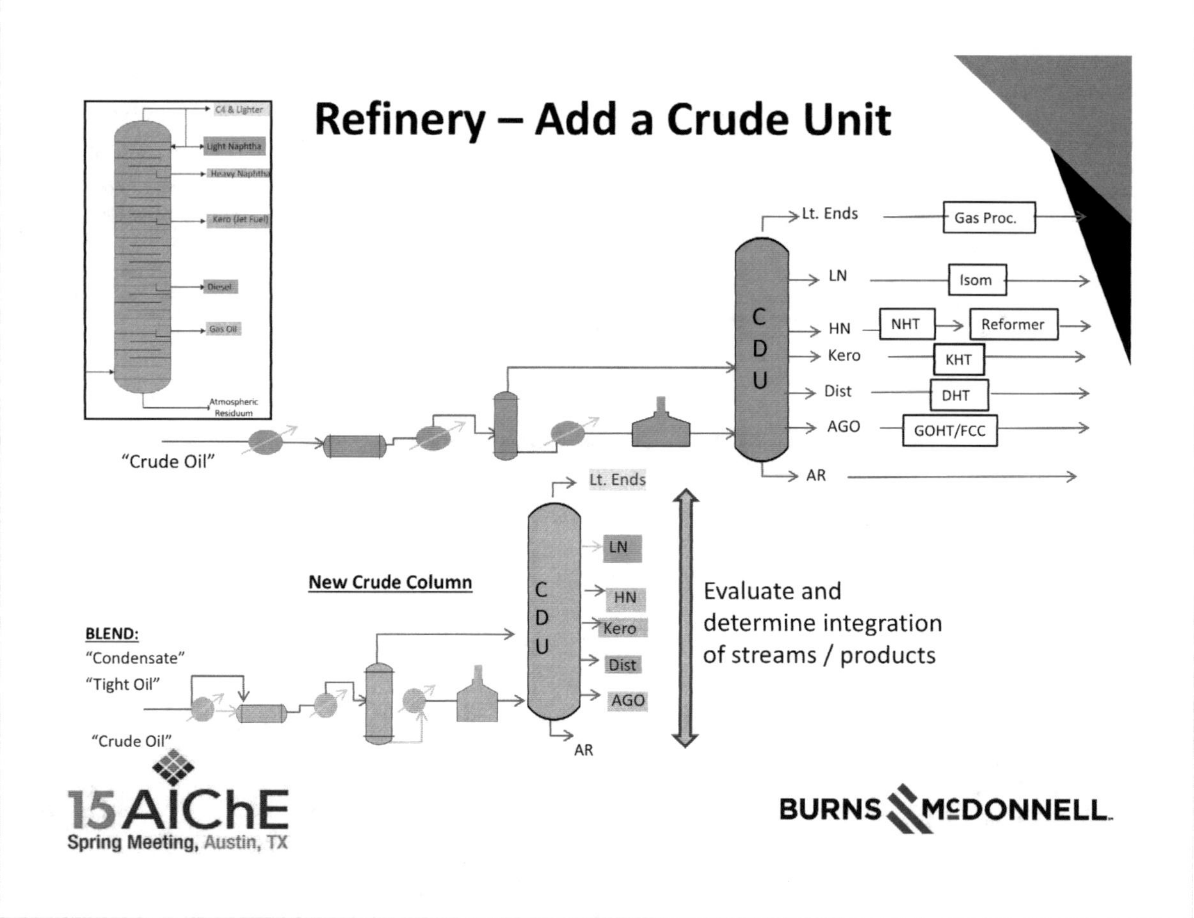
Refinery – Add a Crude Unit
C4 & Lighter
Light Naphtha
Heavy Naphtha
Kero (Jet Fuel)
Diesel
Gas Oil
Atmospheric Residuum
"Crude Oil"
Lt. Ends
Gas Proc.
LN
Isom
HN
NHT
Reformer
Kero
KHT
Dist
DHT
AGO
GOHT/FCC
AR
C
D
U
New Crude Column
Lt. Ends
LN
HN
Kero
Dist
AGO
AR
BLEND:
"Condensate"
"Tight Oil"
"Crude Oil"
Evaluate and determine integration of streams / products
15 AIChE
Spring Meeting, Austin, TX
BURNS McDONNELL

Economics and Logistics

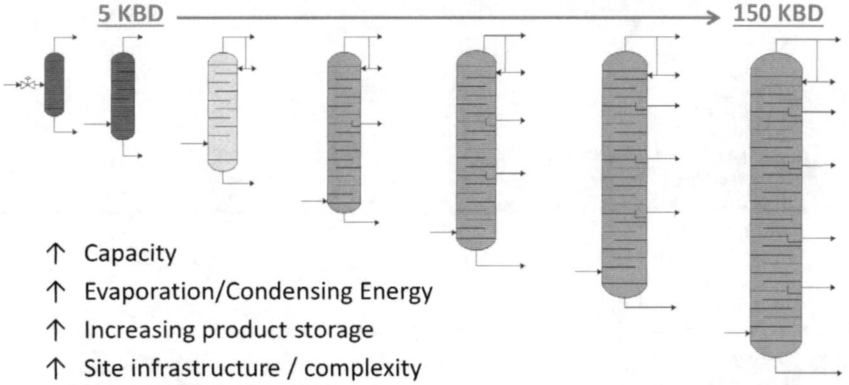

- ↑ Capacity
- ↑ Evaporation/Condensing Energy
- ↑ Increasing product storage
- ↑ Site infrastructure / complexity
- ↑ Supply chain logistics
- ↑ Investment capital

Increasing number of streams separated →

15 AIChE Spring Meeting, Austin, TX

BURNS & McDONNELL

Conclusions – Part 2

- **Location, Logistics and Economics →Upstream, Midstream and Downstream Perspective**
- **Upstream**
  - **At the Wellhead**
  - **Additional distillation implies**
    - **More products, infrastructure, storage, transportation and logistics**
- **Midstream**
  - **Gas Plant Facility**
    - **Condensate (NGL) is exportable**
  - **Condensate Splitter Facility**
    - **Stand-alone facility has high cost:**
      - **Land, infrastructure, utilities, product storage, loading facilities (trucks, ships, rail)**
    - **Best Case - Gulf Coast Location:**
      - **Pipelines, Infrastructure, product logistics – all exist**

15 AIChE Spring Meeting, Austin, TX

BURNS & McDONNELL

Conclusions – Part 2 (cont'd)

- **Downstream**
  - Refinery
    - Add a Condensate Splitter
    - Depending on capacity – may be able to use much of existing infrastructure
    - Incremental capital less – return on investment potentially higher
  - Refinery
    - Add a Crude Unit
    - Depending on capacity – may be able to use existing infrastructure
    - Increasing range of potential products produced
    - Incremental capital less – return on investment potentially higher
- **Final Thoughts:**
  - Distillation requirements more stringent implies investment
  - Moving from upstream to downstream moves to higher capacity equipment and affords the opportunity for more range of products produced
  - Locating in Gulf Coast provides the most infrastructure for supply and product chains
  - Stand-alone ("grassroots") facilities require the highest investment
  - Locating at an existing facility (i.e. Refinery) may lead to the lowest incremental investment with the greatest opportunities to capitalize on existing infrastructure.

15 AIChE Spring Meeting, Austin, TX

BURNS & McDONNELL

Questions and Answers

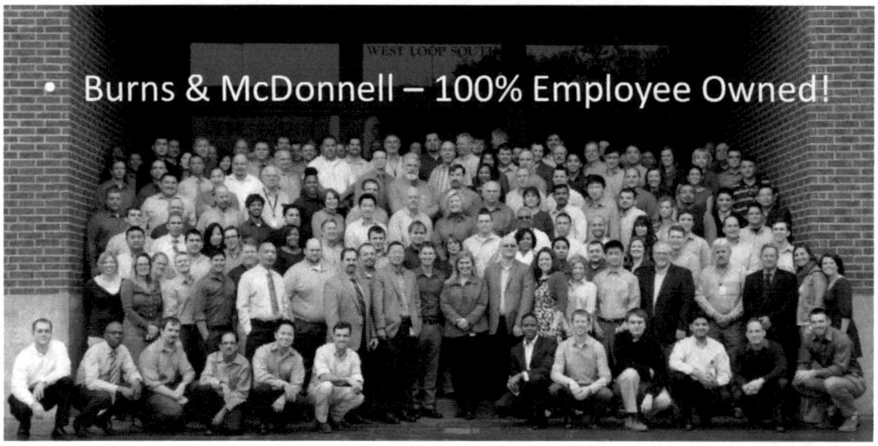

Mark W. Lockhart
mlockhart@burnsmcd.com
832.214.1975

15 AIChE Spring Meeting, Austin, TX

BURNS & McDONNELL

BURNS McDONNELL

15 AIChE
Spring Meeting, Austin, TX

Regulators versus Control Valves Optimizing Selection

Vince Mezzano
FLUOR

Keith Erskine
PUFFER SWEIVEN

15 AIChE
Spring Meeting, Austin, TX

Introduction

- Regulators vs. Control Valves
  - Benefits of each
- Design Considerations
  - User/Site specific requirements
- Applications
  - How will it be used?

15 AIChE
Spring Meeting, Austin, TX

FLUOR
PUFFER SWEIVEN

Regulator vs. Control Valve

Regulator Advantages

- Self Contained
- Fastest Speed of Response
- No Volatile Emissions (no packing)
- Less Complex
- Lower Installed Cost
- Ease of Maintenance
- Process Powered

Control Valve Advantages

- Larger Sizes
- ANSI 900+
- Severe Service Trim Options
- Fluid Versatility
- Remote Feedback
- Diagnostics Capabilities
- Externally Powered

15 AIChE
Spring Meeting, Austin, TX

Regulator Purpose and Types

TO MATCH FLOW DEMAND WHILE MAINTAINING A CONSTANT PRESSURE

Performs the function of a Valve, Actuator, Controller, and Pressure Transmitter in one device

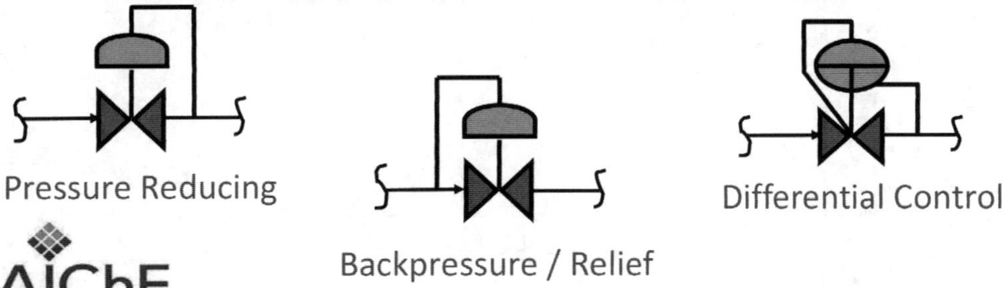

15 AIChE
Spring Meeting, Austin, TX

Process Power vs. Control System

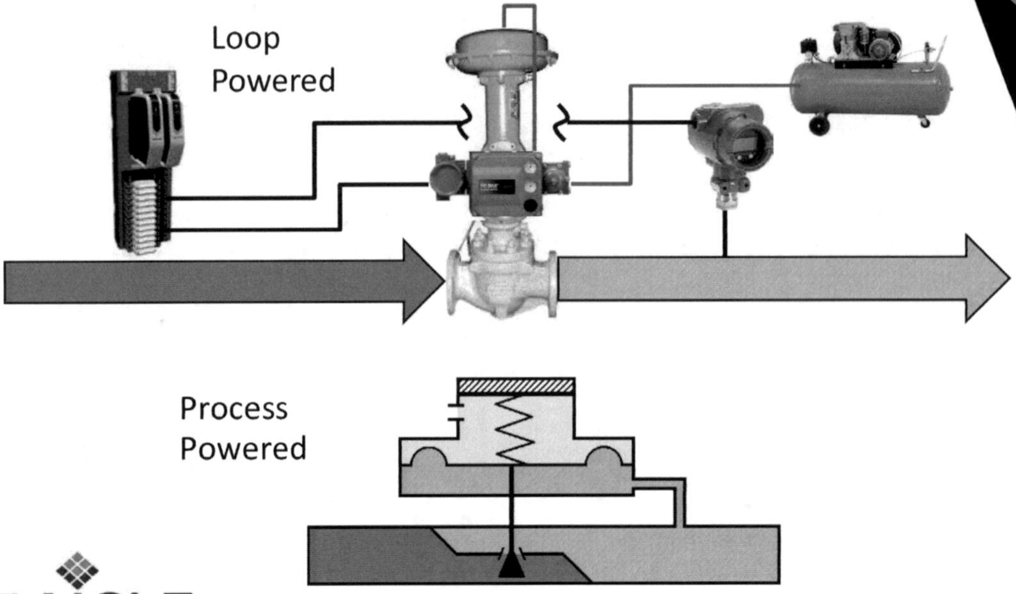

15 AIChE
Spring Meeting, Austin, TX

Capacity vs. Outlet Pressure

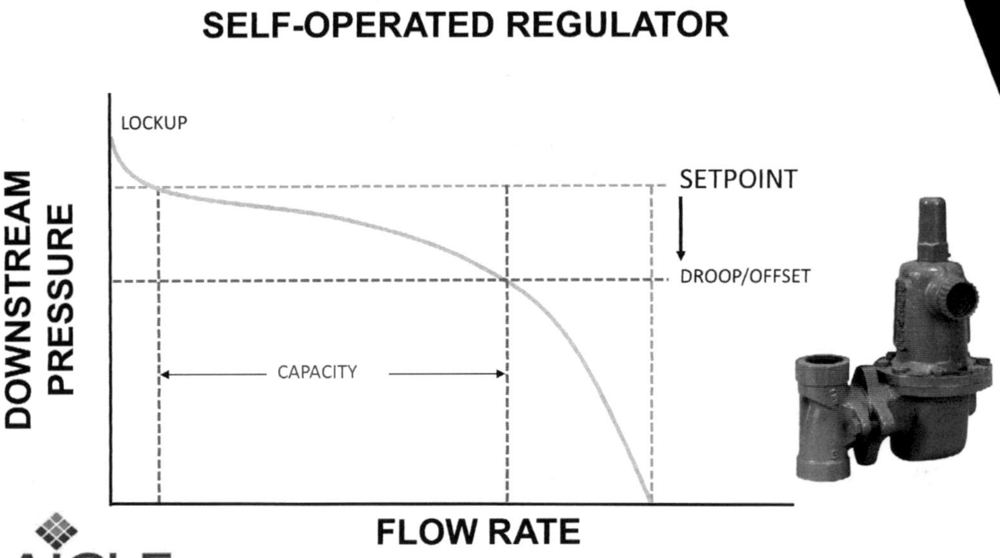

15 AIChE
Spring Meeting, Austin, TX

Capacity vs. Outlet Pressure

PILOT-OPERATED REGULATOR

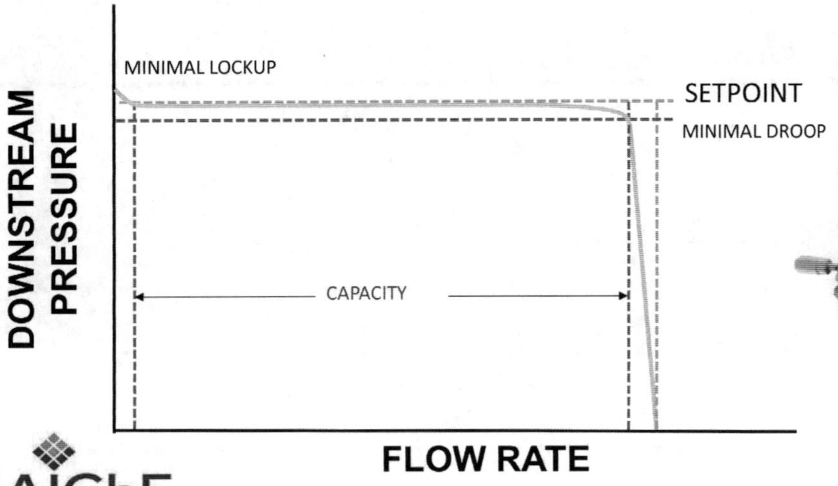

15 AIChE
Spring Meeting, Austin, TX

Capacity vs. Outlet Pressure

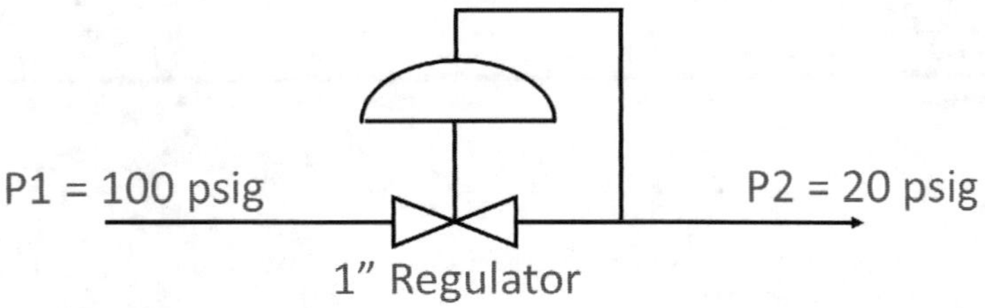

- Self-operated
  - 1″ 630: 11,000 scfh (-4 psi)
  - 1″ MR95H: 15,000 scfh (-4 psi)
  - 1″ 627: 23,300 scfh (-4 psi)
- Pilot-operated
  - 1″ EZR: 72,000 scfh (+/- 1.3 psi)
  - 1″ EZH: 81,000 scfh (+/- 0.5 psi)
  - 1″ 1098-EGR: 89,000 scfh (+/- 0.2 psi)

15 AIChE
Spring Meeting, Austin, TX

Regulator vs. Control Valve

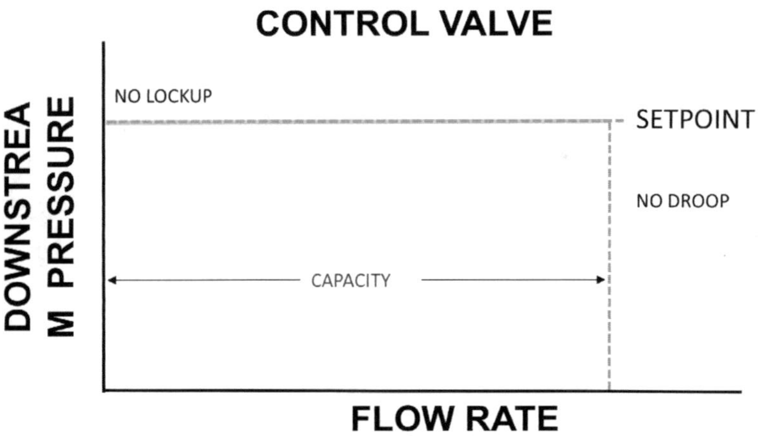

15 AIChE
Spring Meeting, Austin, TX

Regulator Design Pressure Ratings

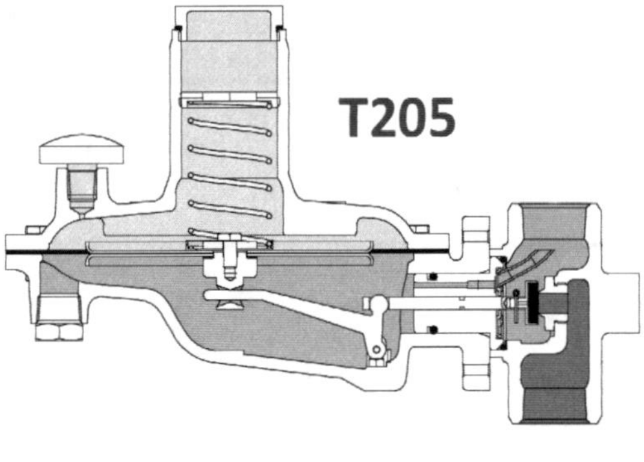

- Maximum Inlet Pressure = 200 psig
- Maximum Outlet Pressures
  - Casing = 75 psig
  - Avoid Internal Parts Damage = 35 psig

- Maximum Inlet Pressure = 1,480 psig
- Maximum Outlet Pressure = 1,480 psig

15 AIChE
Spring Meeting, Austin, TX

Design Considerations

- Criticality of Operation
- Accuracy Requirements
- Diagnostic Monitoring
- Predictive Maintenance
- Design Pressures
- Project Specifications
- Design Temperatures
- Site Standards
- Installed Cost
- Maintenance Cost

15 AIChE
Spring Meeting, Austin, TX

Applications

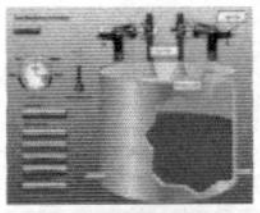
Blanket Gas Control

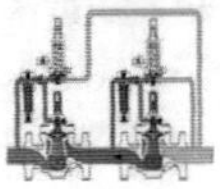
Plant Feed Gas Supply

Severe Service

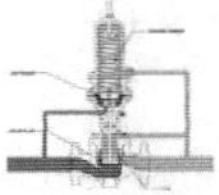
Differential Pressure Control

Boiler Fuel Gas

15 AIChE
Spring Meeting, Austin, TX

Tank Blanketing

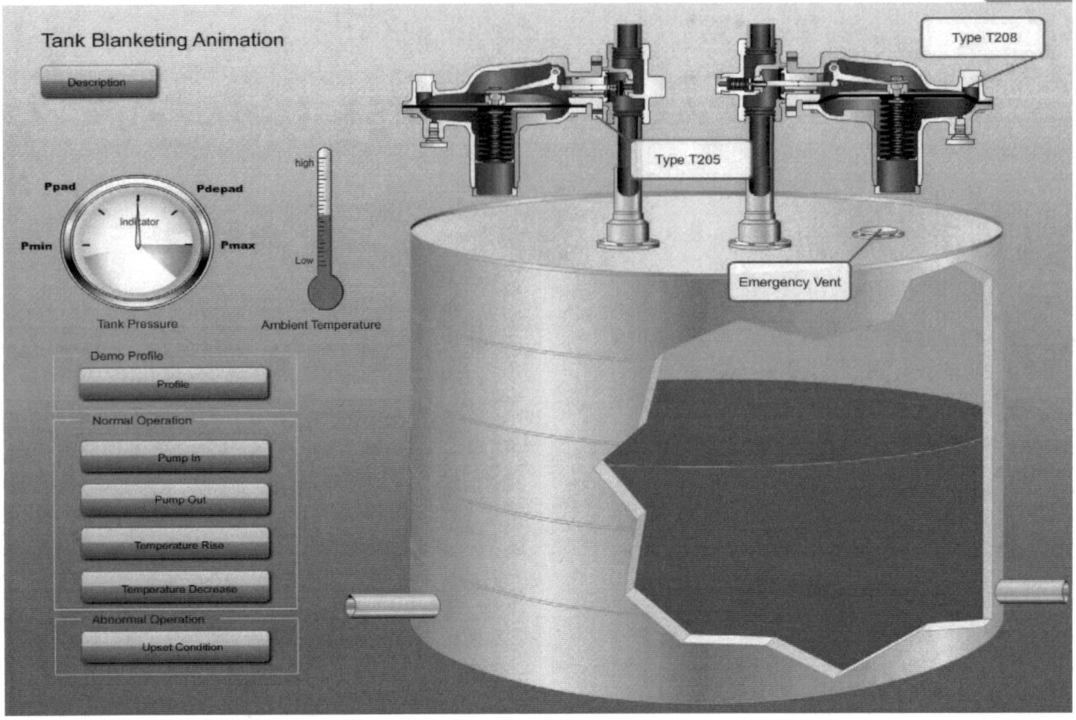

1. Tank Blanketing Gas Control

- Maintain a nitrogen blanket pressure
  - 150 psig N2 supply
  - Blanket control of 14 in w.c. minimum, 20 in w.c. maximum
  - Regulators typically used for fast, automatic response with proportional band of 0.25 in w.c.
  - Diagnostic feedback is not required

T205

ACE95

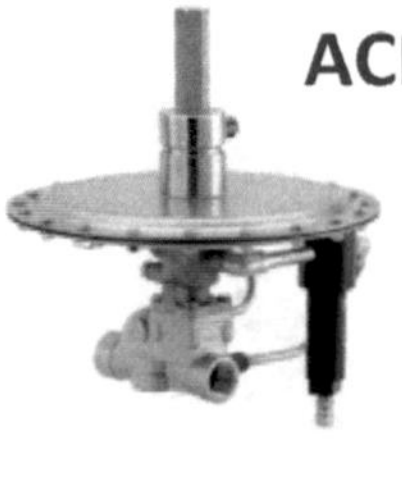

15 AIChE
Spring Meeting, Austin, TX

1. ~~Tank~~ Blanketing Gas Control

Vessel

- Maintain
  - 150 ps
  - Blanke c. maximum
  - Regulı sponse with propor
  - Diagnc

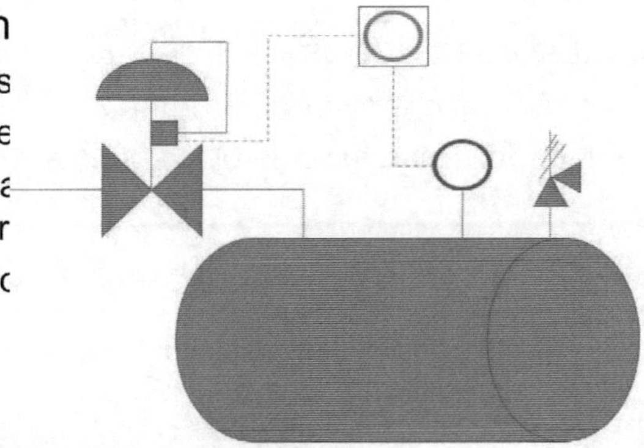

- Vessel MAWP = 600 psig
  - ASME SRV set at 600 psig
  - Regulator pad outlet and depad inlet force a Control Valve

15 AIChE
Spring Meeting, Austin, TX

2. Plant Feed Gas Supply

- Maintain Natural Gas Supply to process unit
  - 350 psig Natural Gas
  - Downstream pressure to be controlled at 75 psig
  - Cannot exceed 85 psig downstream
  - Critical to maintain gas flow

15 AIChE
Spring Meeting, Austin, TX

2. Plant Feed Gas Supply

- Monitor regulators in series
  - Not affected by loss of power or air supply
  - If one regulator fails, the second prevents overpressure

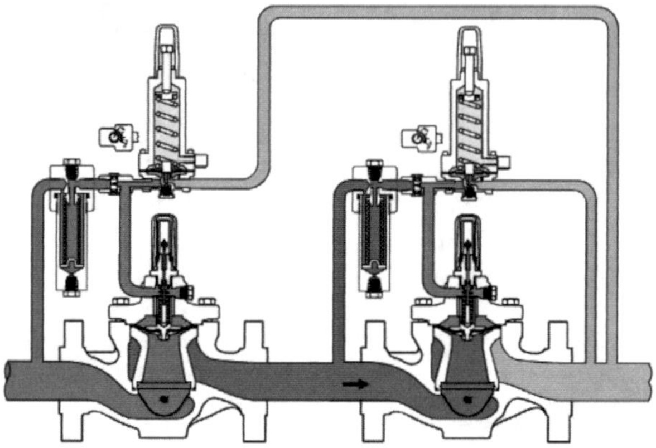

15 AIChE
Spring Meeting, Austin, TX

3. Cooling Water Pressure Control

- Maintain constant pressure on downstream cooling water
  - 300 psig water
  - Downstream pressure to be controlled at 10 psig
  - Critical to maintain pressure
  - Cavitation is predicted for homogenous fluid and 290 psid
  - Control valves offer trim designs to eliminate damaging effects of cavitation, and limit damage from flashing liquids
  - Vapor Pressure = 0.5 psig

15 AIChE
Spring Meeting, Austin, TX

3. Cooling Water Pressure Control

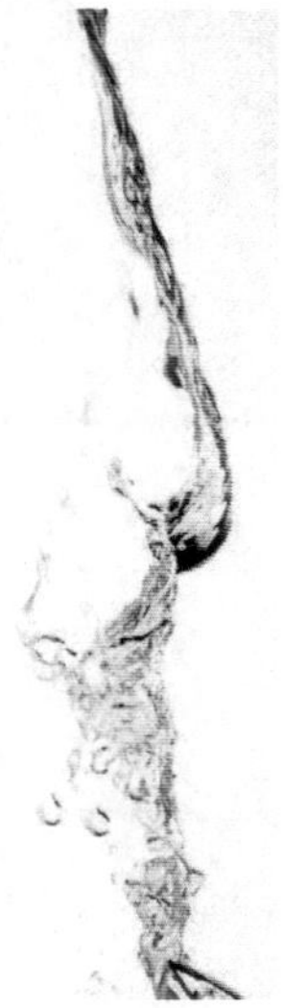

Cavitation Phenomenon

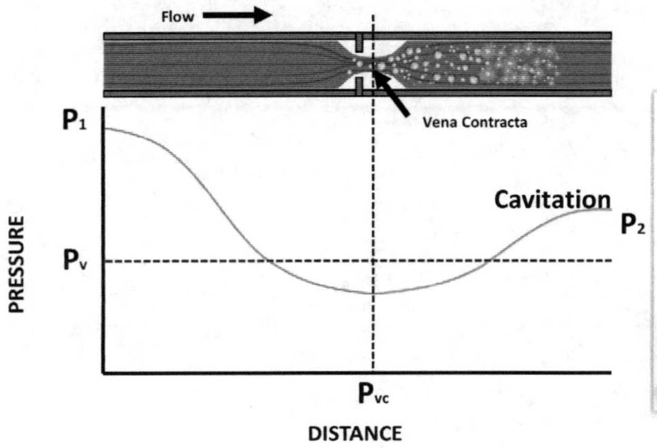

15 AIChE
Spring Meeting, Austin, TX

4. Wellhead to Pipeline Differential Pressure

- Maintain gas pressure exiting separator 5-20 psig higher than pipeline pressure
  - 200 - 800 psig well pressure
  - Pipeline pressure 200-250 psig, MAWP = 300 psig
  - Critical to maintain positive differential
  - Remote areas

15 AIChE
Spring Meeting, Austin, TX

4. Wellhead to Pipeline Differential Pressure

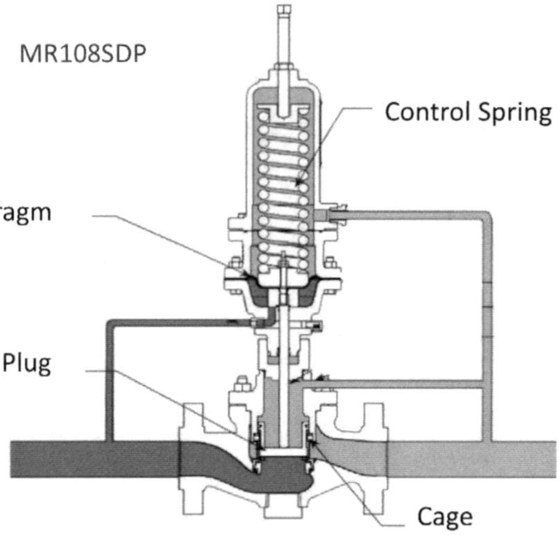

15 AIChE
Spring Meeting, Austin, TX

5. Heater/Boiler Fuel Gas Control

- High capacity fuel gas application with consideration for heater/boiler pilot light maintenance creating a low flow case
  - 120 psig inlet fuel gas pressure
  - 0.8 psig control pressure to burners
  - High capacity required in normal operation
  - Low flow condition when heaters/boilers not in use

15 AIChE
Spring Meeting, Austin, TX

5. Heater/Boiler Fuel Gas Control

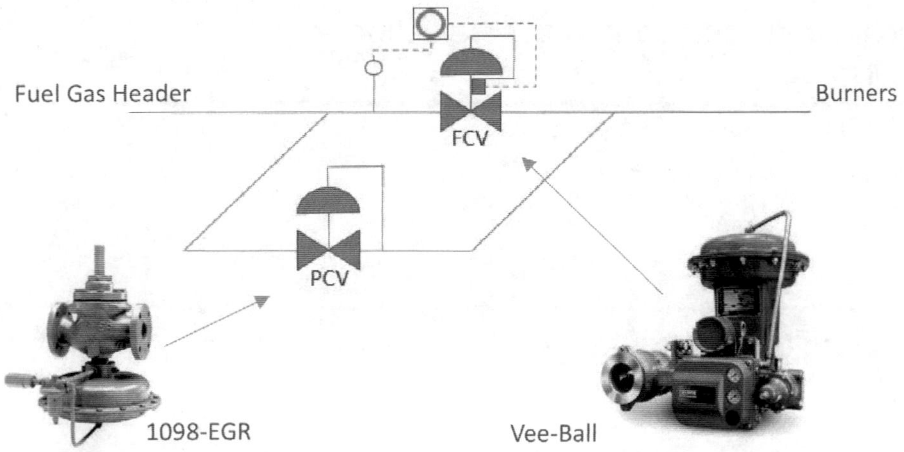

15 AIChE
Spring Meeting, Austin, TX

Results Acheived

- Time reduction
  - Design time
  - Keep the unit on line
  - Understand the application
- Cost reduction
  - Avoid buying more than needed
  - Lower installed costs
  - Lower maintenance costs

15 AIChE
Spring Meeting, Austin, TX

Summary

- Benefits of Regulators and Control Valves
- Design considerations guide selection
- Applications
  - Tailored products fit for use

Feedback is appreciated!

15 AIChE
Spring Meeting, Austin, TX

Where To Get More Information

Visit the source!

- Marshalltown, IA

- McKinney, TX

15 AIChE
Spring Meeting, Austin, TX

Where To Get More Information

- Contact you Local Business Partner

- www.fisher.com
- www.fisherregulators.com

- Regulator Toolkit Software
- Control Valve Handbook
- Emerson Application Guides

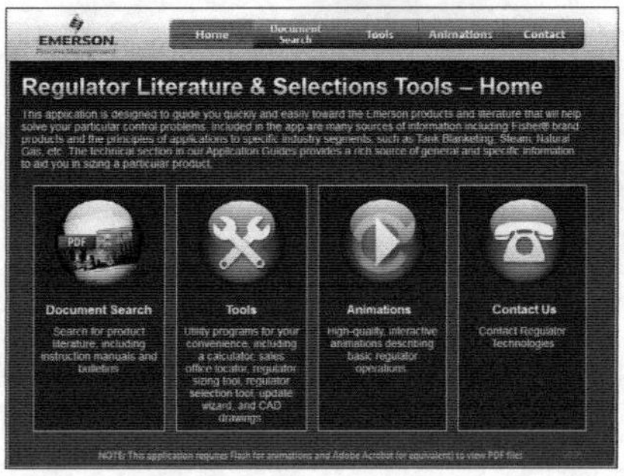

15 AIChE
Spring Meeting, Austin, TX

Questions?

15 AIChE
Spring Meeting, Austin, TX

An Introduction to the Early Planning Process for Plant Automation Projects

Laurie R. Ben

Director, Modernization Business Development

15 AIChE Spring Meeting, Austin, TX

EMERSON. Process Management

Agenda

- The Automation Capital Project Cycle
  - The Problem, Confusion, Where to start?
  - Understanding the Project Cycle
- Why Early Planning? What are the benefits?
  - Independent Project Analysis
  - Construction Industry Institute
  - Definitions & Accuracy
  - References & Websites
- Summary

15 AIChE Spring Meeting, Austin, TX

EMERSON. Process Management

Problem – Allocating funds for best returns

- *Inadequate resources for planning investments*
- *Corporate hurdle rates*
- *Obsolescence will not justify replacement*
- *Scarce expertise in process control*
- *Engineers focused on day-to-day operations*

Allocating scarce capital funds to the projects that will make the company the best returns.

15 AIChE Spring Meeting, Austin, TX

EMERSON. Process Management

Business Objectives Vary

15 AIChE Spring Meeting, Austin, TX

EMERSON. Process Management

Early Project Planning
Where does it begin? Where does it End?

15 AIChE Spring Meeting, Austin, TX

EMERSON. Process Management

Exercise: Understanding The Project Cycle

- Project Example: Kitchen Remodel for wheelchair accessibility

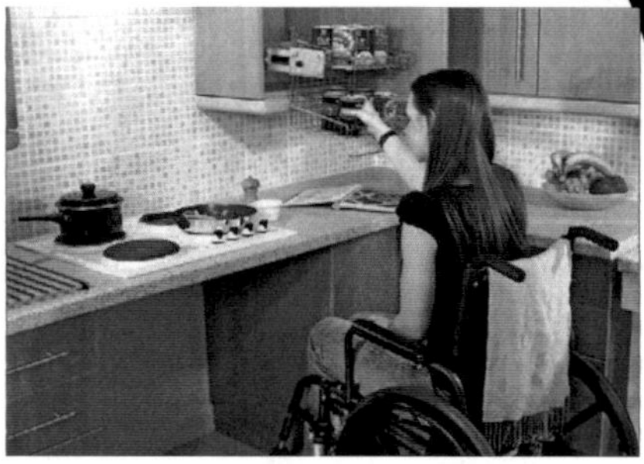

- Project Steps: How do you start?
- Project Risks: What are your concerns?

15 AIChE Spring Meeting, Austin, TX

EMERSON. Process Management

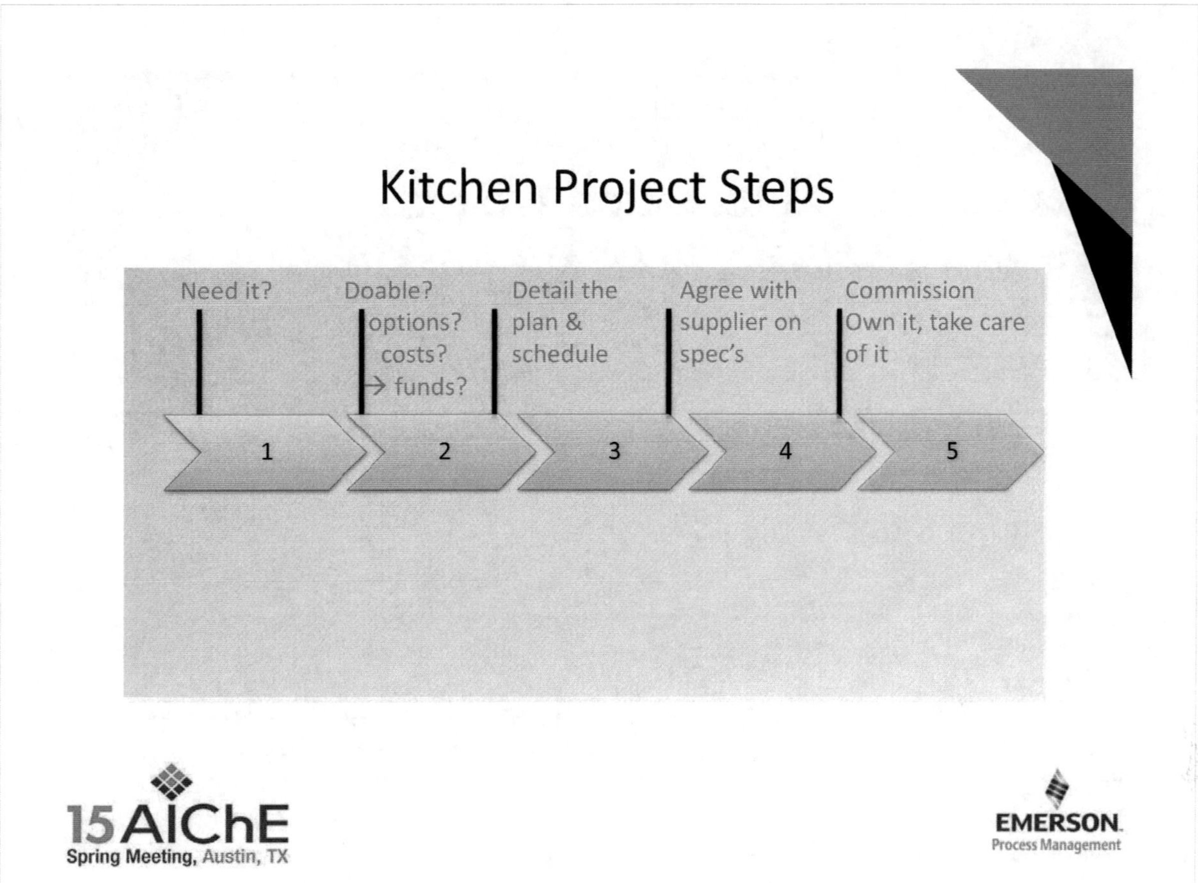
Kitchen Project Steps
Need it?
Doable?
options?
costs?
→ funds?
Detail the
plan &
schedule
Agree with
supplier on
spec's
Commission
Own it, take care
of it
1
2
3
4
5
15 AIChE
Spring Meeting, Austin, TX
EMERSON.
Process Management

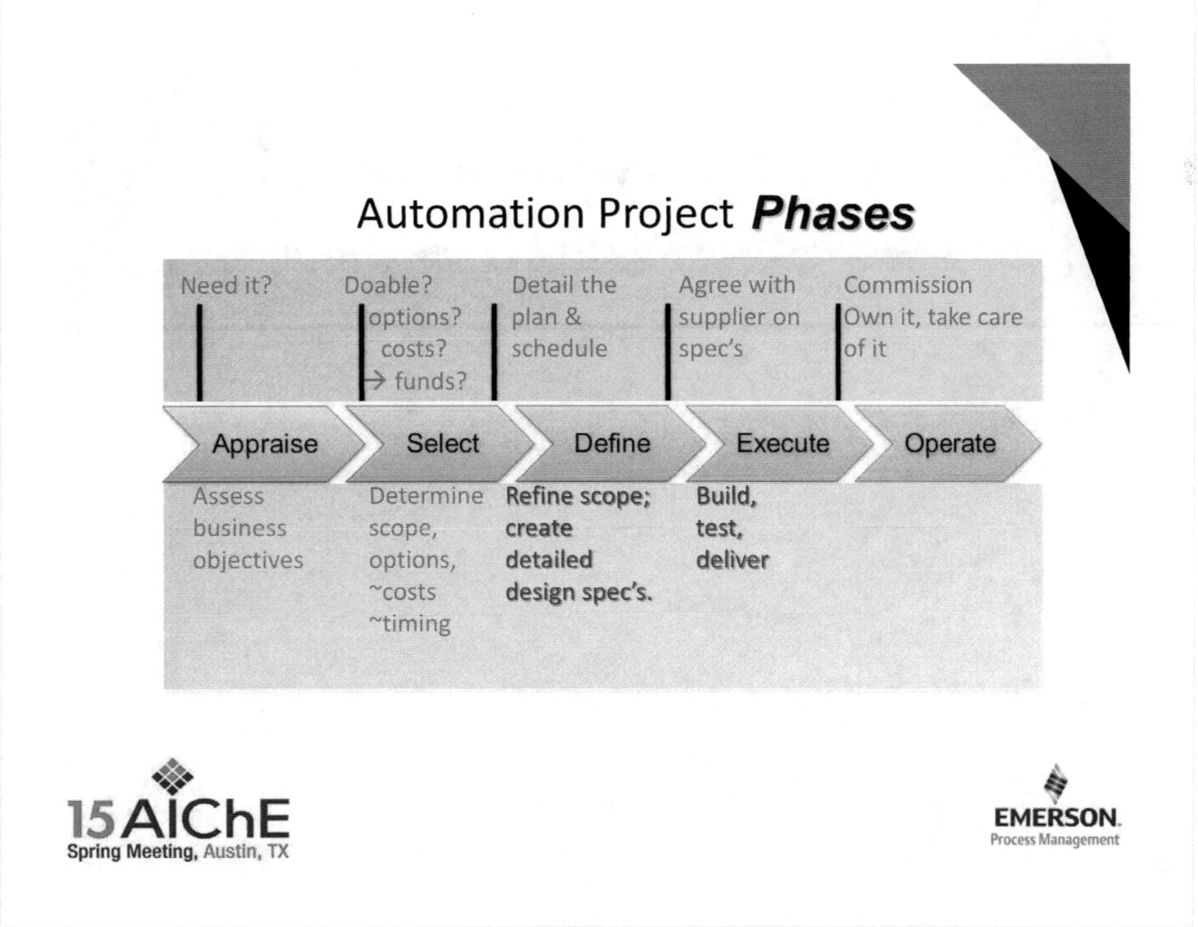
Automation Project Phases
Need it?
Doable?
options?
costs?
→ funds?
Detail the
plan &
schedule
Agree with
supplier on
spec's
Commission
Own it, take care
of it
Appraise
Select
Define
Execute
Operate
Assess
business
objectives
Determine
scope,
options,
~costs
~timing
Refine scope;
create
detailed
design spec's.
Build,
test,
deliver
15 AIChE
Spring Meeting, Austin, TX
EMERSON.
Process Management

Automation Projects

- What is an automation project?
- What drives a company/site to do automation projects?
- What risks are involved?

drivers?

risks?

15 AIChE
Spring Meeting, Austin, TX

EMERSON.
Process Management

EARLY Planning Mitigates Risks

Influence & Expenditures Curve for Project Lifecycle,
Construction Industry Institute CII

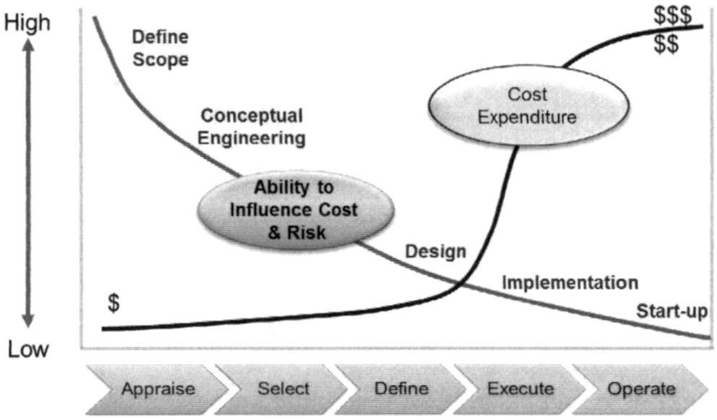

The Decade Rule,
International Business Machines IBM

The Decade Rule	
Phase	**Equivalent Costs**
Appraise	$1
Select	$10
Define	$100
Execute	$1,000
Start-Up	$10,000

15 AIChE
Spring Meeting, Austin, TX

EMERSON.
Process Management

Independent Project Analysis

Appraise → Select → Define → Execute → Operate

Independent Project Analysis (IPA) Model

7.5% increased project ROI
30% decrease in project duration
6% additional capacity

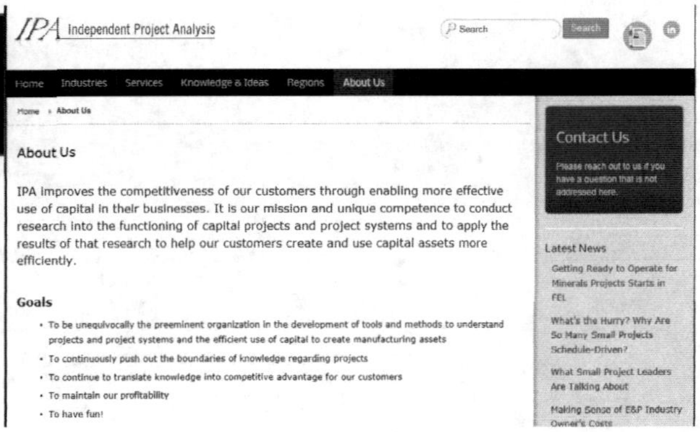

IPA Independent Project Analysis

Search

Home | Industries | Services | Knowledge & Ideas | Regions | About Us

Home › About Us

About Us

IPA improves the competitiveness of our customers through enabling more effective use of capital in their businesses. It is our mission and unique competence to conduct research into the functioning of capital projects and project systems and to apply the results of that research to help our customers create and use capital assets more efficiently.

Goals

- To be unequivocally the preeminent organization in the development of tools and methods to understand projects and project systems and the efficient use of capital to create manufacturing assets
- To continuously push out the boundaries of knowledge regarding projects
- To continue to translate knowledge into competitive advantage for our customers
- To maintain our profitability
- To have fun!

Contact Us

Please reach out to us if you have a question that is not addressed here.

Latest News

Getting Ready to Operate for Minerals Projects Starts in FEL

What's the Hurry? Why Are So Many Small Projects Schedule-Driven?

What Small Project Leaders Are Talking About

Making Sense of E&P Industry Owner's Costs

15 AIChE
Spring Meeting, Austin, TX

CII Construction Industry Institute

- A consortium of leading owners, contractors, and academics working collaboratively to improve the constructed project and the capital investment process.
- An organized research unit of the Cockrell School of Engineering at The University of Texas at Austin

Performance	PDRI Score at end of Front End Planning < 200	PDRI Score at end of Front End Planning > 200
Cost	3% over budget	9% over budget
Schedule	5% behind schedule	21% behind schedule
Change Orders	8% of budget	11% of budget
	(N=25)	(N=83)

Source: CII Project Definition Rating Index – Industrial Projec

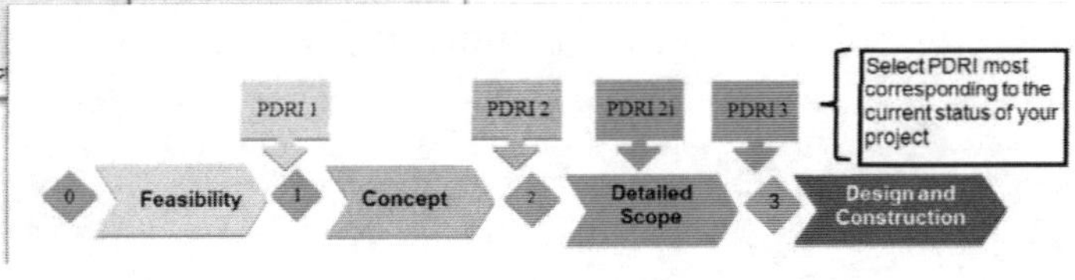

15 AIChE
Spring Meeting, Austin, TX

Gated Process Examples

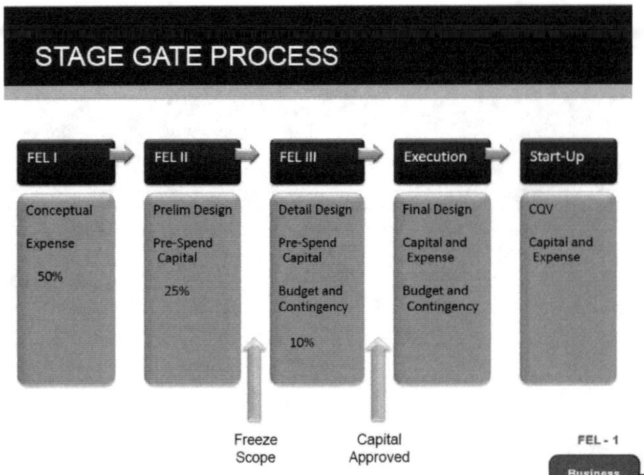

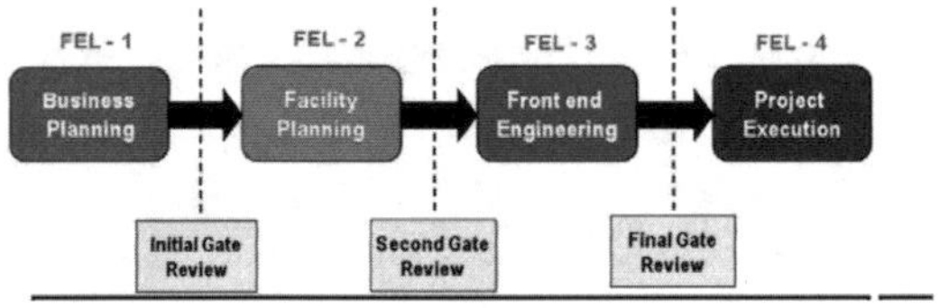

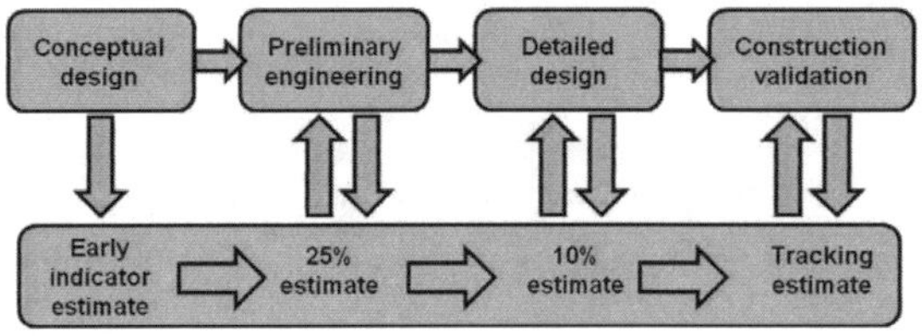

15 AIChE Spring Meeting, Austin, TX

EMERSON. Process Management

Early Planning Definitions What does it mean?

- Feasibility/FEL1:High level analysis of proposed project
  1) Technically & economically feasible
  2) Estimated cost
  3) Profitable (Business Driver/Justification)
  4) Budgetary (order of magnitude) estimate/High Level scope of Work
- Conceptual Engineering/FEL2 : Scope Development
  - generally includes framing the project
  - collecting data, analyzing the data
  - selecting actions/options/alternatives
  - Further definition of scope & Cost
- FEED/FEL 3
  - Basic Engineering for scope of work defined/selected
  - Procurement typically begins
  - Detailed estimates on costs/risks

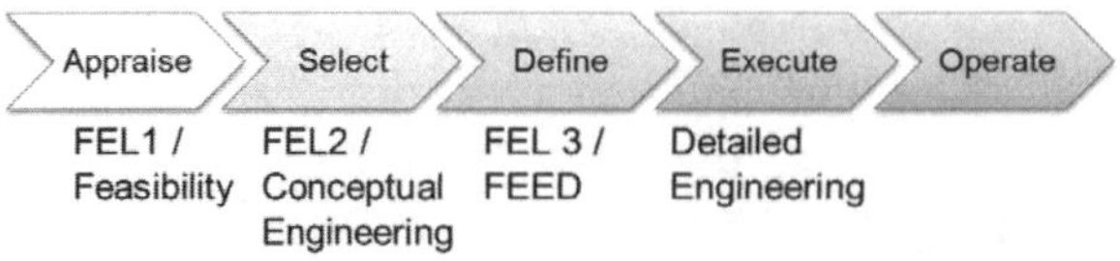

15 AIChE Spring Meeting, Austin, TX

Accuracy Requirements Depend on the Evaluation Stage

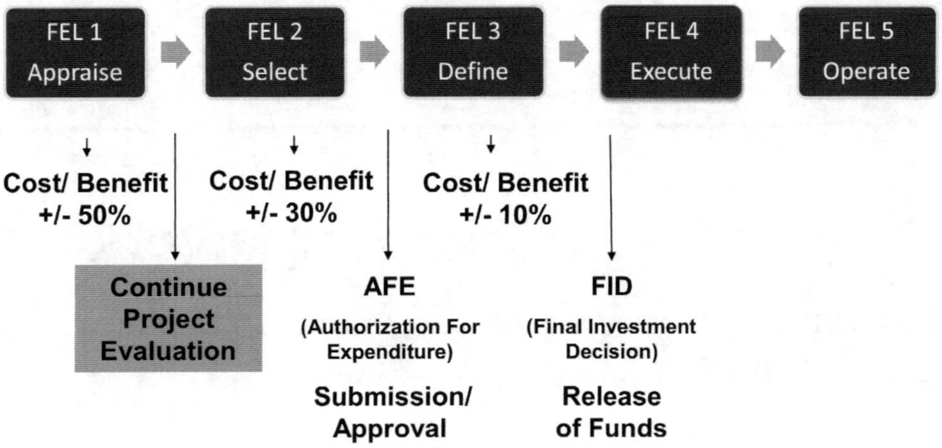

15 AIChE Spring Meeting, Austin, TX

EMERSON. Process Management

Reference Material & Websites

- Independent Project Analysis
  - http://www.ipaglobal.com/
- Construction Industry Institute
  - https://www.construction-institute.org/
- Definitions of FEED, FEL, etc.
  - http://www.epcengineer.com/
  - http://en.wikipedia.org/wiki/Front-end_engineering
  - http://www.fluor.com/epc-services/engineering
- Articles
  - http://www.asee.org/documents/sections/middle-atlantic/fall-2010/01-Teaching-Front-End-Engineering-Design-FEED.pdf
  - http://www.costandvalue.org/download/?id=2098, (Document Project Readiness by estimate Class Using PDRI)

15 AIChE Spring Meeting, Austin, TX

EMERSON. Process Management

Summary - Do Your Homework

- Understand your company's capital project methodology
  - What is your company/site's gated process?
  - What stage are you at in the project evaluation?
  - Do estimating based on current stage
- Understand the drivers for your project
  - Approach justification as a sales process to management
  - Identify key business and operational issues
- Understand the risks for your project
  - Get to know Plant and Operations Management attitude toward automation
- Early Planning is Key to Success
  - Apply Best Practices

15 AIChE
Spring Meeting, Austin, TX

EMERSON.
Process Management

Impact of Control Valve Performance on Control Performance

James Beall
Principal Process Control Consultant
Emerson Process Management

15 AIChE
Spring Meeting, Austin, TX

EMERSON.
Process Management

James Beall

- Principal Process Control Consultant, Emerson Process Management
- 33 Years experience in process control
- 14 Years Emerson Process Management
- 19 Years at Eastman Chemical Company
- Chairman ISA 75.25 Committee - Control Valve Performance Testing
- James.Beall@Emerson.com 903-235-7935

15 AIChE
Spring Meeting, Austin, TX

EMERSON.
Process Management

Agenda

- Importance of Valve Performance
- Standards and References
- Dead Band and Resolution
- Response Time
- Loop Gain
- Specifying Valve Performance
- Benefits

15 AIChE
Spring Meeting, Austin, TX

EMERSON.
Process Management

Control Loops with Excessive Variability

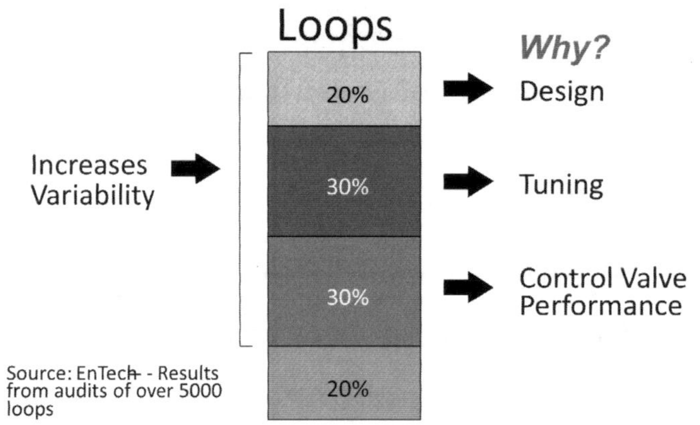

"The undesirable behavior of control valves is the biggest contributor to poor loop performance and the destabilization of product uniformity".

W. L. Bialkowski, President
EnTech Control Engineering

15 AIChE
Spring Meeting, Austin, TX

EMERSON.
Process Management

LNG Plant, Main Gas Feed Flow

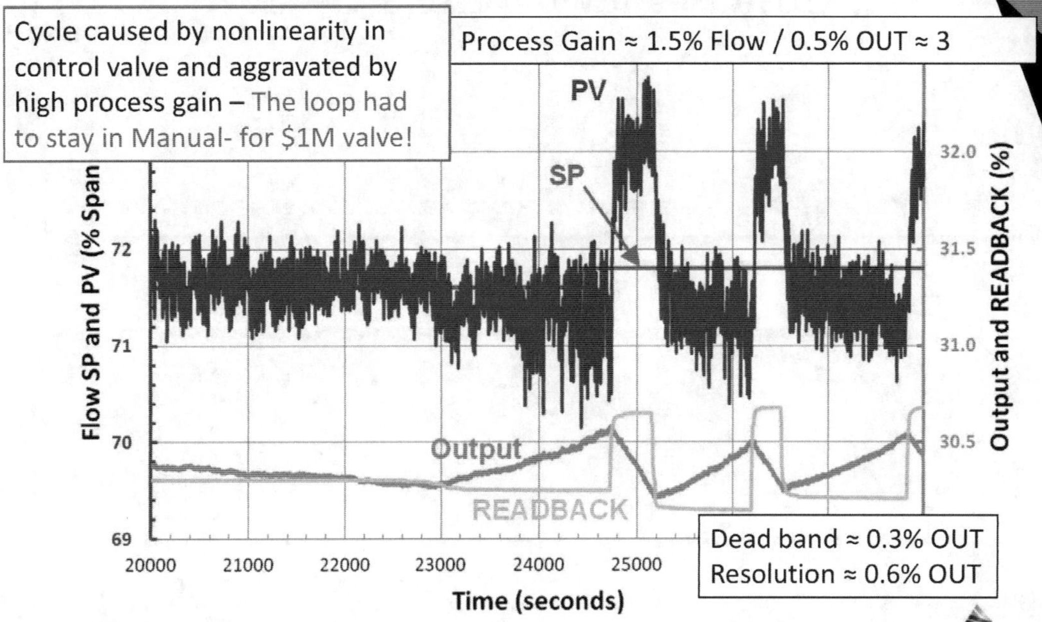

15 AIChE Spring Meeting, Austin, TX

EMERSON. Process Management

Vitamin plant—Steam Valve

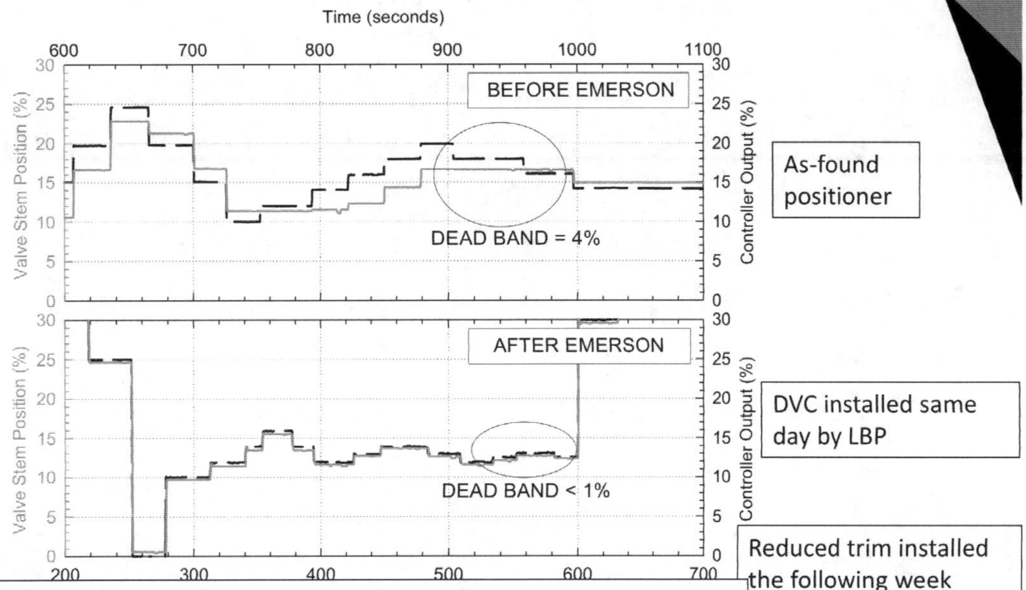

Batch time reduced 27% and steam consumption reduced

15 AIChE Spring Meeting, Austin, TX

EMERSON. Process Management

Chemical Plant – Steam Valve

Flow cycles in automatic

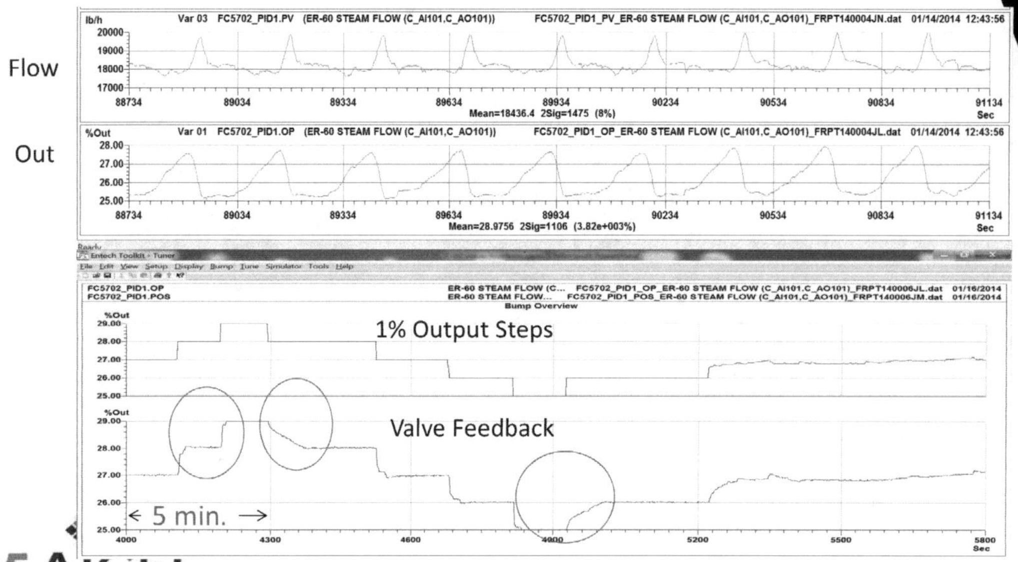

15 AIChE
Spring Meeting, Austin, TX

Step test: 30 – 90 second Response Time!

Control Valve Performance

- ANSI/ISA–S75.25 (Standard) - *Test Procedure for Control Valve Response Measurement from Step Inputs*
  - Defines metrics and test procedure
  - Does NOT specify performance requirements
- ANSI/ISA–TR75.25 (Technical Report) - *Control Valve Response Measurement from Step Inputs*
- *EnTech Control Valve Dynamic Specification V3.0*
  - Version 3 aligned with ANSI/ISA–75.25
  - DOES provide guidance on specifying valve performance requirements

15 AIChE
Spring Meeting, Austin, TX

EMERSON.
Process Management

Some Definitions from ANSI/ISA-75.25

- **Dead band: the range through which an input signal may be varied, with reversal of direction, without initiating an observable change in output signal** [ANSI/ISA-S51.1-1979 (R1993)]. In this standard and ISA-TR75.25.02-2000, it is defined in percent of input span. Note that in some other literature this definition is used for dead zone.
- **Resolution: smallest step increment of input signal in one direction for which movement of the output is observed.** Resolution is expressed as percentage of input span. The term in this document means: the tendency of a control valve to move in finite steps in responding to step changes in input signal applied in the same direction. This happens when the control valve sticks in place, having stopped moving after the previous step change.

15 AIChE Spring Meeting, Austin, TX

EMERSON. Process Management

Dead Band and Resolution

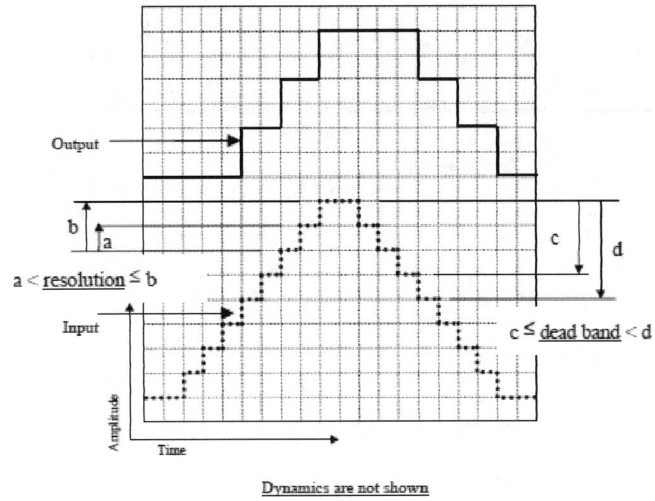

Figure 1 — Dead band and resolution

15 AIChE Spring Meeting, Austin, TX

EMERSON. Process Management

Dead Band

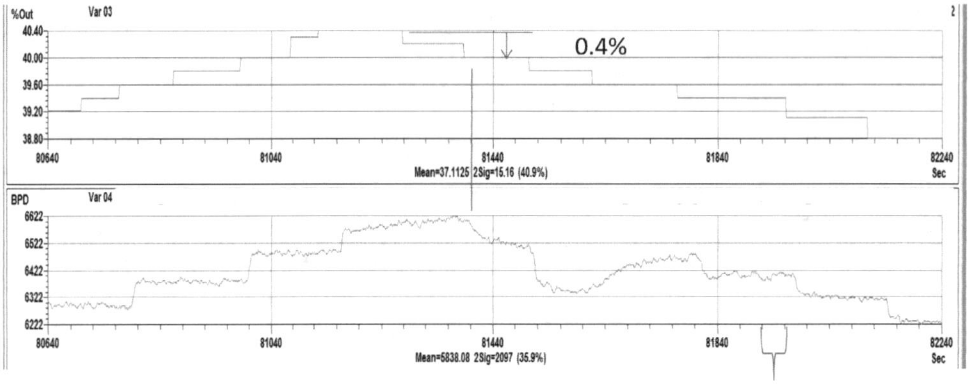

0.2%< Dead Band < 0.4%

40 seconds

15 AIChE Spring Meeting, Austin, TX

EMERSON. Process Management

Resolution

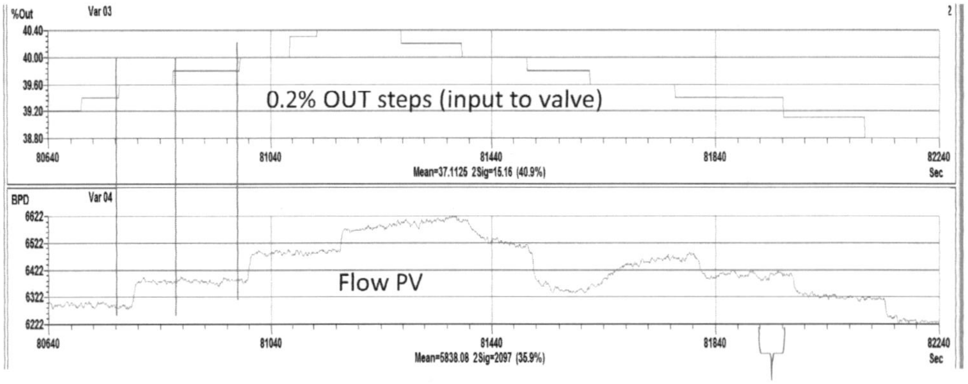

0.2%< Resolution < 0.4%

40 seconds

15 AIChE Spring Meeting, Austin, TX

EMERSON. Process Management

Dead Band and Resolution Cause Limit Cycles

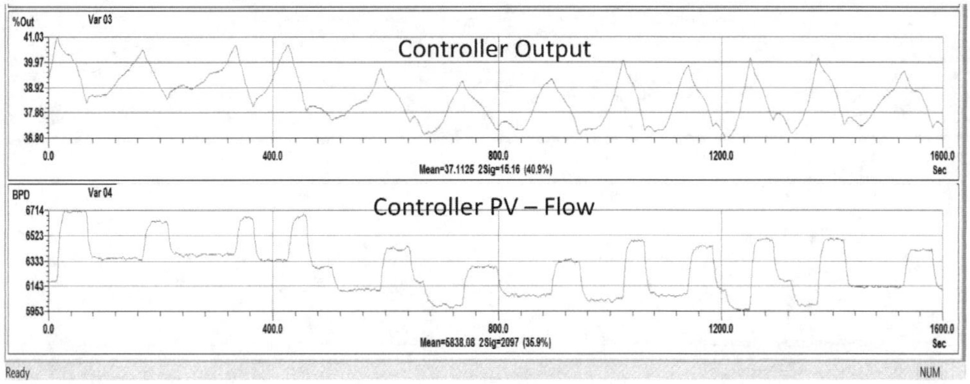

15 AIChE Spring Meeting, Austin, TX

EMERSON. Process Management

Definitions from ANSI/ISA-75.25.01

- **step response time (T86):** the interval of time between initiation of an input signal step change and the moment that the response of a dynamic reaches 86.5% of its full steady state value. The step response time includes the dead time before the dynamic response.

15 AIChE Spring Meeting, Austin, TX

EMERSON. Process Management

Step Response Time (T_{86})

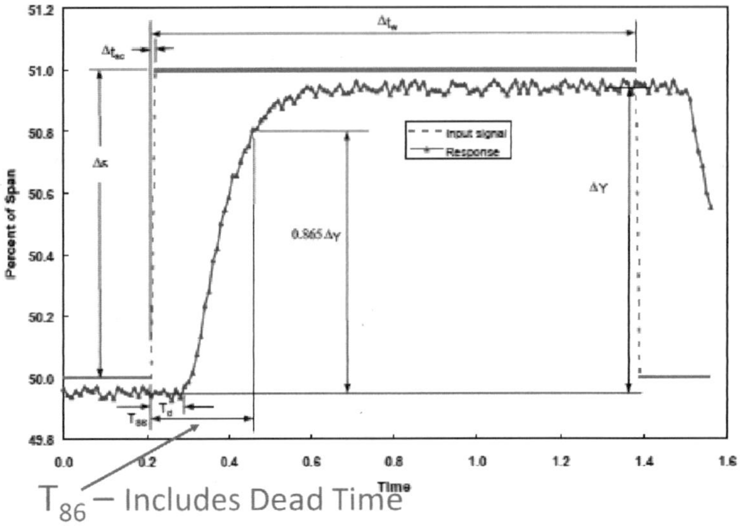

Figure 2 — Typical step change and response without overshoot

15 AIChE Spring Meeting, Austin, TX

EMERSON. Process Management

Step Response Time VS. Step Size

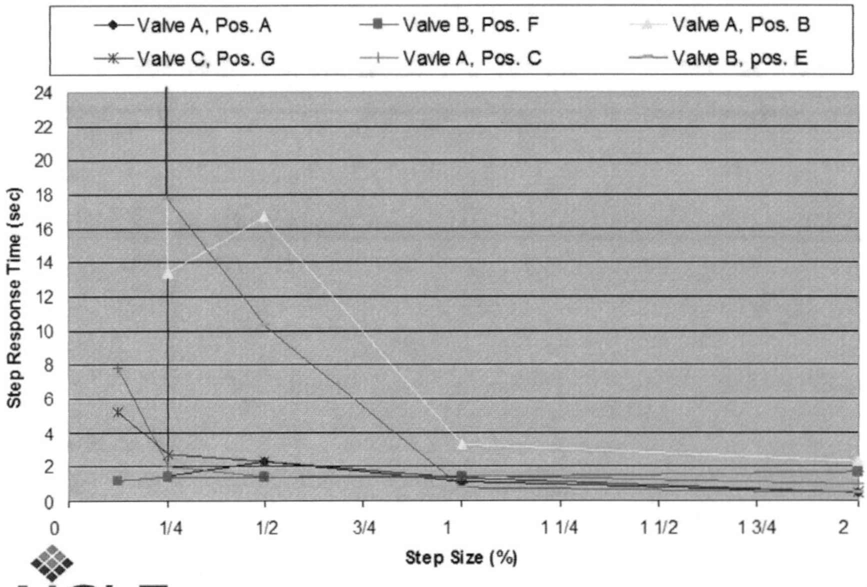

15 AIChE Spring Meeting, Austin, TX

EMERSON. Process Management

From ANSI/ISA–TR75.25

- **Region 1** is defined as small input steps which result in no measurable movement of the closure member within the specified wait time.
- **Region 2** is defined as input step changes which are large enough to result in some control valve response with each input signal change, but the response does not satisfy the requirements of the specified time and linearity.

15 AIChE Spring Meeting, Austin, TX

EMERSON. Process Management

From ANSI/ISA–TR75.25

- **Region 3** is defined as step changes which are large enough to result in flow coefficient changes which satisfy both the specified maximum response time and the specified maximum linearity.
- **Region 4** is defined as input steps larger than in region 3 where the specified magnitude response linearity is satisfied but the specified response time is exceeded.
- *EnTech Control Valve Dynamic Specification V3.0* is similar, using A, B, C, D instead of 1, 2, 3, 4.

15 AIChE Spring Meeting, Austin, TX

EMERSON. Process Management

Step Response Time VS. Step Size

- Region 1 and 2 are bad for control performance
- Region 3 performance provides the most linear operation of the valve and hence the best control performance
- Region 4 performance is typically inconsequential to control performance (large step, velocity limited)
- So, minimize the size of Region 1 and 2, maximize the size of Region 3

15 AIChE Spring Meeting, Austin, TX

EMERSON. Process Management

Limit Cycle Due to Region 2 Slow Response

Region 2 T_{86} = 43 seconds; Region 3 T_{86} = 1 second

Limit Cycle

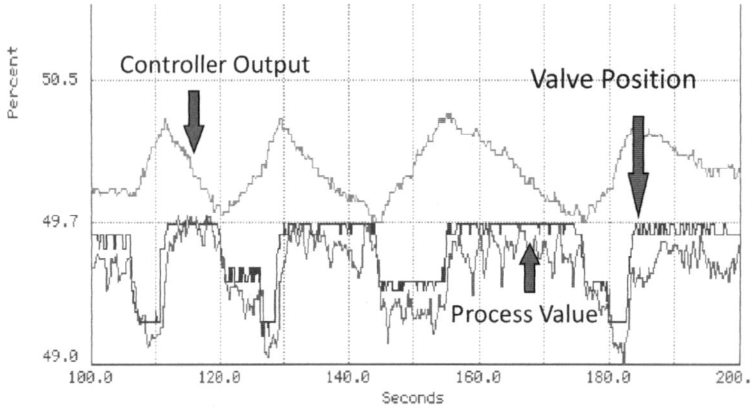

15 AIChE Spring Meeting, Austin, TX

EMERSON. Process Management

Loop Gain

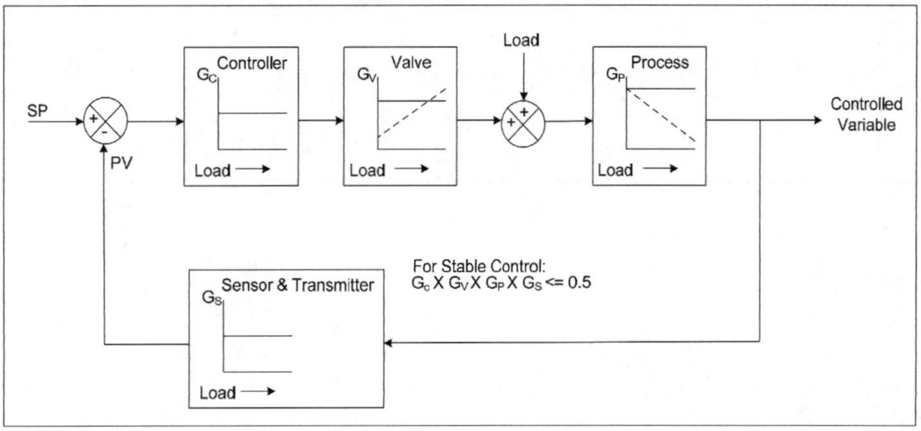

15 AIChE Spring Meeting, Austin, TX

EMERSON. Process Management

Total Loop Gain & Installed Valve Gain

- Total loop gain is equal to the product of the various gain components within the loop.
  - **Total Loop Gain = (Controller Gain) x (Valve Flow Gain) x (Process Gain) x (Transmitter Gain)**
- **Installed** Valve Flow Gain

$$ValveFlowGain = \frac{\Delta Flowrate(\%)}{\Delta StemPosition(\%)}$$

$$Flowrate = Cv\sqrt{\Delta P / SG}$$

  - Combined representation of the valve gain and the fluid process gain under **operating** conditions
  - Accounts for the valves true characteristic and subsequently changes the gain equation to reflect stem position (%) rather than valve capacity or % of full open Cv.

15 AIChE Spring Meeting, Austin, TX

EMERSON. Process Management

Control Valve Flow Characteristics (Inherent)

Control Valve Inherent Flow Characteristics :

-The relationship between flow through the valve & valve stem position at constant pressure drop

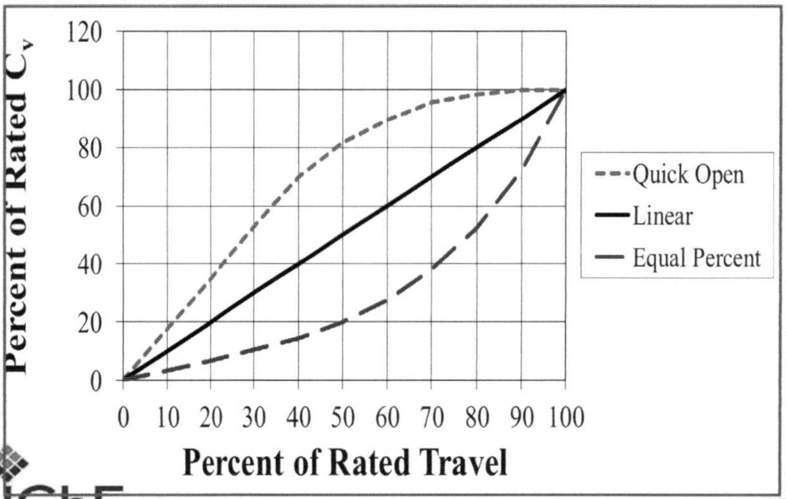

15 AIChE
Spring Meeting, Austin, TX

EMERSON.
Process Management

Objective of valve trim selection example:

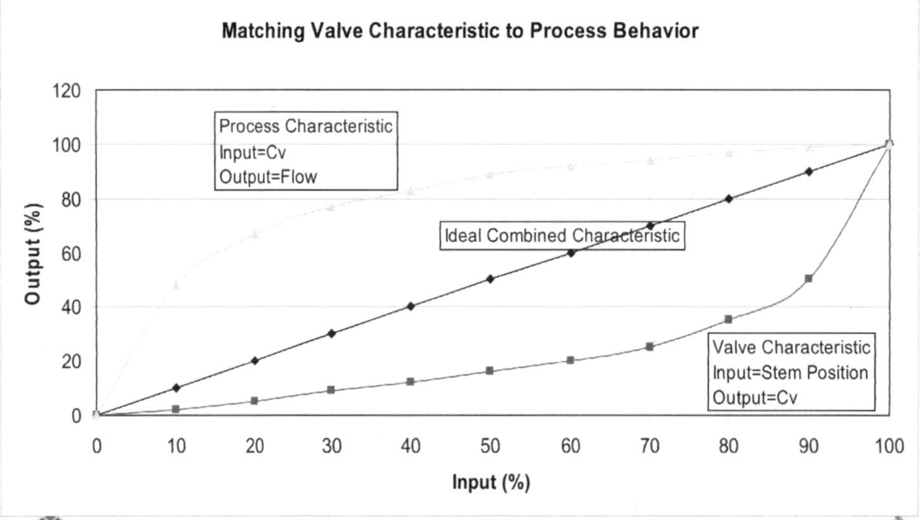

15 AIChE
Spring Meeting, Austin, TX

EMERSON.
Process Management

Installed Gain – Goals and Selection Criteria

Gain ≤ 2.0 (to avoid oscillatory and destabilizing behavior)

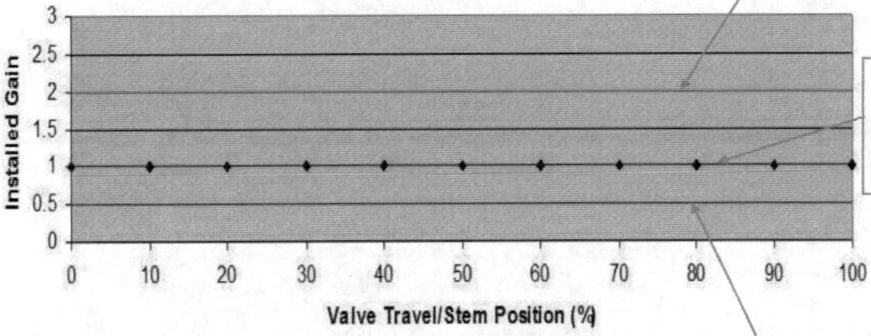

Gain as close to 1 as possible and as constant as possible

Gain ≥ 0.5 (to avoid sluggish / non-responsive behavior)

Generally, plotting installed gain for available trim choices will indicate the best choice. Valve manufacturers can provide one-off trim designs for special applications as well…

15 AIChE Spring Meeting, Austin, TX

EMERSON. Process Management

Example: installed characteristics and process gains

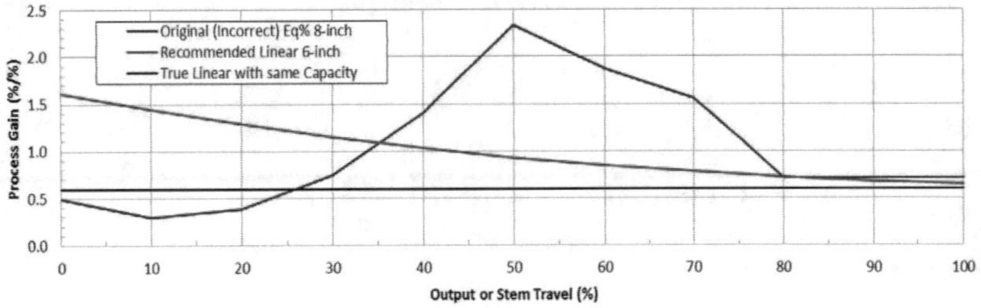

Installed gains

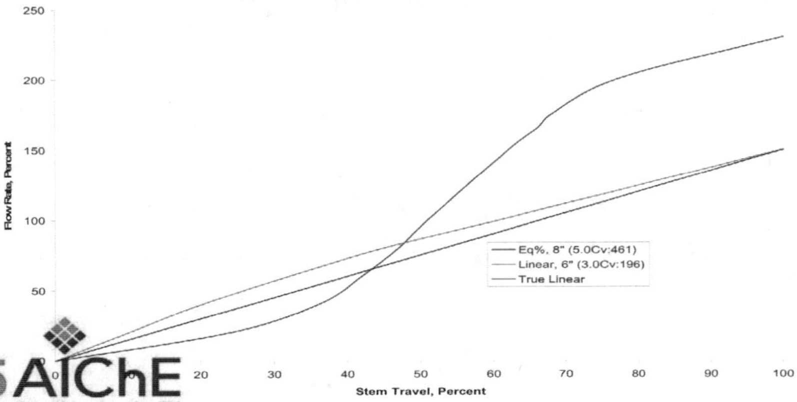

Installed characteristics

15 AIChE Spring Meeting, Austin, TX

EMERSON. Process Management

Bad Performance from Gain Change

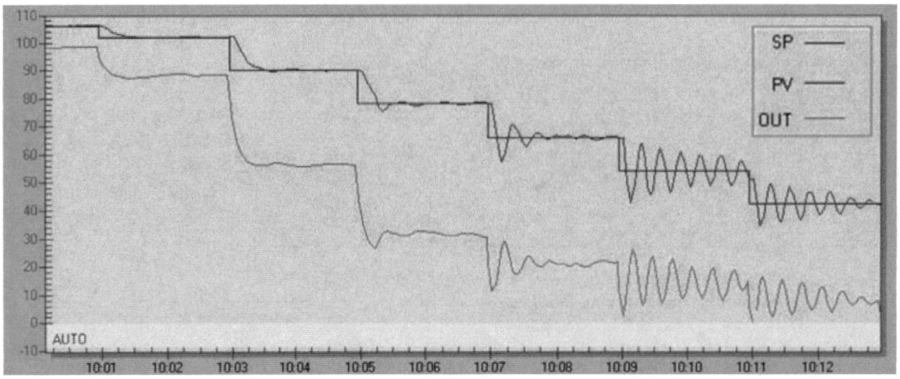

Valve Gain increases as controller output decreases, causing loop instability.

15 AIChE Spring Meeting, Austin, TX

EMERSON. Process Management

Specifying Valve Performance

- Important "Static Performance" Parameters
  - Dead Band
  - Resolution
- Important "Dynamic Performance" Parameters
  - Response Time (T_{86})
  - Size of Region 1- (~=dead band) - bad, minimize
  - Size of Region 2 – bad, minimize
  - Size of Region 3 – wide as possible but over 10% probably not required
  - Size of Region 4 – not of great consequence to control
- *EnTech Control Valve Dynamic Specification V3.0* is a good guide

15 AIChE Spring Meeting, Austin, TX

EMERSON. Process Management

Benefits of Good Valve Performance

- Performing the front end engineering work to properly select the control valves provides major incentives
  - Reduced oscillation of control loops enables smoother plant startups and reduced maintenance costs
  - During commissioning stage, controller tuning controllers is simplified and accelerated, given that the total loop gain is nearly constant
  - Process control performs better over a wider range of conditions when the total loop gain is nearly constant

15AIChE
Spring Meeting, Austin, TX

EMERSON.
Process Management

Summary

- Many operational problems can be prevented, and process control significantly improved, by proper selection of the control valve
  - Considering the control loop stability and performance during procurement!
  - Valve gain, dead band, resolution, response time regions are clearly defined parameters
  - Severe control valve nonlinearities **cannot** be corrected in the controller
- These principles apply across all process industries.
- Questions?

15AIChE
Spring Meeting, Austin, TX

EMERSON.
Process Management

400181 Impact of Tight Oil on Refining

Abstract:

The impact of tight-fracked hydrocarbons has been one of the major contributors to what marginal economic highlights our country has had in the last 6 years. But what an unfolding event it has been. We now have reserve levels on the order of Saudi Arabia, and are close to energy independence, at least within the bounds of the NAFTA agreement of 1996.

At the same time not all tight hydrocarbons are the perfect match to each purchasers needs. Every well, each local formation and production region, in reality, produce products of strongly site-variant constituents, contaminants, and final user qualities and properties.

Product markets for the well outputs range from natural gas, NGL's, and naphtha condensate, to full range crudes (with or without resid-type end points) with all sorts of petrochemical industry implications derived from these sources.

This presentation will discuss two distinct areas :

1. The effect of drilling, recovery, bulk transportation and final treating/conversion facilities on the gas and midstream sales qualities of products and on the physical infrastructure from well to final fuel tap.

1. The changes that will likely be needed in existing or new conventionally-planned refineries to

make using this "wonder crude" wonderful

We will call upon experiences primarily in the Eagleford, Barnett, and Bakken formations to help compare and contrast just some of the current production and ultimate marketing realities of this resource.

Copyright © American Institute of Chemical Engineers. All rights reserved.

The availability of cheap natural gas has made the United States an attractive location for manufacturing projects. Between 2012 and 2020, announced manufacturing projects in the U.S. will result in nearly $122 billion of economic investment and over 150,000 permanent jobs. The analysis on which this presentation is based broke down the country into eight regions– Gulf Coast, Northeast, Southeast, Mid-continental/Southwest, Midwest, Rockies/ Great Plains, Northwest, and West Coast –by shale basins to analyze the geographic distribution of shale-gas related manufacturing investments and their potential economic impacts. Detailed analysis for each geographical region includes granular data on investment dollars and job creation for selected manufacturing sectors – including chemicals, iron & steel, plastics & rubber, and petroleum refining.

The presence of existing capital (both physical and human) was found to be the most significant driver for regional economic investment, greater even than the proximity to accessible shale resources. Recoverable shale gas in the Northeast is more than 4-fold that in the Gulf Coast, however announced investments in the Gulf Coast is nearly 10-fold that of the Northeast. The Gulf Coast accounts for 66% of all U.S. manufacturing investments with over $80 billion identified in the region, of which $72 billion is in the chemicals industry. Improving the infrastructure in shale resource rich regions and mitigating the environmental concerns surrounding shale extraction techniques will support the continued growth of manufacturing across the U.S. Additionally, manufacturing investments in sparsely populated regions – such as those surrounding the Bakken –may be contingent on the ability to attract and retain workers.

This presentation is based upon work in a forthcoming research paper performed under contract for Oak Ridge National Laboratory.

11TH GLOBAL CONGRESS ON PROCESS SAFETY

Fire from the Cascading Failure of an Oxygen Supply System

Delmar "Trey" Morrison, Ph.D., P.E.
Exponent, Inc.
4580 Weaver Parkway, Warrenville, IL 60555 USA
tmorrison@exponent.com

Sean J. Dee, Ph.D., P.E., and Brenton Cox, Ph.D., P.E.
Exponent, Inc.

Vijay Kumar
EQUATE Petrochemical Company
PO Box 9717, Ahmadi, 61008, Kuwait
kumarv@equate.com

Mohammad Al-Shamary, Abdulqader Al-Qabandi
EQUATE Petrochemical Company

Prepared for Presentation at
American Institute of Chemical Engineers
2015 Spring Meeting
11th Global Congress on Process Safety
Austin, Texas
April 27-29, 2015

UNPUBLISHED

AIChE shall not be responsible for statements or opinions contained
in papers or printed in its publications

Fire from the Cascading Failure of an Oxygen Supply System

Delmar "Trey" Morrison, Ph.D., P.E.
Exponent, Inc.
4580 Weaver Parkway, Warrenville, IL 60555 USA
tmorrison@exponent.com

Sean J. Dee, Ph.D., P.E., and Brenton Cox, Ph.D., P.E.
Exponent, Inc.

Vijay Kumar
EQUATE Petrochemical Company
PO Box 9717, Ahmadi, 61008, Kuwait
kumarv@equate.com

Mohammad Al-Shamary, Abdulqader Al-Qabandi
EQUATE Petrochemical Company

Keywords: fires, explosions, oxygen enriched atmosphere, case study

Abstract

This paper discusses the investigation, findings, and root cause analysis of a fire due to the loss of oxygen containment in an ethylene oxide process. This is one of many types of processes that may contain a high pressure oxygen supply system. In such systems, the pressure and velocity of the oxygen stream are related to the potential for ignition of system components. Thus, measures are taken in design, operation, and maintenance to utilize ignition resistant materials of construction.

While more ignition resistant than carbon steel, stainless steel is still ignitable under certain process conditions in oxygen enriched atmospheres. In this incident, unanticipated conditions led to ignition of internal parts and stainless steel components in the high purity oxygen system. The resulting loss of oxygen containment caused a fire that destroyed the oxygen injection system and allowed a significant release of a combustible process gas stream. While automated safety instrumented safeguards shut down the oxygen supply to the system, the ensuing fire involving the combustible process gas stream continued. Consumption of both steel and ignition-resistant alloy components in the system during the oxygen enriched fire masked the underlying causes of the system failure. No injuries were sustained in the incident, but the process unit sustained major damage and required a lengthy outage for repair.

Given the level of damage to the oxygen injection system, the failure analysis required rigorous and systematic investigation by a multidisciplinary team. Through evaluation of the process data, metallurgical analysis of the damaged components, and understanding of the process dynamics, detailed hypotheses were developed, tested and refuted leaving a few most likely hypotheses explaining the cause of the incident. The lessons learned from the incident were used to develop a system redesign and maintenance guidelines for the rebuilt unit to mitigate the likelihood of future failures.

1 Introduction

A fire occurred in the oxygen supply system in an Ethylene Glycol Unit of the Shuaiba, Kuwait facility of EQUATE Petrochemical Company (EQUATE), which led to significant damages, unit shutdown, but no injuries. Since the fire involved an oxygen system, it escalated rapidly, but emergency shutdown systems quickly isolated oxygen flow. The piping system was breached at two locations: a carbon steel piping flange upstream of an automated block valve and a manual stainless steel gate valve downstream of check valve in the oxygen piping. The remainder of the process gas feeds was isolated at the unit's battery limits during the emergency shutdown, but the inventory of flammable process gas partially blew down through the breached oxygen system. This contributed to a large local fire that heavily damaged the oxygen system, process unit structure, and surrounding equipment.

Given the destructive nature of the fire, a detailed investigation was undertaken to identify the cause of the initial loss of containment in the oxygen system. The oxygen system failure analysis required rigorous and systematic investigation by a multidisciplinary team. Through evaluation of the process data, metallurgical analysis of the damaged components, computational fluid dynamics (CFD) analysis, and understanding of the process dynamics, detailed hypotheses were developed, tested and refuted leaving a few most likely hypotheses explaining the cause of the incident. The lessons learned from the incident were used to develop a system redesign and maintenance guidelines for the rebuilt unit to mitigate the likelihood of future failures.

2 Background on Pure Oxygen Systems

Pressurized oxygen systems pose significant hazards due to the acceleration of combustion of materials when exposed to an oxygen-enriched environment. Industrial experience has shown that oxygen enhanced fires rapidly accelerate in intensity since even apparently non-combustible materials can be consumed [4]. The ignition of both metallic and nonmetallic components is thus a concern in O_2 enriched atmospheres (i.e., greater than 23.5% O_2). Components that would not normally burn in ambient air (i.e., 21% O_2) may burn vigorously in an O_2 enriched environment. This can lead to loss of containment of high purity oxygen and rapid escalation of fire and explosion incidents. High oxygen concentration combined with a fuel and ignition source can result in a breach in oxygen containment systems, rapid loss of containment of pressurized oxygen, rapid fire growth, and rapid fire extension to other areas.

Purity and concentration of oxygen in the system directly affect the dominant design considerations. Oxygen enriched systems are defined as containing at least 23.5% O_2 by volume with the remaining volume being inert [1]. Oxygen-enriched systems can be classified into a number of categories depending on purity, including [1]:

- Low purity oxygen – contains less than 35% O_2 by volume
- Standard purity oxygen - equal or greater than 99.5% O_2 by volume
- Ultra high purity oxygen - greater than 99.999% O_2 by volume

Industrial standards and the underlying research efforts by industrial oxygen users have led to the development of safety and design guidelines and standards including ASTM G94-05, ASTM G63-99, ASTM G88-05, CGA G-4.4, NFPA 50, and NFPA 53.

2.1 *Understanding Potential Modes of Ignition*

Initial ignition of a component or particle within an oxygen system may lead to a kindling chain mechanism of fire growth where the energy released from an initial, smaller, easily ignitable fuel source causes ignition and combustion of a more ignition-resistant material [Reference [2], Section 5.2.14]. The kindling chain may quickly lead to fire growth within a system and overcome (i.e., ignition and continued combustion) the combustion resistance of the components in a piping system leading to loss of O_2 containment. Ignition and combustion resistance of materials are related to the material composition, thickness, configuration, and environment (e.g., temperature, pressure, and velocity). Several potential ignition mechanisms for metal alloys have been identified in the literature [1]:

- Particle impact
- Mechanical friction[1]
- Adiabatic compression
- Promoted ignition
- Electrical arcing (e.g., static electric discharge, lightning strike, electrical system fault)

The collision of high velocity particles with stationary surfaces can generate localized heating of the particles due to the conversion of kinetic energy to heat at the impact point. If the heat generated is sufficient, ignition and combustion of the particle may occur which could then ignite the impacted material or other nearby fuels (e.g., polymeric gaskets). Both impinging (impact face-on in the direction of flow), and non-impinging (impact obliquely or parallel to the direction of flow) impact with internal piping system surfaces may result in ignition of either the particle or the surface material [Reference [2], Section 5.2.6]. Particles may be introduced from the attrition of surface scale deposits from corrosion in piping, mechanical wear of system internals, or other sources of particle contamination. The nature of the particle along with the velocity and other system conditions will affect the likelihood of ignition of the particle. For example, silica particles are non-ignitable, but rust particles are ignitable.

Mechanical frictional heating may result from metal-to-metal rubbing, which can be experienced in moving valve internals. This rubbing mechanism can wear protective coatings from surfaces as well as generate small particles [Reference [2], Section 5.2.8.1]. Adiabatic compression of oxygen can occur when a valve is opened to a dead end space in a pipe with O_2 at a lower pressure. The resulting mechanical compression of the gas in the dead space as it is pressurized to the pipeline pressure can result in heating sufficient to ignite non-metallic valve seats or gaskets [Reference [2], Section 5.2.7]. These could, in turn, ignite metallic components leading to a system fire.

---

[1] While frictional heating may occur from gaseous flow against a pipe wall or through a small opening, flow frictional heating causing ignition of metals is generally only supported by "unverifiable anecdotes" in high pressure oxygen systems (e.g., greater than 500 psi) [ASTM G88-05, 5.2.12 – 5.2.12.1]. These pressures are significantly higher than those in the system discussed in this manuscript. Hence, frictional heating is more appropriately discussed from a standpoint of surface friction induced by internal moving components in valves, pumps, or other devices with moving parts.

Several design steps can be taken to limit the potential for ignition, or to stop the kindling chain before the system is breached. The use of filters can reduce the presence of particles, but it is infeasible to remove all particles for an oxygen system [Reference [2], Section 5.2.6.1]. The likelihood of ignition from both frictional heating and particle impact can be limited by selecting combustion-resistant materials of construction [Reference [2], Sections 7.6.3 and 7.8.3]. The likelihood of ignition by adiabatic compression can be limited by system design and throttling controls, for example to limit the rate of valve opening/closure or ignition resistance of materials in the adiabatic compression zone.

2.2 Metal Alloys in Oxygen Service

Metal alloys for use in O_2 service are classified as either combustion-resistant or exempt, based upon the alloy, configuration, temperature, pressure, and gas flow velocity. A combustion resistant material at the given conditions may undergo ignition, but the combustion process will quench with limited propagation away from the point of ignition. An exempt material has no thickness or pressure limitation for use in O_2 systems, based upon the standards [1]. Although the alloys listed are referred to as "exempt" from velocity limitations, this does not mean that the metals are immune to oxygen-enhanced combustion. Sufficient evidence exists in the literature and from practical experience to indicate that damage can occur to these materials from repeated combustion events without causing a propagating combustion front.

The kindling chain may quickly lead to fire growth within a system and overcome the burn resistance of the components in a piping system leading to loss of O_2 containment. Ignition and burn (i.e., combustion) resistance of materials is related to the material composition, thickness, configuration, and environment (e.g., temperature, pressure, and velocity). For example, carbon steel alloys are unsuitable for use in oxygen systems at concentrations greater than 35% by volume due to their reduced resistance to combustion [Reference [1], Section 4.2.3.2].

Metals are more difficult to ignite than non-metals due to their higher ignition temperature, higher thermal conductivity, and ability to generate ignition-resistant oxide coatings [Reference [3], Section 5.4.2]. If sufficient energy can be added to initiate combustion (i.e., ignition), then a given material may start to burn. However, if the oxygen concentration, density, and velocity are not within certain bounds, the combustion event will extinguish. In practical terms, repeated particle impact may be able to eventually cause a hole in a non-exempt material part. Various modeling approaches have been discussed that can be used to estimate the rate of consumption of a metal containment system [4],[6],[7],[8]. Estimates of the propagation rate of the combustion front can be used to design containment systems (e.g., wall thickness) and the timing for automated shutdown systems.

Examples of metallic alloys used in O_2 distribution systems are provided in Reference [1], Appendix C and D. The table lists some common stainless steel alloys and nickel alloys. Carbon steel alloys are typically unsuitable for use in oxygen systems at concentrations greater than 35% by volume due to their reduced resistance to combustion [Reference [1], Section 4.2.3.2]. The exemption status of a given metal is determined based upon a maximum observed consumed length of 30 mm in some standardized tests, or maximum consumed lengths of 10 cm to 15 cm, in others [Reference [1], Appendix B]. For a material in a thickness less than that tested, the standards indicate that it should be treated as flammable at pressure and velocity above the exemption curves as displayed below.

Table 1. Example O_2 oxygen system alloy exemption pressures and minimum thicknesses.

Material	Nominal Composition	Min. Thickness	Exemption Pressure
MONEL® 400	67Ni, 32Cu, 1Fe	None Specified	21 MPa (3000 psig)
MONEL® K-500	66.5Ni, 30Cu, 3(Al, Ti)	None Specified	21 MPa (3000 psig)
Hastelloy C-276	56Ni, 12Cr, 13.5Mo, 4Fe, 3W, 2.5Co	None Specified	5.3 MPa (750 psig)
INCONEL® X-750	74Ni, 15.5Cr, 7Fe, 2.5Ti, 1Al	None Specified	6.9 MPa (1000 psig)
Stainless Steel, cast, CF-8M	19Cr, 11Ni, 3Mo, 1.5Si, 1.5Mn, Bal. Fe	3.18 mm (0.125 in) 6.35 mm (0.250 in)	1.4 MPa (200 psig) 2.0 MPa (290 psig)
Stainless Steel, wrought, 316	17Cr, 12Ni, 2Mn, 3Mo, 1Si, Bal. Fe	3.18 mm (0.125 in) 6.35 mm (0.250 in)	1.4 MPa (200 psig) 2.0 MPa (290 psig)
Stainless Steel 317 [5]	19 Cr, 13 Ni, 2 Mn, 3.5 Mo, 0.75 Si, ~Bal. Fe	Not listed	Not listed
Gray Cast Iron	3C, 2Si, 0.8Mn, Bal. Fe	3.18 mm (0.125 in)	0.27 MPa (25 psig)

For non-exempt materials (i.e., combustion-resistant), velocity limitations for impingement and non-impingement oxygen flow are shown in Figure 1 and Figure 2 respectively [Reference [1], Sections 4.4.2 and 4.4.3]. For each graph, systems containing oxygen-contact stainless steel alloy components should be designed to keep the pressure and velocity below and to the left of the curve to limit particle kinetic energy [Reference [2], Section 7.6.2]. The velocity limit as a function of pressure is lower for the impingement curve than the non-impingement curve because a theoretical particle impact face-on to a surface will generate more energy than an oblique impact.

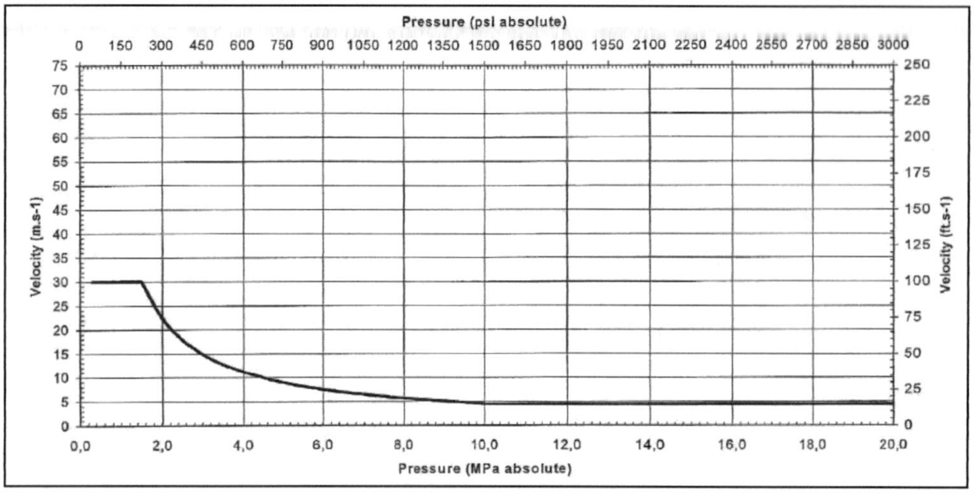

CGA G-4.4; Figure 1, p12

Figure 1. Impingement velocity curve.

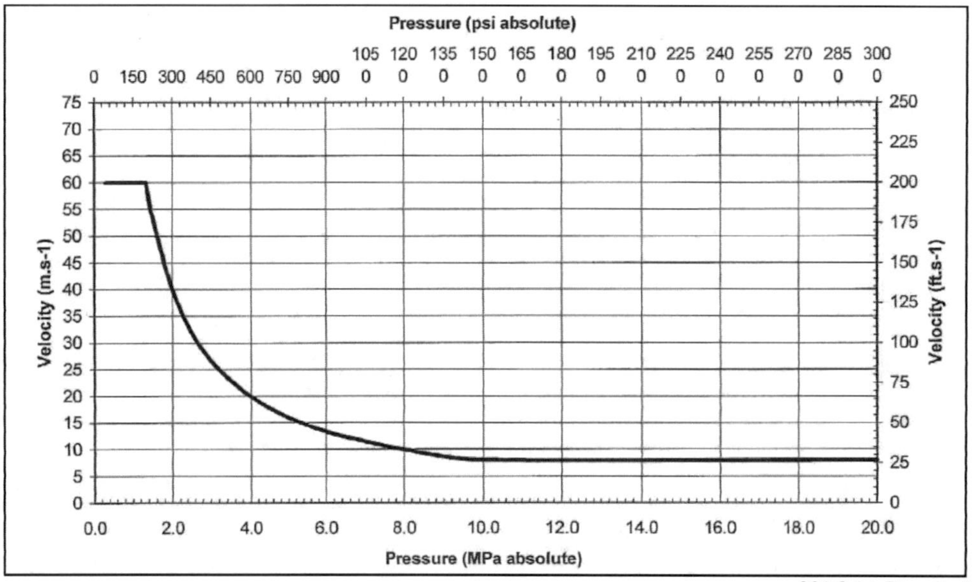

CGA G-4.4; Figure 2, p13

Figure 2. Non-impingement velocity curve.

3 Failure Analysis of Oxygen System

In undertaking an investigation, the scientific method of analysis is an essential framework for completing a thorough and objective analysis. Figure 3 provides a flowchart that outlines the approach. This is a hypothesis-driven approach where scientifically defensible answers are sought. The key objectives of the investigation are to gather the facts, perform an objective and systematic analysis, and determine the most likely causes of the incident. After these most likely causes have been identified, evaluated, and understood, the investigation can reliably move toward providing useful lessons learned and recommendations for preventing a similar incident from recurring.

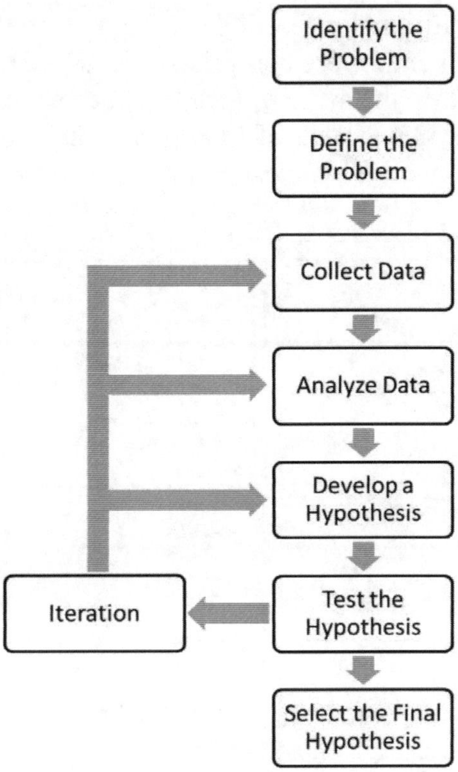

Figure 3. Flow chart for the scientific method of investigation.

3.1 Overview of the Oxygen System

The oxygen system was installed in approximately 2000, and the piping diameter was increased from 4-inches to 6-inches sometime after the initial installation to meet operational requirements. The system was equipped with an emergency shutdown system that tripped based on temperature or pressure deviations outside the acceptable operating envelope. In approximately 2005, the emergency system actuated. The root cause investigation determined that the most likely causes involved fugitive particles in the system. The system operated without incident for several years until the loss of containment incident in 2012.

The system supplied oxygen at approximately 27 bar (400 psig) and 13 kg/s (29 lb/s) to the process. The oxygen feed was filtered before passing into the piping control system. A piping isometric describing the system is provided in Figure 4. Components made from ignition resistant alloys are colored green in Figure 4, while stainless steel alloys are red. In the oxygen handling system, high purity oxygen was passed through two sets of filters before it was fed to the O_2 Isolation Section. In the O_2 Isolation Section, oxygen flow to the process was modulated through a MONEL® flow control valve (6). Two automated stainless steel block isolation valves (5 and 7) were tied to the emergency shutdown system to isolate and stop oxygen flow in the event of an emergency. Oxygen then flowed through a 6-inch tee connection (8) to the N_2 Purge Section. Oxygen reverse flow back into the N_2 Purge Section was restricted by a stainless steel swinging check valve (10). MONEL® piping connected the O_2 Isolation Section to the O_2 Process Section through a tee at this location. In the O_2 Process Section, oxygen passed through a MONEL® dual disk check valve (3) with internal components made of either MONEL® or INCONEL® alloys. Directly downstream of the dual disk check valve was a stainless steel manual gate valve (2) that connected to additional MONEL® piping before the oxygen was fed to the process unit. Within the system, stub ends were comprised of MONEL® with non-ignition resistant flanges. Additionally, gaskets within the system were found to have stainless steel inner rings and MONEL® coils. Thus, in the oxygen system, the oxygen contact surfaces were found to be composed of ignition resistant materials, with the exception of any exposed surface of the gaskets' inner ring.

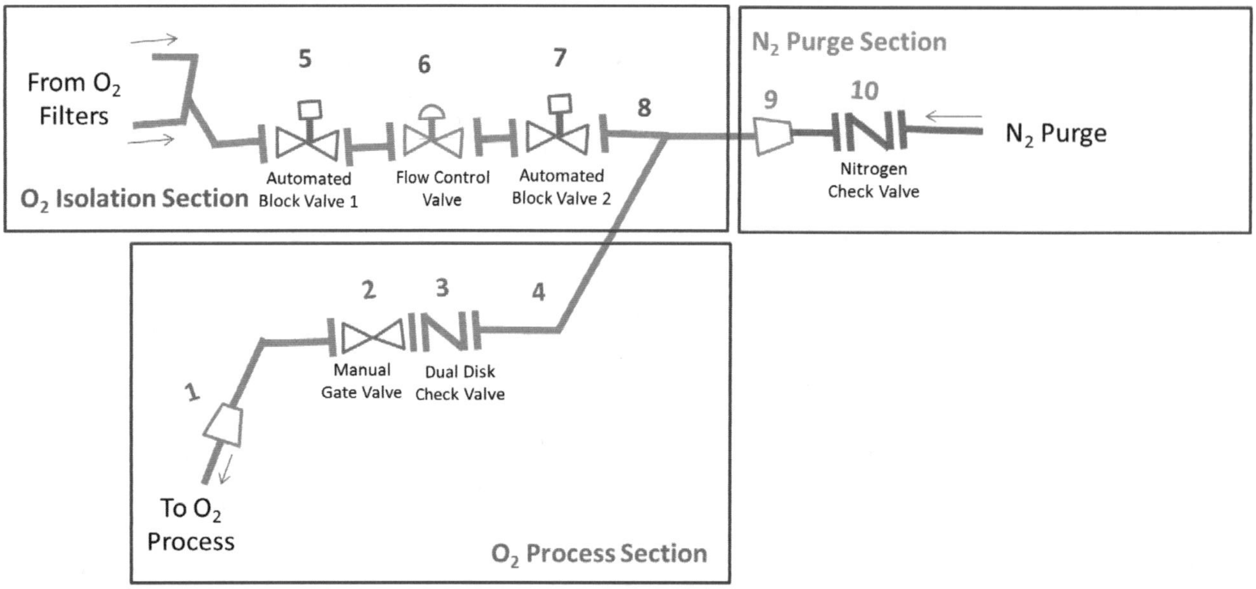

Figure 4. Piping isometric diagram of the O_2 system.

A pre-incident image of the O_2 system is provided in Figure 5. The O_2 filters are located out of the image to the left with flow from the left to right matching the isometric figure. The blue actuators correspond to the automated isolation block valves (5 and 6). The green actuator corresponds to the flow control valve (6). The hand wheel shows that the manual gate valve (2) is in the open position. A post-fire photograph of the O_2 system is depicted in Figure 6 indicating that most of the piping system components partially survived the fire.

Figure 5. An aerial view of the O_2 system prior to the incident.

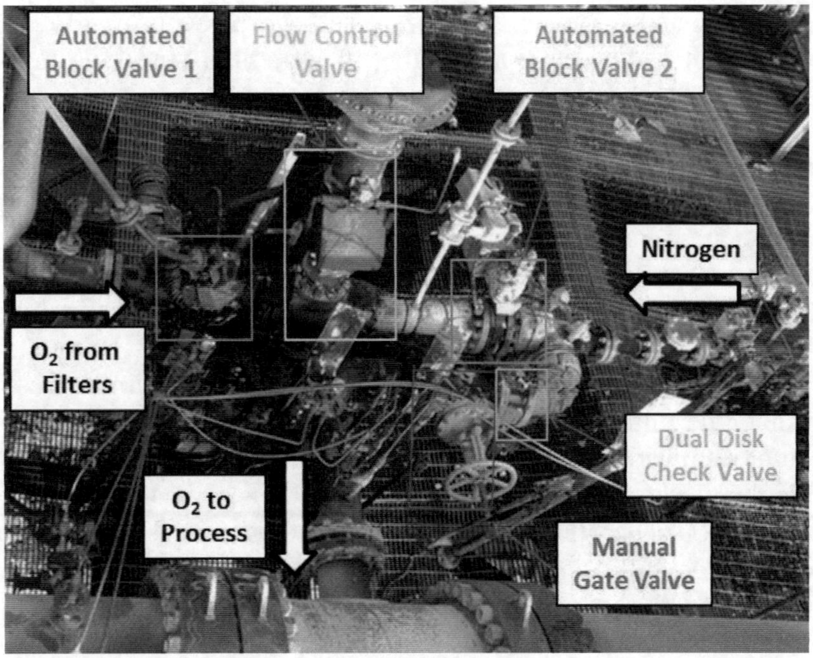

Figure 6. Overhead view of piping from oxygensystem fire.

3.2 Determining the Initiating Event

Two potential areas were identified that could correspond to the initial loss of containment incident: the carbon steel flange upstream of automated block valve (5) and the stainless steel manual gate valve (2). Both components were found to contain significant material loss due to oxygen-enhanced combustion. The upstream flange was refuted as a potential source of the initial LOC due to the evidence of automated isolation system action and a delayed pressure blowdown observed in the recorded process data. Thus, the investigation focused on determining the cause of the breach at the manual gate valve. The most likely potential initial ignition modes for this high purity oxygen system were determined to be frictional heating and particle impact. To characterize their relative likelihood as the potential cause, the average operating conditions based on process data for the oxygen system leading up to the incident are summarized in Table 2.

Table 2. Average operating conditions of O_2 system prior to incident.

Pressure bar (psi)	Temperature (°C)	Mass Flow Rate $kg\ s^{-1}$ ($lb\ s^{-1}$)	Calculated Volumetric Flow Rate $m^3\ s^{-1}$ ($ft^3\ min^{-1}$)	Calculated Nominal Velocity $m\ s^{-1}$ ($ft\ min^{-1}$) 6-inch Diam.
27.9 (396)	22.5	13 (29)	0.4 (850)	20 (3900)

An important aspect of designing systems and processes that utilize high purity oxygen is conducting an oxygen compatibility assessment (OCA) [9]. In a typical OCA, oxygen-wetted materials are evaluated under their potential worst case operating environment to assess their potential flammability. If a material is deemed to be flammable in the operating environment, then the potential ignition mechanisms are evaluated and scored according to their likelihood to promote ignition of the flammable material. Finally, the potential for a kindling chain or reaction effect is evaluated with respect to other nearby materials.

In this particular incident, OCA was used as a tool to assist in evaluation of potential origin and cause hypotheses. Within the oxygen system, five metals were evaluated for oxygen compatibility at 400 psi at ambient temperature: MONEL® 400, MONEL® K-500, INCONEL® X-750, 316 Stainless Steel, and 317 Stainless Steel. The MONEL® alloys are considered ignition resistant at 400 psi and ambient temperature because published literature has demonstrated that these alloys can withstand pressures up to 10,000 psi without failing a promoted ignition test. 300 Series stainless steel has been shown to burn at pressures as low as 111 psi, and fail promoted ignition tests at 200 psi. Therefore, the stainless steel alloys in the oxygen system would be considered flammable at 400 psi. INCONEL® X-750 has not been as extensively tested in the literature as MONEL® or steel alloys. The available data has only demonstrated that INCONEL® X-750 fails the burn length criteria at 1,000 psi, and has not been tested at lower pressures. At 400 psi, the flammability of INCONEL® X-750 is undetermined.

The classification of materials as "non-flammable", as opposed to "ignition resistant", must be approached with caution from a fire and explosion hazard assessment. Traditional classification of a material as "non-flammable" is based on the material exhibiting less than 1 inch of combustion consumption of a 0.125 inch diameter rod in a promoted ignition test [Reference [1], Appendix B]. Thus, materials classified as "non-flammable" can still exhibit metal combustion and consumption but "pass" the promoted ignition test. Promoted ignition and combustion of these "non-flammable" materials may still occur. For example, in the oxygen system under investigation, MONEL® may be considered "non-flammable" in an OCA, but particle impact, frictional heating from metal/metal rubbing, or kindling chain reactions may still result in combustion of the alloy, albeit in a prolonged fashion.

The limited available data for special alloys, can also lead to confusion in material selection related to OCA. In the example above, INCONEL® X-750 is listed as being exempt up to 1,000 psi, but the available data demonstrate that at 1,000 psi INCONEL® X-750 failed the promoted ignition tests. Further testing data was not available at lower pressures. It is of critical important that consideration of material exemption pressures as a safety limit or threshold be supported with test data.

4 Evaluating the LOC Hypothesis

The most severe damage and metal consumption occurred on the underside of the stainless steel manual gate valve and on the flange between the check valve and gate valve. Once ignited in the O_2 environment inside the piping system, combustion of the stainless steel components would be anticipated to occur rapidly leading to a breach in the valve body and release of O_2 pressure from the piping system. There are several distinct features of the damage that were used to identify the most likely point of origin for the fire inside the assembly.

4.1 Explaining Physical Damages (Metal Consumption)

Based on the literature, a representative rate of consumption for 316 stainless steel rods of 1 to 1.2 cm/sec was estimated for the valve body at 2.76 MPa (400 psi) [6]. Although this rate cannot be directly correlated to the more complex geometries involved, it does give a frame of reference of a few seconds to reach the degree of metal consumption observed [4].

The stainless steel gate valve body was consumed internally on the bottom side of the valve, but did not contain internal wall thinning on the top inside of the valve. The valve body consumption was focused around two holes on the bottom side located on either side of the valve seat. Based upon the differential consumption and metal wall thickness, it appears that these two holes were a consequence of the thinner wall thicknesses at these locations in the valve body casting when compared to the thicker cross section near the valve seat. The high pressure O_2 release caused the outer valve body to be consumed in jagged, irregular patterns. A few bolts in the valve flanges and the valve bonnet area were also partially or totally consumed during the O_2 release.

Given the nature of enhanced O_2 combustion of metal, the greatest amount of consumption can be related to the duration of combustion at that location. For a thin wall thickness specimen with continued O_2 flow across the surface, the hole will grow quickly in diameter with the perimeter of non-consumed metal moving quickly away from the point of origin. After O_2 flow through the hole stops (i.e., O_2 system isolation), consumption of the metal will likely stop immediately since ambient O_2 concentration is insufficient to support the metal combustion and a process gas-fueled fire is insufficient to cause stainless steel consumption in an ambient oxygen environment.

The damage to the gate valve and adjacent check valve is most severe at the connection of the two components as evidenced by the thickness of the original flange, which was completely burned through, and is located at one of the bottom bolts connecting the gate valve flange to the check valve body (see Figure 7). Not only is the gate valve-to-check valve flange consumed through its full thickness, but the adjacent gate valve bonnet flange is also externally consumed near this location. The damage patterns, both internally and externally, point back to this location as the origin of the initial high pressure leak in the piping system. The damage, and thus the combustion of the valve, extends away from this point of origin.

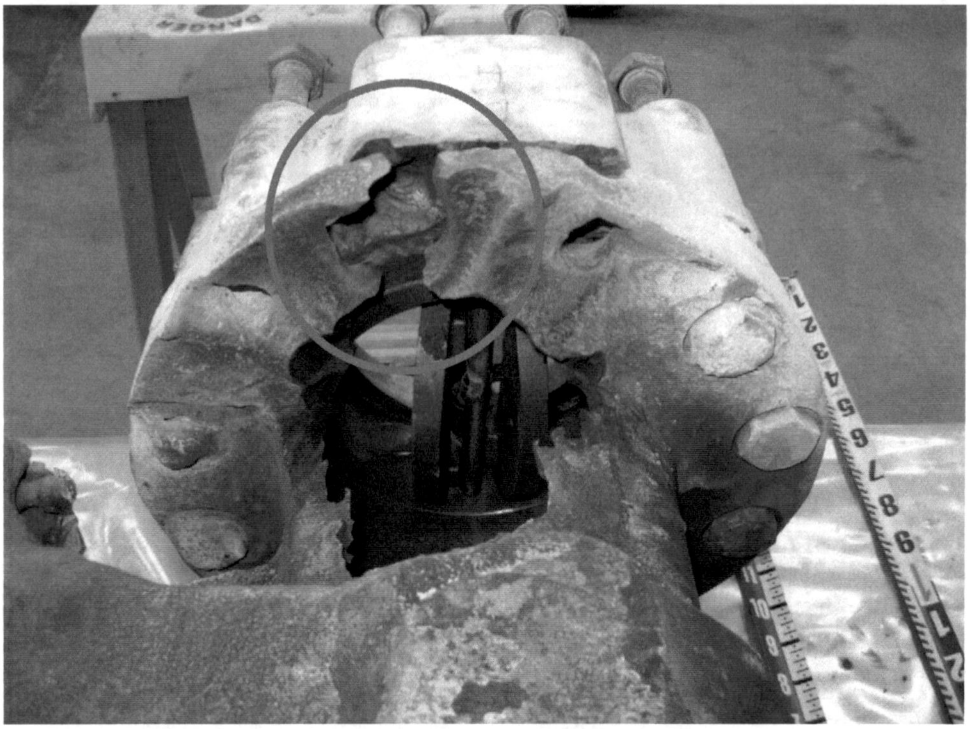

Figure 7. Localized consumption of stainless steel gate valve flange adjacent to check valve. (Installed such that the circled region was facing downward.)

The section of valve flange shown in Figure 7 was unique. No other flange thickness was completely compromised. The degree of consumption is consistent with this being the point of the initial loss of containment; thus, this it was most likely near the origin of the internal fire in the stainless steel valve body. Examination of the spiral wound gasket between the valve body and check valve indicated that the inner ring, which was in oxygen contact, was composed of stainless steel. The inner ring, spiral winding, and outer ring were consumed on the downward-facing edge, consistent with the area of greatest damage to the valve body and gate valve flange.

4.2 Cause of Ignition

Given the high consumption rate for combustion of the stainless steel components, the point of ignition must have been very close to the origin of the damage in the valve assemblies (see Figure 7). There are two likely modes of ignition that could cause ignition in the oxygen system either individually or when combined: frictional heating and particle impact.

The flow control valve was not being stroked at the time of the incident, and the seat was largely intact. However, evidence of particle creation and particle impact upstream in the O_2 system was noted during the investigation (see Figure 8). During disassembly, that upstream gasket of the flow control valve was found to be sloughing off layers of corrosion products that could be carried in the O_2 stream.

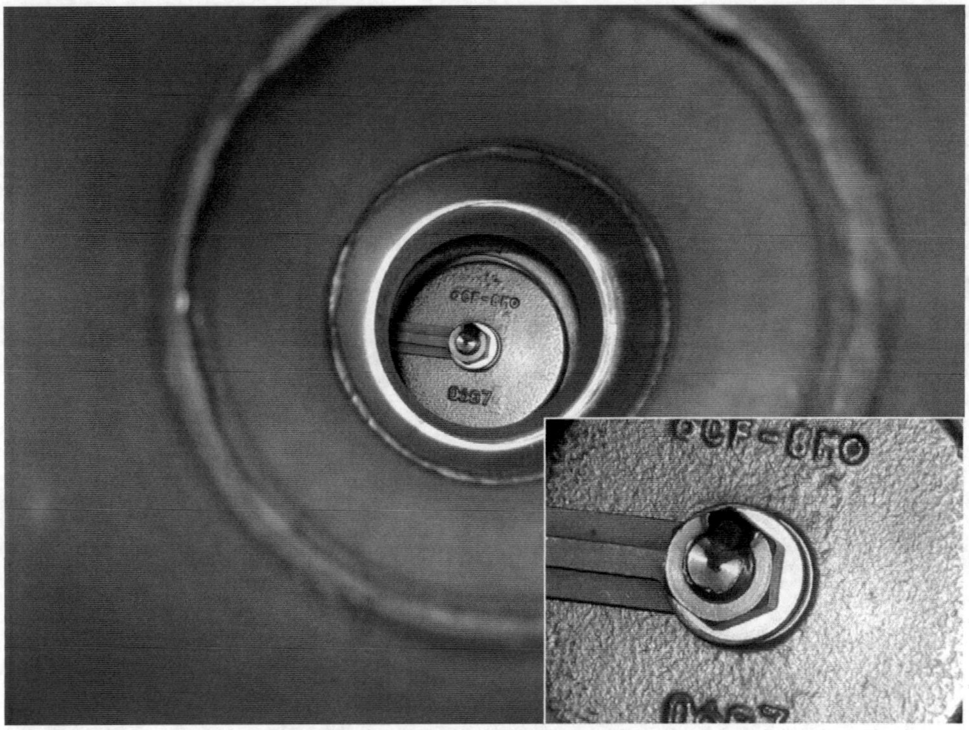

Figure 8. An apparent particle impact point with non-propagating combustion of the stainless steel check valve component was observed on the O_2 side of the N_2 check valve.

Two types of damage were found to the MONEL®/INCONEL® check valve internals just upstream of the gate valve: mechanical wear and melting. The mechanical wear is due to relative motion of the internal parts. The melting is either due to frictional heating or due to heat from nearby non-propagating combustion of the components. Movement of the parts is required for either potential mechanism. Given the high oxygen flow velocity and evidence of molten metal droplets being carried downstream in the piping, the check valve damage was not likely caused by the metal fire moving upstream from the gate valve to the check valve against the flow of O_2.

The wear and melt-over-wear damage patterns to the components as shown in Figure 9 were consistent with prolonged oscillation of the valve disks. The dual disk check valve opened and closed each disk with independently operating torsion springs. O_2 flow pushed on the disks to open the valve, but in the event of a loss of O_2 flow, the torsion springs were intended to shut the disks and prevent reverse flow. After one or both springs were compromised, the disks were no longer restrained, and the oscillations could increase in frequency, thereby accelerating the mechanical wear process and resulting in potentially significant frictional heating. The washers displayed evidence of wear on their internal circumference and some washers actually cut through the hinge pin.

Figure 9. View of the check valve internals. The two worn and melted gouges on the stop pin (top pin) are consistent with the location of the torsion spring restraints (see Figure 11). The stop pin was determined to be a MONEL® alloy. The grooves on the surface of the center of the hinge pin (bottom pin) are consistent with mechanical wear from the torsion spring.

The check valve internal components were combustion-resistant MONEL® and INCONEL® alloys, thus a single particle impact could not be wholly responsible for causing the observed damages to the check valve. Particle impact may have damaged one or both INCONEL® alloy torsion springs, which may have accelerated mechanical wear within the check valve due to a loss of the restraining force for the disks. However, mechanical wear over time due to disk oscillation without particle impact may also be sufficient to create the observed damage. Thus, mechanical wear alone could not be refuted as the root cause of the failure.

A view of the internal components prior to disassembly is depicted in Figure 10. The internal damage on the base of the check valve flows downstream and connects with damage to the gasket and gate valve body. Based on the flow dynamics, the configuration of the components, and the localized fire damage, the most likely cause of the internal fire damage at the MONEL® check valve is due to frictional heating at the bottom hinge-to-check valve body interface. Additional sprays of molten (black) metal alloy are visible on the disks along with erosion of the stop pin.

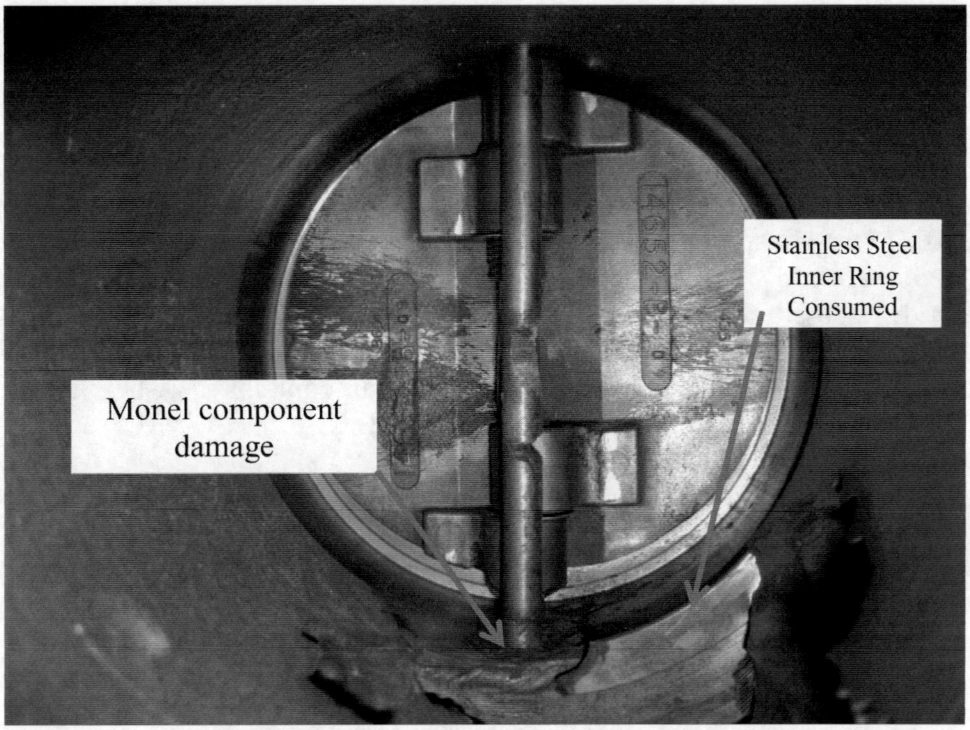

Figure 10. View through the gate valve at the downstream side of check valve. The area at the bottom of the valve that includes the stop pin and hinge contains surface melting and MONEL® consumption up to the mating surface of the partially consumed stainless steel inner ring for the spiral wound gasket.

Although M-400 is a velocity exempt material and is considered to be combustion resistant, there was evidence of non-propagating, localized combustion and melting on many of the internal components. The area of damage at the bottom of the check valve body around the contact area for the bottom hinge included the stainless steel inner ring of the spiral wound gasket. The thermal damage to the bottom of the MONEL® valve body appears to have been sufficient to cause ignition of the stainless steel inner ring of the gasket. Combustion of the inner ring led to combustion of the adjacent stainless steel gate valve body, which then propagated through the gate valve further downstream. Failure of the thin gasket likely led to the high-pressure release and noise identified by witnesses prior to the louder sound of the deflagration. The gasket failure, combined with a high-velocity oxygen release, also led to consumption of the local full thickness of the gate valve flange and external damage to the nearby valve bonnet flange.

4.3 Review of Preventative Maintenance

The check valve was initially thought to have been in service since approximately 2001. However, the investigation discovered that the check valve was scheduled to be replaced in 2010, but a defect in the replacement valve body led to just the internals being replaced. No detailed report was included in the maintenance file assessing the state of the original internals.

A few photos were provided from the 2005 emergency shutdown investigation, which depicted the check valve's internal components after five years of service. This is the same check valve body that was in service during the incident.

In hindsight, two photos indicated the progressive nature of the mechanical wear damage to the valve internal components. Figure 11 shows the downstream side of the valve with the disks held closed by the pair of torsion springs. In the figure, indications of wear-related failure are apparent for the bottom two washers. There is no washer between the bottom two hinges, which indicates one of three possibilities: (1) no washer was originally installed, (2) the washer has been worn through and its remains passed downstream, or (3) the washer has severed the shaft and passed downstream. Given the fact that the top hinge pair contains a washer, possibility (1) is unlikely. No evidence of heating or ignition of the MONEL® or INCONEL® alloy components is apparent in the photo.

In the slightly different view of the image in the right side of the figure, additional mechanical wear damage is visible. The center washer is missing, and a groove is worn in the hinge pin where the washer was originally located. This indicates that the washer was worn through and passed downstream. Additionally, the top spring in the photo has surface damage on the coil immediately adjacent to the missing washer. Although the 2005 investigation was searching for particles within the O_2 piping system, they did not find the two missing washers. In the current incident, three washers were unaccounted for, but no remains were found. The reason for the disappearance of washers could not be explained, but it is possible that they were consumed in the fire, or overlooked or lost during equipment disassembly.

Figure 11. Check valve internals during 2005 examination.

4.4 Oxygen Flow Dynamics

Design calculations for the O_2 system indicated that the bulk average flow velocity of the system was below the limiting velocity curve. However, the actual piping geometry including the 6-inch dual disk check valve, which varied from a 5.56-inch inlet diameter to a 6.75-inch outlet diameter, caused local regions of higher velocity flow. The check valve disks should open when oxygen flow starts and close when it stops. If the flow rate through the valve is lower than the minimum velocity required to keep the valve open, then the valve disks may oscillate ("flutter") with perturbations in the local gas flow. The available documentation for the valve did not provide a minimum velocity to reduce or eliminate disk oscillation [10]. The minimum velocity to pin the disks in the open position was estimated as 35.6 m s^{-1} based on calculations derived from the listed "cracking" pressure to cause the disks to start to open. The process data indicated that the mass flow rate supplied to the O_2 process during the hours prior to the incident fluctuated from 12.4 kg s^{-1} to 13.1 kg s^{-1}, corresponding to a nominal velocity of 19.1 m s^{-1} to 20.2 m s^{-1}. This velocity range was therefore about half that estimated for pinning the disks in the open position to prevent oscillation.

A computational fluid dynamics (CFD) model demonstrated that the local geometry had a dramatic impact on the flow velocity through the dual disk check valve. The check valve geometry created streamline velocities as high as 40 m s^{-1} on the outside port of the valve and much lower on the inside port (relative to the upstream elbow). The unbalanced streamline velocities at the check valve face (from ~20 m $s^{-1}$to 40 m s^{-1}) likely also contributed to oscillation of both disks (i.e., fluttering) when coupled with the force required to open the disks. The installation manual from the manufacturer of the dual disk check valve recommended placement of the valve at least 5 pipe diameters downstream of any piping elbow or pump discharge "to reduce the likelihood of turbulent flow through the valve, which could shorten valve life due to component vibration."[10] Additionally, 40 m s^{-1} is significantly higher than the velocity limitations suggested for both impingement (16.3 m s^{-1}) and non-impingement (29 m s^{-1})[2] surfaces in oxygen service as shown in Figure 12. Thus, the flow rate acceleration caused by the internal components of the check valve led to a velocity increase that not only increased the wear of the internal components (potentially creating particles), but also left the stainless steel gate valve susceptible to particle impact ignition when compared against the velocity limitations for the stainless steel components.

Oscillation of the disks in the check valve likely provided the mechanism for mechanical failure of the valve. This mechanical failure caused parts of the valve such as the washers to pass downstream into the stainless steel gate valve. It also appears to have caused non-propagating local ignition of the combustion-resistant MONEL® and INCONEL® alloys. Although the damage to these alloy components was limited, the adjacent stainless steel components were not exempt from velocity limitations in high pressure O_2 systems.

---

[2] At an assumed operating pressure of 400 psi.

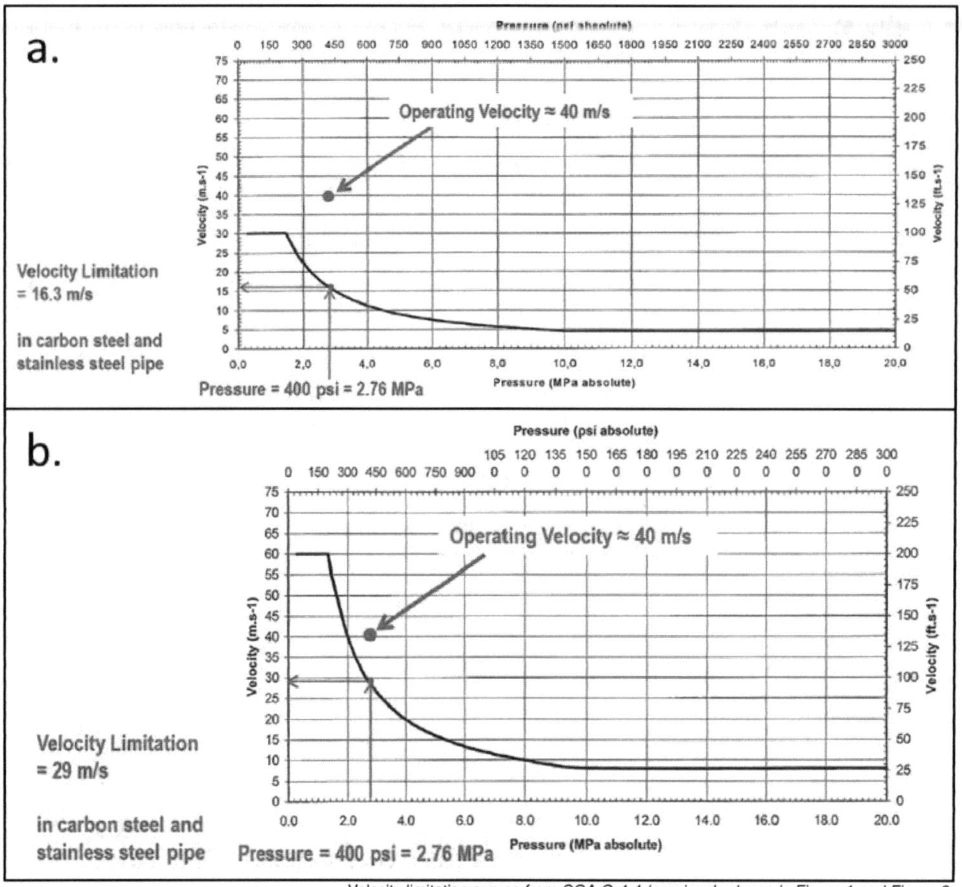

Velocity limitation curves from CGA G-4.4 (previously shown in Figure 1 and Figure 2.

Figure 12. Velocity limitations for impingement (a) and non-impingement (b) for oxygen flow at 400 psi. Note: Suggested velocity limitation for O_2 system pressure (400 psi) is indicated on the plots.

5 Conclusions and Recommendations

Based on the result of the investigation, several lessons were learned. The investigation identified complexities of oxygen flow and dual disk check valve performance that were not apparent to the original designers. Although the nominal bulk velocity for O_2 through the system fell below the recommended threshold for particle impact ignition, the complex internal geometry of valves drastically increased the local velocity. Additionally, the necessary velocity for maintaining the spring-loaded dual disk check valve pinned in the open position exceeded the maximum velocity limitation for stainless steel in oxygen service.

The system contained a combination of ignition resistant alloy components and stainless steel components. Even though ignition resistant alloys were used, the high velocity oxygen flow still facilitated non-propagating ignition of these alloys. Unanticipated and undocumented mechanical wear in the dual disk check valve components also provided the most likely source of ignition energy for the incident. Frictional heating at the bottom hinge-to-valve body interface and particle generation in the check valve are the most likely ignition sources to have caused the fire. Excessive wear to the internal dual disk assembly in the oxygen check valve was likely caused by fluctuations in flow velocity through the valve and operation below a minimum velocity required to pin the disks in the open position. Ignition of the stainless steel appeared to be a direct result of non-propagating ignition of adjacent MONEL® or INCONEL® alloy components.

As a result of the investigation findings, the preventative maintenance for the oxygen system was modified to include more frequent examination and replacement of internal components that were prone to wear. Technicians were also trained to identify and address signs of particle impaction inside the oxygen system during maintenance activities. More detailed analysis of the effects of piping and valve geometry on the oxygen flow dynamics was also implemented as part of the Management of Change program.

The following specific lessons were implemented to prevent future recurrence:

- Replace the SS316 gate valve with an INCONEL® gate valve.
- Replace all gaskets in oxygen service with gaskets that have a MONEL® inner ring for oxygen contact.
- Careful management, identification and control of materials of construction in critical services such as oxygen are vital. It is important to manage the materials originally selected for use, verify installation of those materials, and to manage later changes during subsequent maintenance activities. The special materials and their usage in oxygen systems must be documented and shared during training and certification programs for maintenance staff.
- Management of change, understanding what constitutes a change and the use of this system to prevent unwanted material, geometry, and operating changes is critical. Whenever a change is done it needs to be evaluated how this change will impact the system according to the MOC process. For example, in this case even small change such as allowing the purchase of a stainless inner ring gasket had significant impacts long after the change occurred.
- Careful documentation is required when working on, inspecting or repairing critical oxygen systems. The equipment owner must review and approve what was done (repair/ change/ no finding). This includes clear documentation of what was found, what repairs were made, what specifications were used, and verification of materials of construction. This will help ensure incorrect materials of construction/ specifications are not installed.

We would like to acknowledge the dedicated staff of EQUATE Petrochemical Co. and Exponent for conducting the detailed root cause investigation and sharing the learnings with the industrial fraternity for improvement of process safety and operations.

6 References

[1] CGA G-4.4, "Oxygen Pipeline and Piping Systems," Compressed Gas Association, 2012.

[2] ASTM G88-05, "Standard Guide for Designing Systems for Oxygen Service," ASTM International, November 2005.

[3] ASTM G94-05, "Standard Guide for Evaluating Metals for Oxygen Service," ASTM International, October 2005.

[4] Cox BL, Dee SJ, Hart RJ, Morrison DR. Development of a steel component combustion modelfor fires involving pure oxygen systems. Process Safety Progress 2014; 33(3):299–304.

[5] Sandmeyer Steel Company Specification for 316, 316L, 317L Stainless Steel. http://www.sandmeyersteel.com/images/316-316L-317L-Spec-Sheet.pdf. Note: Specification for 317L. Averages were calculated where ranges were reported in the specification.

[6] Benz, F. J., Shaw, R. C., and Homa, J. M., "Burn Propagation Rates of Metals and Alloys in Gaseous Oxygen," *Flammability and Sensitivity of Materials in Oxygen-Enriched Atmospheres: Second Volume. ASTM STP 910.* M. A. Bennings, ED., American Society of Testing and Materials, Philadelphia, 1086, pp. 135-152. Higher pressure consumption rates are listed in ASTM G94-05, Table X1.1, p 15.

[7] Slockers, M. J., and Robles-Culbreth, R, "Ignition of Metals at High Temperatures in Oxygen," Journal of ASTM International, Volume 3, NO. 9, 2006.

[8] Sato, J. and Hirano, T., "Fire Spread Limits Along Metal Pieces in Oxygen," Flammability and Sensitivity of Materials in Oxygen-Enriched Atmospheres: Third Volume, ASTM STP 986, D. W. SchroU, Ed., American Society for Testing and Materials, Philadelphia, 1988, pp. 158-171.

[9] Beeson H. D., Smith, S.R., Steward, W.F., Safe Use of Oxygen and Oxygen Systems: Handbook for Design, Operation, and Maintenance. ASTM International, 2nd edition, Baltimore, MD 2007.

[10] Crane Brochure for Duo-Chek II High Performance Check Valves, 2002

Lessons Learned from an Incident at a Cryogenic Gas Processing Facility

Adrian Pierorazio
BakerRisk
304-5515 North Service Road,
Burlington, Ontario, Canada
AdrianP@BakerRisk.com

Prepared for Presentation at
American Institute of Chemical Engineers
2015 Spring Meeting
11th Global Congress on Process Safety
Austin, Texas
April 27-29, 2015

UNPUBLISHED

AIChE shall not be responsible for statements or opinions contained
in papers or printed in its publications

Lessons Learned from an Incident at a Cryogenic Gas Processing Facility

Adrian Pierorazio
BakerRisk
304-5515 North Service Road,
Burlington, Ontario, Canada
AdrianP@BakerRisk.com

Keywords: Incident investigation, cryogenic, hot work, welding, coke oven gas

Abstract

In September 2009, a fatal incident occurred in a cryogenic gas processing plant. The investigation of the incident indicated a number of potential issues that may have contributed to or caused the event. These issues include hot work procedures, electrostatic discharge, electrical conduit sealing, convective "breathing" due to multiple vents, equipment age, maintenance, and worker training. This paper provides an overview of the process and facility, a timeline of events, a summary of the investigative process, and a discussion of the lessons learned from this event.

1 Introduction

Incidents that occur in industry often provide key learnings for other operators. While many of the specifics may change from site to site, it is valuable to consider how the lessons learned at one facility may apply to others, even if the processes are quite different. The incident being considered here is part of a coke oven system but illustrates a number of issues that have wider application. In this case, the process is a sublimation process to extract the fuel gases from the toxic and inert components of coke oven gas. The cryogenic and sublimation aspects are interesting, but should not detract from the key lessons learned. This paper starts with an overview of the process and then presents the facts about the activities underway at the time of the incident before moving on to a narrative of the event and an analysis of the technical causes of the incident. Some of the interpretations of the events may differ between the experts who investigated this incident, but the purpose here is to promulgate the lessons learned.

2 Process Description

The system was designed, built, and commissioned in the 1960s to convert coke oven gas to fuel gas for other parts of the complex. The system uses a cryogenic process to convert coke oven gas into three streams:

- Downriver gas
- Underfire gas
- Sub gas

The downriver and underfire gases are used as fuels in other areas of the complex. One part of the facility is downriver, giving rise to the "downriver" name. The underfire gas is a lower pressure fuel and is used as part of the fuel for the coke ovens. The sub gas contains the components that have been taken from the coke oven gas to produce the two product fuel streams. The sub gas is processed further in another area to remove more fuel gas and further concentrate the heavier components out of the stream.

The processing of the gas is performed using cryogenic separation. The coke oven gas is flowed over a bed of cold stones in a series of vessels called "regenerators." The stones are cold enough (as low as -258°F) to cause the gases to liquefy or freeze. The inlet flow is then turned off and the pressure is dropped. The control of pressure and the use of sweep gas cause the components to sublimate, allowing the capture of the three product streams by switching the outlet at the appropriate time.

The regenerators are clustered into groups of four, called "quads." Each quad has its own insulated "cold box" enclosing the bottom of the regenerators and the valves and piping there. These quad cold boxes are constructed with an insulated double wall with a void for the pipes and valves. The outer wall is made of steel plate that is seal-welded to hold it in place. The inner skin is made of thin metal sheets that are clipped in place. Mineral wool insulation is installed between these skins and along the floor and roof to help maintain the cryogenic process temperatures. Each quad is numbered, with a total of seven quads being installed in the main regenerator area.

The piping and valves that connect the quads together are housed in cold boxes at the end(s) of each quad. There are three such cold boxes and these are designated using letters A through C. These lettered cold boxes differ from the quad cold boxes in that they are packed with insulation rather than having an insulated double wall around a hollow space.

The lettered cold boxes (A through C) are connected through the quads and also through a "cross-tie" box that permits gas to bypass the quads and flow between the A and B Cold Boxes. In this report, the cross-tie box is referred to using the letters of the cold boxes at the ends of the cross-tie (i.e., "AB").

The cold boxes are designed to keep the space under the Quad 3 regenerators cold. This is done by attempting to seal the cold box to prevent air from being introduced into the cold box and to prevent gas from inside the cold box from leaking out.

The numbered cold boxes each have a nitrogen supply and backpressure regulator system. The quad boxes each have a single nitrogen injection point consisting of a 2-inch nominal diameter pipe penetrating the double-wall of the box. There are two 10-inch nominal diameter vents on the quad cold boxes, each equipped with a flapper-style relief valve to relieve overpressure and to allow air to be drawn in to relieve vacuum to prevent the cold box from collapsing in the event of a substantial vacuum forming in the cold box. One of the vent lines on each quad box is also fitted with a backpressure regulator with a separate open-ended vent line that is designed to maintain a positive pressure in the cold box. By contrast, the lettered cold boxes receive nitrogen from a sparger pipe and have a single vent pipe that runs up to roof level.

3 Summary of Facts

The following sections provide a summary of the key facts obtained during the investigation.

3.1 Activities at the Time of the Incident

Extensive maintenance activities had been underway since 2005 and were ongoing at the time of the incident. Quads 1, 2, and 3 and Cold Box A were in operation and work was progressing on the other equipment. Cold Box B was in the process of being rehabilitated, with the east half (i.e., the half connected to the south end of Quad 3) having been completed and returned to service weeks before the incident. The west half of the Cold Box B was being stripped of insulation at the time of the incident, in preparation for repair, re-insulating, and return to service. The insulation was being removed by two vacuum trucks that were parked outside of the main regenerator area with hoses running into the cold box.

Work was being performed in the main regenerator area on the day of the incident. Rock wool insulation was being vacuumed from the west side of Cold Box B. Grinding and welding repairs were being made at the south end of the AB Cross-tie Box adjacent to the north face of Cold Box B, on the south face of Cold Box B, and on top of the Quad 5 Cold Box. Hot work was being performed in the control room 2 building around the time of the incident. This included the insulating work in Cold Box B as well as welding and grinding work that was being performed to repair the wall plates of the cold boxes (Cross-Tie AB, Cold Box B and Quad 5 being worked on the day of the incident). There were numerous portable gas meters in use in the west half of the B Cold Box, under the AB Cross-tie Box, and on top of Quad 5 at the time of the incident.

Two work crews were in the area at the south end of the AB Cross-tie Box (i.e., on the north wall of the B Cold Box near the junction of B Cold Box and Quad 3). A welding crew was welding a plate at the intersection of Cold Box B and the AB Cross-tie Box, within a few feet of Quad 3's cold box. Another crew was working at ground level, preparing to mix concrete to refurbish a footing for Cold Box B at the time of the incident. It was reported that the welding crew on the north side of Cold Box B (approximately 50 feet from their return lead) had just struck an arc on the AB Cross-tie.

Quads 1, 2, and 3 were in operation, along with Cold box A and parts of the Cross-tie AB box and Cold Box B. At the time of the incident, one of the control valves in Quad 3 was switching.

3.2 Equipment Condition Prior to the Incident

The main regenerator equipment was being repaired to allow it to return to service and some cold boxes had been out of service for some time.

During the refurbishment project, the flapper valves at the top of the vent stacks were being inspected. Some of the flapper valves were repaired and/or replaced. Following the incident, some of the flapper valves were found to have eroded and the seals were missing.

The welding crew on the north side of Cold Box B was using one of the group of four individual Witherup welding machines that were stacked and/or bolted together near the road on the west side of the Control Room 2 area and shared a piece of angle welded to the opposite side of Cold Box B for their return lead location. The crew had welded a support to Cold Box B and had used a bottle jack to lift a member on the bottom of Cross-tie Box AB in order to restore a support clip for it back to the Cold Box B structure. The original connections had failed and new connections were being made as part of the refurbishment project.

During the investigation, one expert opined that parts of the system were beyond their lifespan. Whether this was true or not was subject to debate regarding the assumptions in the lifespan model. In the end, the company had implemented an inspection program to find and address cracks and any other flaws in the equipment and repair these when found. Many chemical processing facilities are of a similar vintage, or older, and rely on such inspection/repair programs to extend the life of their assets, so this is a common condition.

3.3 Safety Procedures

Standard operating procedures in the regenerator area required workers to use personal gas monitors to monitor for flammable gas, oxygen, and carbon monoxide as well as area monitors and periodic checks by operations with a hand-held monitor. A review of the logs and the results of testing of the personal monitors collected after the incident confirms that none of these gas detectors sensed alertable levels of flammable gas or carbon monoxide or an oxygen deficiency in the area prior to the incident.

There was a specific requirement for patching the live cold boxes to be performed as "cold work" with a gas sampling port and a nitrogen purge port to allow the atmosphere behind the plate to be sampled and then purged before hot work (i.e., welding) could commence under a separate hot work permit. The welding crew was performing hot work at the junction of the High Box and the Cross-tie Box, specifically on the Cross-tie Box, using hot work, with no separate permit, no purge, and no continuous monitoring of the sample port.

3.4 Damage Sustained

On September 3, 2009, an "explosion" was reported in Quad 3, resulting in one fatality and one worker being sent to the hospital for treatment. The fires that followed the event burned for a number of hours and were concentrated in the area around and under the Quad 3 regenerators.

Inspections after the event showed that the Quad 3 Cold Box floor had been deformed outward to the point that it impacted the concrete below, and the walls of the Quad 3 Cold Box had become detached and displaced/deformed outwards on both sides of the cold box to the point where they were no longer enclosing the space underneath.

A double wythe brick infill wall to the east of Quad 3 failed and fell into the axi compressor building, away from the Quad 3 Cold Box.

The damage was generally confined to the immediate vicinity of Quad 3. The only significant structural damage observed in Cold Boxes A and B were to the walls that separate them from Quad 3. This damage resulted in wall movement that moved the pipes running between these lettered boxes and Quad 3, resulting in some mechanical damage to the valves and piping in the lettered boxes. The walls were deflected away from the Quad 3 Cold Box.

The damage patterns are consistent with an "explosion" having occurred inside the Quad 3 Cold Box.

4 Interpretation of the Event

The investigation concluded that the incident that occurred on September 3, 2009 was a confined vapour cloud explosion (VCE). The limited extent of the damage indicates that the energy involved was similarly limited. The damage is consistent with a relatively small vapour cloud, and the intensity of the overpressure also indicates that the VCE was a deflagration rather than a detonation.

A vapour cloud explosion is a combustion event and, as such, requires three elements to be present in sufficient quantity:

- Fuel
- Oxidizer
- Ignition

Each of these elements must exist in balance with the others. The fuel must exist in a concentration between the lower and upper flammable limits, the oxidizer (oxygen from the air in this case), must exist at above the minimum oxygen concentration (MOC) required to support combustion, and the ignition source must be in the correct location and of sufficient strength to produce a self-sustaining combustion. Each of these elements is discussed in the sections below.

4.1 Fuel Source

The underfire, downriver, and sub gas lines run through the Quad 3 Cold Box. These lines run at cryogenic temperatures. When these lines are serviced and the quad is brought on-line, the pipes are cooled to below -250 °F. The stresses due to cryogenic processes can cause flanges and other connections to leak. When these leaks occur, they are generally small. Even small leaks in the relatively sealed space of the cold box can produce accumulations of gas and the concentration can increase over time. After the incident, the piping was purged with nitrogen and maintained at a slight positive pressure of nitrogen. As part of their investigation, handheld quad gas meters were used to detect locations with low oxygen readings. These low oxygen areas were identified as areas where nitrogen was being dispersed from the piping. Several such areas were identified and it was concluded that, as expected, there were small leaks in the piping that could allow gas to escape. While flow rates were not measured, it was concluded that these were a credible source of fuel for the event.

4.2 Oxygen

As designed (and as in operation on the day of the incident), the nitrogen is introduced at a single point from the cross-tie alley (i.e., on the west of the odd numbered quads and on the east of the even numbered quads) at the south end of each quad cold box. With the erosion of the vent flapper valves, the nitrogen supply was unable to provide a significant positive pressure in the cold box. The leak in the flapper valves could result in the venting of the incoming nitrogen through the nearest vent (located just south of the C regenerator) and the nitrogen would not sweep through the cold box to purge the air.

With hydrogen-rich gas (i.e., underfire or downriver gas) leaking from the process piping, the gas mixture would have a lower density (i.e., be "lighter") than normal air. The result of this was that the gas would rise out of the vents creating a convection loop in which the light gas rises out of the cold box and denser oxygen-laden air entered the cold box to replace it and providing the oxygen for combustion to take place. Calculations show that this mechanism is a credible source of oxygen to support combustion in the Quad 3 cold box, if sufficient fuel is present along with a competent ignition source.

4.3 Ignition Source

With fuel and oxygen present in the cold box the only remaining required element is ignition. A number of potential ignition sources were considered including the following:

- Electrical equipment or wiring inside the quad 3 cold box
- Electrical equipment or wiring outside the quad 3 cold box
- Static electric sparks from the vacuum truck operations
- Static electric sparks from within the quad 3 cold box
- Ignition due to hot surfaces in the quad 3 cold box
- Ignition due to deformation/failure of metal components in the system
- Ignition due to stray currents induced by process equipment in the control room 2 area
- Ignition due to stray currents induced by the welding in operation at the time of the incident

The following sections provide a summary of how each potential ignition source was evaluated.

4.3.1 <u>Electrical equipment or wiring inside the quad 3 cold box</u>

One of the first hypotheses considered was that there could be a short or malfunction of electrical equipment or wiring in the quad 3 cold box. The only wiring that was found to run inside the cold box itself was thermocouple wiring to measure the process temperatures. The voltage expected from a thermocouple under normal operating conditions in the cold box would be less than 10 mV and would be a static voltage; this is well below what would be required to ignite a flammable gas cloud.

4.3.2 Electrical equipment or wiring outside the quad 3 cold box

As a corollary to the scenario above, there is an amount of electrical equipment outside of the quad 3 cold box that does operate at voltages of 120 volts or more. Since the wiring and equipment are in an electrically classified area, the wiring is routed inside of conduit and junction boxes intended to prevent potentially flammable gas mixtures from being present where significant voltages and/or currents were present. To investigate this, the junction boxes near the cold box were opened and inspected. It was observed that the seals between the quad 3 cold box were of an elastomeric caulking-like material, rather than the cast solid material that would normally be used in classified conduit. It was further revealed that some of the locations where cast or poured seals would be expected (i.e., where conduit is connected to a junction box) were missing seals or had gaps in the seals.

The conduit was analyzed to determine the minimum gas concentration that would produce a flame that could propagate through the conduit that connected to the thermocouple junction box. The pressure produced by the deflagration of this minimum gas concentration was then predicted and compared to the pressure capacity of the conduit and junction boxes. The analysis shows that any flammable gas concentration that can propagate through the conduit would cause failure of the junction boxes and/or conduit. Since these components did not fail due to overpressure and since there were no other signs of combustion inside the conduit and junction boxes and since there were signs that there were functional seals between the cold box and the electrical system outside the cold box, it was concluded that this was not a credible ignition source.

4.3.3 Static electric sparks from the vacuum truck operations

Work crews were working on west side (quad 4 side) of Cold Box B at the time of the incident. As part of this work, they were using a vacuum truck to suction the old insulation away to clear the area for rehabilitation. The vacuum hose ran east under quads 5 and 6, along the east side of quad 5, then under cold box B to the work location. The hose was not conductive hose and so there is a potential for static buildup and discharge from the hose. Static discharge is a high voltage/low current phenomenon which occurs locally where a static source with high voltage potential creates an arc to ground. In this case, the potential ignition locations would be inside the pipe (charged particle to hose) or along the path of the hose. Since there was no report of fire or ignition at the work site or vacuum truck, it can be surmised that there was no deflagration propagating in the hose. Since the hose was generally lying on the ground or near quad 4/cold box b, any local discharge from hose to structure would be unable to ignite a flammable gas mixture in quad 3. As a result, ignition from static discharge due to vacuum operations was ruled out.

4.3.4 Static electric sparks from within the quad 3 cold box

Potential sources of static electric discharge inside the quad 3 cold box were sought. These included moving parts that were well insulated from any metallic structure, and any particles or gases that can flow in such as was as to develop an electrical charge. None of these potential sources were found and so this source of ignition was ruled out.

4.3.5 Ignition due to deformation/failure of metal components in the system

It has been postulated that the ignition may have occurred as a result of deformation or failure of a component. The only mechanism for such an event to occur inside the quad 3 cold box is a gross failure of the process equipment. When the process equipment was inspected, the damage was consistent with mechanical failure due to the rupture of the cold box under positive pressure. The only evidence of any component that may have failed prior to combustion is the bellows at the bottom of regenerator 3a. However, this failure appears to be inwards and not outwards, indicating that it is more likely due to impact during the failure of the cold box, than due to rupture prior to the incident. In any event, the deformations in this area are relatively small and not credible as being sufficient to heat the metal close to the autoignition temperature of the fuel.

4.3.6 Ignition due to stray currents induced by process equipment in the control room 2 area

With ignition due to electrical, chemical, and mechanical systems inside the quad 3 cold box ruled out, attention turned to how operations outside the cold box could produce ignition inside the cold box. The most logical of these would be for ignition to occur as a result of electrical current flowing through the cold box from external sources.

Elementary electrical theory holds that the current between parallel paths will divide as the reciprocal of the resistances of the various paths. Stated more simply: current will flow through all available paths from a source to a sink (often ground) with low resistance paths getting a higher portion of the flow than high resistance paths. The current that flows through alternative paths such as along structural elements and process piping is referred to as "stray currents".

The cold boxes run at cryogenic temperatures, making the resistance of paths though the steel of the structure and piping very low. This makes paths through the cold boxes the preferred path for currents flowing through the structure. A search was undertaken to find electrical equipment that either did not have a viable local ground connection (to cause currents to need to go through the structure), or was capable of providing high enough currents to the system that even a small fraction of the current would be significant enough to cause ignition.

In the process area, there are a number of compressors which use high voltages and currents. These are all a fair distance from the cold boxes. The compressors were tested found not to be shorted and had reasonable connections to ground to account for stray currents produced locally. These were then ruled out.

The lighting and electrical switches in the regenerator area are much closer to the cold boxes (implying lower resistance to the cold boxes), but none were found to have shorted to the structure or otherwise lack a proper electrical return path. As a result, these were ruled out as likely ignition sources.

4.3.7 Ignition due to stray currents induced by the welding in operation at the time of the incident

In evaluating electrical sources outside of the cold boxes themselves, the welding activity being conducted at the time of the incident was also considered as a potential source of appreciable stray currents. In welding operations, currents as high as 100 amperes, or more, flow through the components being welded together. The current then flows through the metal to the return lead to complete the circuit. At the time of the incident, a crew was welding on the north side of the B cold box at the west side of the intersection with the crosstie box (i.e., approximately two stories up, very close to quad 3). The return lead (colloquially, if inaccurately, referred to as the "ground clip"), was located near the middle of the south wall of the B cold box near ground level. This means that the full welding current was being passed through a large part of the cold box structure as a result of the way the return clip was located. The combination of high current source, low conductivity path through the quad 3 cold box, and the long path to the return lead, indicates that there is a high potential for substantial currents to be flowing throughout the quad 3, crosstie box, and cold box b areas.

Witnesses reported that the ignition occurred at virtually the same time that a welding arc was struck on the AB Cross-tie Box. Electrical testing post-incident revealed that welding being performed in the plant at that time was producing stray electrical currents that were strong enough to interfere with electrical testing being performed. This indicates that the welding practices in place at that time were transmitting energy to areas away from the work location.

Locating the ground return lead on the far side of the B Cold Box approximately 50 feet from the welding location, increased the energy sent to other areas of the system. Since it was shown that stray current paths existed in the cold boxes, that the welding was being performed in a way that would encourage the creation of significant stray currents along parallel paths in the cold boxes, that there was a correlation in time between the start of the welding near Quad 3 and the incident, and since all other ignition mechanisms considered were ruled out, it is concluded that the welding is the most probable cause of ignition.

5 Summary and Lessons Learned

While the processes described in this paper are not typical of wider chemical industry, there are some lessons learned that can be applicable to other processes. As with any incident, the root cause is often a failure of the PSM system and this incident was no exception.

5.1 Training and Contractors

Despite documentation and training to the contrary, the welders (contractors) did not seem to be aware of the hazard of welding on an enclosure that could contain flammable gas. Their personal gas monitors would not be able to detect a flammable gas mixture inside the cold box from their location outside of the box and the welding current was able to penetrate the skin of the cold box easily. The safety personnel were responsible for monitoring the gases in the area and did not note this discrepancy in the sample location.

5.2 *Hot Work*

This incident appears to be a failure of the hot work permitting system in that the sampling and welding practices did not prevent ignition. Safe procedures were in place, but not followed.

5.3 *Mechanical Integrity*

As with many aging facilities, erosion, wear, and fatigue were present in this facility. While active testing and repair programs corrected many issues, the erosion of the flapper valves was critical to the safety of the operation.

Post-incident inspections and testing demonstrated that the grounding was disconnected in some places. While this was not likely a cause of the incident, it is valuable to note that this electrical issue may occur in other process areas and may go undetected unless specifically sought out. While it is not specifically mechanical, the observation that the seals in the classified electrical conduit and junction boxes were incomplete, missing, or sealed in a manner that is not consistent with good practices (i.e.,. using an elastomer rather than a castable seal) is a valuable lesson as well. The conduit and junction boxes had been installed many years prior and, it seems, had not been inspected to determine if all seals were in place. This may have been found if the seals were inspected.

5.4 *Human Factors*

The flapper valves were located at the top of exhaust stacks, well above the roof of the process building. To access these valves required a crane and basket and so the equipment was not regularly inspected or maintained. As such, personnel were unaware of the erosion of the valves, which led to the presence of air in the cold box.

5.5 *Hazard Identification*

The design, as received from the technology licensor included two large vent stacks. The presence of two stacks, rather than one, allowed a convection loop to form. The designer and operator failed to identify this hazard and modify the design appropriately.

The nitrogen purge system, coming in at a point rather than being distributed by a sparger, allowed pockets of air to remain. This hazard was not identified prior to the incident.

The welders were not aware of the hazards of welding on live cryogenic equipment which may contain flammable gases. There was a belief that their flammable gas monitors would have alarmed if flammable gas was present in the area.

The welders were unaware of the path taken by their welding current. While there was general understanding that the current would follow the path of least resistance, this was interpreted as the shortest distance through the steel without considering parallel paths or the decreased resistance of steel at low temperatures.

5.6 Incident Investigation

The cold box had a much smaller event four years prior to this incident. Investigators at that time concluded that it was caused by lightning strike, and so recommended lightning rods be installed to prevent a second strike. While there is debate about the veracity of this cause of ignition, the earlier investigation could have identified the fuel and air sources and recommended ways to control them. Had the original investigation found the eroded flapper valves and identified convection as a means of introducing air into the cold box, the valves could have been replaced or the vent piping re-configured to prevent this from occurring and so could have prevented the event.

11TH GLOBAL CONGRESS ON PROCESS SAFETY

1996 Amines Plant Explosion

Jerry Forest
Celanese
1601 W. LBJ Freeway, Dallas, TX 75234
Jerry.Forest@Celanese.com

Prepared for Presentation at
American Institute of Chemical Engineers
2015 Spring Meeting
11th Global Congress on Process Safety
Austin, Texas
April 27-29, 2015

UNPUBLISHED

AIChE shall not be responsible for statements or opinions contained
in papers or printed in its publications

1996 Amines Plant Explosion

Jerry Forest
Celanese
1601 W. LBJ Freeway, Dallas, TX 75234
Jerry.Forest@Celanese.com

Keywords: Incident, PSE, explosion, Lessons Learned

Abstract

Process safety management systems should be in place to ensure repeatability in operations. A good system consists of higher level procedures that describe what to do. At the plant level, the management system describes who does the activity and how it is done. Following procedures leads to predictability in operations. In 1996, Celanese had established a rigorous management system that consisted of 21 elements of process safety. However, failure to rigorously apply several elements led to an explosion in our Cangrejera facility resulting in 2 fatalities. This paper discusses the management system failures of conduct of operations elements, safe operating limits, standard operating procedures and training, and critical safety equipment management and how each failed due to a common weakness of management systems.

1. Introduction

In 1987, Celanese had an event that fundamentally changed the way we manage process safety. An explosion at our Pampa facility resulted in three fatalities and 39 injuries. Not only were the lives of those directly involved severely impacted, the company made a substantive shift in how we define and practice process safety.

Corrective action from this incident led to the development of a comprehensive 21 element process safety model; similar to the one that the CCPS later and independently developed with 20 elements. A strong driver for Celanese in the development and implementation was the recognition that leaders drive change and create the culture.

Despite having implemented such an aggressive approach to process safety, an explosion occurred at our Cangrejera Mexico facility nine years later that resulted in two fatalities. While the management systems were in place, they were not rigorously followed. As with many process safety incidents there were several opportunities for existing systems and controls to prevent the explosion. However, each of them failed. There was a mode failure of our process safety culture, as it allowed partial bypasses in the process safety management system.

2. The Incident

Among other chemicals, Celanese produces amines at our facility in Cangrejera Mexico. This unit uses ammonia and methanol as raw materials. The reaction that produces the amines is endothermic and relies on an electrical heat exchanger to provide the necessary heat required to initiate and maintain the reaction.

On June 7, 1996, the facility experienced a brief electrical power outage. The temporary interruption caused a loss of cooling water and a loss of power to the electrical heat exchanger that provided the necessary heat for reaction for the amine unit. During the course of the outage a pressure relief device on a nearby distillation column lifted and failed to reseat. The amines unit therefore had to be shut down rather than start-up from warm, in order to re-seat the pressure relief device.

Standard operating procedures were in place to cut power to the reaction section electrical heat exchanger during a shutdown. However, due to a communication error between the outside operator and the board operator, this did not occur. During the course of the outage, power and therefore heat, continued to the unit exceeding the maximum allowable working temperature of the heat exchanger resulting in a rupture in the shell wall.

Reactor contents consisting of ammonia and methanol were released. The resulting vapor cloud found an ignition source from a nearby welding operation. The ensuing explosion fatally injured the welder and fire watch. Ignition of the reactor contents propagated back to the heat exchanger source resulting in a secondary jet fire.

3.0 Primary Causes

As is typical with many process safety incidents that have severe consequences, there were several layers of protection that failed allowing this incident to occur. They include a failure of conduct of operation elements; failure to recognize a safe operating limit challenge; failure to follow standard operating procedures and training on those procedures; and failure to manage critical safety equipment.

3.1 *Conduct of Operations*

Conduct of operations (COO) concerns how work is performed in a manufacturing unit with the overall objective of producing consistent, repeated and desired results. COO reduces the likelihood of human error. The term operating discipline is often used to describe how to bring a consistent approach to various operations to minimize errors. [1]

Proper communication between and among operators, crafts, and supervisors is inherent to a good deal of how work is performed in a unit. Communications should be crafted in a standardized format that includes positive verification and acknowledgement. [2] The Celanese COO management system includes requirements for closed loop communications and defines the standard format that leads to proper communication for operators.

At the time of the unit shutdown described above, a communication error occurred between the outside and inside operators regarding shutting off power to the electric heat exchanger. The COO closed loop communication requirement in our management system failed. Had the communication taken place as intended, this incident would not have occurred.

3.2 ***Safe Operating Limits***
A second level of protection from this type of incident is found in defining true safe operating limits (SOL) that allow for pre-determined action by the operator to prevent an incident. Associated with the SOL are alarms and operator training on the consequence if exceeded and the actions that should be taken to prevent the consequence from occurring.

Since power was on the electrical heat exchanger during the course of the outage, the temperature in the process exceeded the maximum allowable working temperature of the heat exchanger. Failure to set a proper SOL with an associated alarm, and failure to train operators on the consequences of deviation and predetermined actions voided this layer of protection from preventing this incident.

3.3 ***Standard Operating Procedures***
The third level of protection is standard operating procedures (SOPs). SOPs should be followed as written each time an activity is performed. At the time of the Cangrejera incident, the unit shutdown SOP did define that power should be turned off to the electrical heat exchanger to prevent the consequence of vessel rupture by exceeding the maximum allowable working temperature. However, this step in shutdown did not occur.

3.4 ***Management of Critical Safety Equipment***
Communications, SOLs, and SOP layers of protection are human interactions that prevent process safety incidents. Engineered controls are also in place to prevent them.

Specifically, a typical design for this type of heat exchanger includes both a low flow and high temperature interlock. Either interlock would remove power from the electrical heat exchanger in this shutdown situation and prevented the incident. However, in this instance the temperature interlock was removed due to poor reliability without a management of change. At the time of the shutdown, the low flow interlock was bypassed – again, without a management of change.

3.5 ***Management System Failure***
How can so many levels of protection fail at the same time? As is seen with so many major process safety incidents, the answer lies in failure of the overall process safety management system. In this case, we had the process safety systems in place: COO, SOLs, SOPs, and properly designed interlocks. The common failure mode was lack of a process safety culture that insured two-way communication was always followed, safe operating limits were set and challenges investigated, safe operating procedures were always followed, and engineered controls were operable.

4.0 Response to the Incident

As a result of this incident Celanese took corrective action to update and reinforce our COO program. In particular, COO now includes a process safety officer (PSO). Each site identifies process safety critical activities that need special attention. Examples include critical startups and shutdowns where we might be more vulnerable to a process safety incident. A PSO is assigned and has the following duties:

- Present during safety critical startup, shutdowns, and other operating modes identified in the PSO reference manual
- An in-depth knowledge of the hazards of the process
- A working knowledge of unit operations
- An understanding of all applicable SOPs
- **Have no other duties that distract from their PSO responsibilities**

As applied to the amines plant explosion, the PSO would both observe two-way communication to ensure proper transmission and understanding and verify that the communicated activities are properly executed. The PSO would also verify the SOP was followed, monitor the SOLs, and understand the effect of any safety critical equipment bypasses and alternate mitigations.

Another key action taken is that all SOPs are risk ranked using our risk scoring tool. The ranking identifies how frequently tasks are performed in an SOP and the possible process safety consequences that could result if the tasks are done incorrectly or not followed. SOPs are then classified as critical, reference, and guideline.

Critical and reference SOPs must be followed as written. Critical SOPs are written in a checklist format, must be present and in hand during execution and each step is signed off by the operator. When complete, a responsible second party such as PSO or front line supervisor independently validates that all activities are complete.

Finally, MOCs are required for critical equipment bypassing, and these bypasses must be posted on a bypass board in our control rooms. In-service by-pass is not allowed in Celanese unless there is another variable that can be monitored with the same intent of the one being bypassed. More recently, Celanese has recognized that knowledge of past incidents has a definite life. If we are to prevent the mistakes of the past, we must somehow ingrain both knowledge of the incident and the corrective actions that were put in place. Thus, our process safety lessons learned program has been put in place to reinforce our process safety culture.

5.0 About Celanese Process Safety Lessons Learned Program

The fundamental premise of this program is to go beyond incident communication and deeply embed and institutionalize knowledge of incidents and the mitigations that resulted from them in the organization. This is done by a constant reinforcement of the message or to state another way, by "retelling our story". This concept of defining and strengthening culture comes from two sources. First, Trevor Kletz is often quoted that "organizations don't have memory, people do." [3] That combined with reference materials from Loyola University on how group cultures are formed and strengthened indicates that in order to define and build a culture, the story must be defined, presented and repeated. This therefore is the basis of our Lessons Learned program.

In order to achieve maximum effectiveness, our program is divided into five parts.

1. Global Lessons Learned – The major incidents in Celanese that have defined how and why we do process safety

2. PS Rules (Process Safety Rules) – Technology specific process safety rules that define the critical few incidents that could occur and what mitigations are in place to prevent
3. Event Sharing – A comprehensive catalog of all Tier 1 and Tier 2 process safety incidents that have occurred at all Celanese sites since 2010.
4. Industry Learnings – A catalog of resources with learnings from industry incidents that are applicable to Celanese
5. PS BITS (Process Safety Basic Information to Share) – short informal videos submitted by employees that encourage employee participation in the learning process.

5.1 *Global Lessons Learned*

The intention of the Global Lessons Learned is to define for each manufacturing employee why we do process safety by telling the story of the major incidents that have occurred at our sites from our past. We do this by producing high quality videos (similar to CSB) that are shown to each manufacturing employee each year. Our intention is to limit this portion of the program to a few videos so that we do not reduce the impact of each. To date, we have produced:

1. Pampa Plant Explosion, 1987 – The result of this incident was the development of our comprehensive 21 element process safety program.
2. Cangrejera Amines Plant Explosion, 1996 – Reinforces the need for strong process safety management systems to prevent incidents
3. Hot Work, Shelby 1999, and Ocotlan 2014 – Reinforces fundamentals of life critical hot work procedures and hazard recognition
4. Organic Peroxide Explosion, Edmonton – Reactive hazards (in production)

Each video has subtitles in the languages for the various countries in which we operate. Some are overdubbed in the native language depending on those countries wishes to do so. The videos use animation, historic video clips, pictures, and interviews to convey the story of the incident, the causes, the key learnings and follow-up actions from each.

5.2 *PS Rules (Process Safety Rules)*

Celanese has several groups of similar technologies that in general, have similar risks and mitigations of risks. The PS Rules Lessons Learned is an attempt to target operators with the key hazards of the process by developing a short, memorable "rule." In order to be effective and follow the premise of Lessons Learned, the PS Rules have background training material that describes the hazard and the engineered and administrative controls in place to prevent. To ensure that operator can commit the PS Rules to memory:

- There are no more than 5 rules per technology
- They are written in such a way so that they are memorable using "Always do x", or "Never do y".

An example rule for our ethylene vinyl acetate technology is "Never mix vinyl acetate with initiators". The training materials then include the possible hazard created by this mixing and what controls are in place to prevent an incident. The PS Rules become part of refresher training for operators.

5.3 *Event Sharing*

This portion of the Lessons Learned program follows industry's traditional incident sharing program. For each Tier 1, Tier 2, and serious near miss, a single Power Point incident flash is produced and distributed to each site. The purpose of the flash is to raise awareness that an incident has occurred and provide an opportunity for each site to determine if action should be taken at their site. The corporate process safety department determines if a global teleconference call is needed to broadly communicate the learnings of the incident.

What might separate our event sharing program from our peers is the way we catalog the incident flashes on our Lessons Learned portal. They are listed by technology and site and have several uses. A primary use of the portal is to prepare for PHA by identifying and analyzing incidents of interest to the PHA team. We also encourage use of the event sharing in our New Leader Process Safety Certification Program which requires a review of historic incidents for certification. Other programs where this portal is regularly accessed include the pre-work requirements for our front line supervisor (FLS) certification program, MOC hazard reviews, supplemental operator training, and toolbox meeting topics for FLSs.

5.4 *Industry Learnings*

As stated above, our PHA process requires review of incidents that are related to the processes that we operate, both internal and external to the company. Furthermore, each site is encouraged to stay current with industry incidents in order to determine applicability for our sites and lessons that can be applied to prevent similar incidents.

This portion of Lessons Learned is a convenient repository of reference materials, pictures, videos, and investigation reports for external incidents that might have applicability for our manufacturing sites. Examples include Bhopal and Texas City.

In addition to those incidents that have helped define process safety, we publish a quarterly event sharing newsletter. The newsletter highlights similar incidents that have recently occurred in industry and compares and references them to similar incidents that have occurred at Celanese. A major source of industry event information is the AFPM Advancing Process Safety Event Sharing Database. Each manufacturing site is expected to consider applicability of learnings at their site.

5.5 *PS BITS*

PS BITS, or process safety basic information to share, is the newest addition to our overall Lessons Learned program. PS BITS engages employees by having them produce short video segments on process safety topics. These clips are less than 5 minutes, can be recorded with a smart phone and edited with common tools such as Microsoft Movie Maker.

The key to this portion of Lessons Learned is that employees are telling the story in their own words for other employees.

6.0 Conclusion

Good process safety management systems are in place to understand and manage the risks of the process. By understanding the risks, we define the layers of protection to prevent consequences.

Managing the risks includes assurance that both the human controls, and engineered controls are in place and working.

The common element to understanding and managing risks is the commitment to process safety. This commitment takes the form of a leadership driven culture of process safety and ensuring that process safety management systems are in place and are not compromised. Feedback to the commitment comes in the form of learning from process safety incidents and acting to improve.

The explosion at the Celanese amines plant illustrates how these four pillars of process safety introduced by CCPS must be in place to prevent incidents. [4]

Reference

[1] J. Forest, Management Discipline, Process Saf Prog XX (2014)

[2] Center for Chemical Process Safety, Conduct of Operations and Operational Discipline, American Institute of Chemical Engineers, Newark, NJ, 2011.

[3] T. Kletz, Learning from Accidents in Industry, Butterworths, London, 1988

[4] Center for Chemical Process Safety, Guidelines for Risk Based Process Safety, American Institute of Chemical Engineers, Hoboken, NJ, 2007

U.S. Chemical Safety Board's Final Investigation Report on the Chevron Richmond Refinery Pipe Rupture and Fire

Mark Wingard
Chemical Safety Board
Denver, CO
mark.wingard@csb.gov

Steve Cutchen
Chemical Safety Board
Denver, CO
Steve.Cutchen@csb.gov

Lauren Grim, Dan Tillema, Amanda Johnson
Chemical Safety Board
Denver, CO

REPORT NO. 2012-03-I-CA
JANUARY 2015

Presentation at
American Institute of Chemical Engineers
2015 Spring Meeting
11th Global Congress on Process Safety
Austin, Texas
April 27-29, 2015

U.S. Chemical Safety Board's Final Investigation Report on the Chevron Richmond Refinery Pipe Rupture and Fire

Mark Wingard
Chemical Safety Board
Denver, CO
mark.wingard@csb.gov

Steve Cutchen
Chemical Safety Board
Denver, CO
Steve.Cutchen@csb.gov

Lauren Grim, Dan Tillema, Amanda Johnson
Chemical Safety Board
Denver, CO

Keywords: enter a few keywords here to aid searching

Abstract

On August 6, 2012, the Chevron U.S.A. Inc. Refinery in Richmond, California (Chevron Richmond Refinery) experienced a catastrophic pipe rupture in the #4 Crude Unit. The incident occurred from piping referred to as the "4-sidecut" stream, one of several process streams exiting the refinery's C-1100 Crude Unit Atmospheric Column. The pipe rupture occurred on a 52-inch long component of the 8-inch 4-sidecut line. The rupture of the 4-sidecut piping was the result of the 52-inch component being extremely thin due to unidentified, accelerated sulfidation corrosion in the low-silicon piping component.

The ruptured pipe released a flammable, 640°F light gas oil, which then partially vaporized into a large, opaque vapor cloud that engulfed 19 Chevron employees. Approximately two minutes after the release, the flammable portion of the vapor cloud ignited. Eighteen of the employees safely escaped from the cloud just before ignition; one employee, a Chevron refinery firefighter, was inside a fire engine that was caught within the fireball when the vapor cloud ignited. Because he was wearing full body fire-fighting protective equipment, he was able to make his way through the flames to safety. Six Chevron employees suffered minor injuries during the incident and subsequent emergency response efforts. The leak resulted in a large plume of vapor which traveled across the Richmond, California, area. The ignition and subsequent burning of the hydrocarbon process fluid created a large cloud of particulates which also swept across the Richmond area. In the weeks following the incident, approximately 15,000 people from the surrounding communities sought medical treatment at nearby medical facilities for ailments including breathing problems, chest pain, shortness of breath, sore throat, and headaches. Approximately 20 of these people were admitted to local hospitals as inpatients for treatment.

This report details the CSB's technical findings on sulfidation corrosion and inspection techniques, as well as delves into the Chevron organizational and Chevron emergency response factors that contributed to the incident.

http://www.csb.gov/assets/1/19/Chevron_Final_Investigation_Report_2015-01-28.pdf

11TH GLOBAL CONGRESS ON PROCESS SAFETY

Bhopal 30 Years Later

Ronald J. Willey, PE, PhD
Dept. of Chemical Engineering, Northeastern University
313 SN, 360 Huntington Ave., Boston, MA 02115
r.willey@neu.ecu

Daniel A. Crowl
Dept. of Chemical Engineering, Michigan Technological University
1400 Townsend Drive, 203 Chem Sci & Eng. Bld. Houghton, MI 49931
crowl@mtu.edu

All rights reserved by the authors

Prepared for Presentation at
American Institute of Chemical Engineers
2015 Spring Meeting
11th Global Congress on Process Safety
Austin, Texas
April 27-29, 2015

UNPUBLISHED

AIChE shall not be responsible for statements or opinions contained

in papers or printed in its publications

Bhopal 30 Years Later

Ronald J. Willey, PE, PhD
Dept. of Chemical Engineering, Northeastern University
313 SN, 360 Huntington Ave., Boston, MA 02115
r.willey@neu.ecu

Daniel A. Crowl, PhD
Dept. of Chemical Engineering, Michigan Technological University
1400 Townsend Drive, Houghton, MI 49931
crowl@mtu.edu

Keywords: Bhopal, Layers of Protection, Process Safety Management, Management of Change

Abstract

On the early morning of December 3rd 1984 a large release of methyl isocyanate (MIC) gas occurred in a pesticide production plant located in Bhopal, India. The plant was partially owned by Union Carbide (USA) and its affiliate Union Carbide India Limited (UCIL). Thousands of people died and approximately 200,000 were exposed to the MIC gas in that early morning.

This paper reviews the prevailing accident theories, the chronology of events that occurred after the accident, the industry and regulatory response after the accident, and the current state of the Bhopal site.

1. Introduction

For those us practicing chemical engineering in 1984, it is hard to believe that 30 years has passed. Worldwide, every practicing chemical engineer was influenced by the process safety incident that unfolded in Bhopal, India on December 2 and 3, 1984. Over 2,500 lives were lost due to inhalation and exposure to methyl isocyanate vapor released from a Union Carbide India Limited pesticide plant. 200,000 people suffered various levels of exposure that ranged from death, to blindness, to nausea [1]. As the horrifying photographs of the chaos reached the news media in the United States, the US public was drawn in. Our profession was challenged. Are we really the premier engineers that our salaries command? Or, are we another group that puts production above safety? The incident woke up our political system. Can it happen here? What regulations are in place to prevent a Bhopal from happening again and especially within the USA? The outcome: the establishment of the Center for Chemical Process Safety within the AIChE [2], the enactment of OSHA 29 CFR 1910.119 [3] Process Safety Management (PSM) of Highly Hazardous Chemicals, and EPA 40 CFR Part 68 [4] Sub Part G Risk Management Plan (RMP). Specific governmental requirements were enacted on chemical plants to demonstrate to the public that they can operate safely. The U.S. Chemical Safety Board [5] (CSB) an independent governmental agency was formed. Its mission is to investigate major chemical industry incidents, and share the outcomes such that future incidents are prevented. There is not one chemical engineer reading this manuscript who has not been affected by the incident in Bhopal and the resultant actions taken to prevent such again. This manuscript reviews the technical aspects of the accident and discusses what has transpired over 30 years. The intended audience is the safety and health practitioner who joined the profession after 1985.

2. The Accident

2.1.Exothermic Reactions in Closed Vessels

Explanation of this accident begins with chemistry, the storage of a very reactive chemical inside a 15,000 gallon, 40 foot long, 304 SS storage tank, and contamination by water. When chemical reactions form more stable molecules from more energetic molecules the result is the release of bond energy as heat. The energetic molecule involved: methyl isocyanate, or MIC as it has become abbreviated in many articles. Methyl isocyanate has two adjacent double bonds as shown in Figure 1 [6].

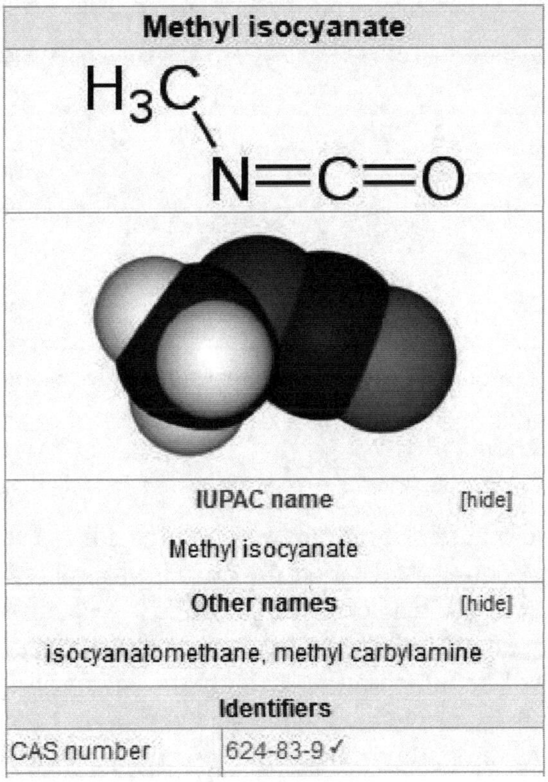

Figure 1. Molecular formula of MIC (ref. 6)

This concentration of electrons make it very reactive. It is a combination used in isocyanate chemistries to form higher molecular weight nitrogen containing organic chemicals that range from insecticides to insulation. It is an organic chemist's answer to a prayer. It is a chemical engineer's answer to a nightmare. The reason is the heat released when this chemical reacts with nearly any compound. If we had one chance to define a runaway reaction, this event is the example. When the organic chemist works on the bench scale, he is using test tubes, with high surface areas, creating plenty of cooling surface. When this reaches the application stage at 1,000's of gallons, heat transfer is much more challenging. Specifically, the combination of MIC with water releases up to 1,360 kJ/kg (585 BTU/lb) of MIC, or about 8,600 kJ/kg (3,700 BTU/lb) of water [7]. Recall that the heat of vaporization of water is approximately 2,200 kJ/kg (1,000 BTU/lb). There is enough energy available to raise the temperature of a mixture above the boiling point of the mixture and generate significant vapor pressure within a closed container. The reaction also forms carbon dioxide as a gas and this adds pressure to a closed system such as a storage tank. It was the entry of an estimated 1,000 to 2,000 lbs of water that initiated the runaway [8]. Abnormal amounts of chloroform, 1,500 to 3,000 lbs, was also present in Tank 610 creating a source for a chloride ion that attacked the stainless steel tank walls[8].

Several exothermic reactions happened in parallel and series. Three key reactions are shown in Figure 2 [9,10]. They are ordered based on the analysis of core samples collected after the accident and the initiator reaction, MIC reaction with water. The major product found was a trimer formed from 3 molecules of MIC (Rxn-1). This reaction is accelerated in the presence of a catalyst. In this case the catalyst was iron. Reported to be 0.18 to 0.26% or approximately 20 to 40 lbs within the tank [8].

Main residue, Trimethyl isocyanurate (TMI) ~40 to 55%, formed by a trimerization reaction:

$3\ CH_3N{=}C{=}O \rightarrow$ Trimethyl isocyanurate $-\Delta H=36$ kJ/mol
MIC trimer (a ring structure) Rxn-1

1,3-Dimethylisocyanurate (DMI) ~ 13 to 20%

$2\ CH_3N{=}C{=}O + HNCO \rightarrow$ Dimethylisocyanurate $-\Delta H=??$ kJ/mol
MIC Isocyanic Acid DMI Rxn-2

Initiating reaction

$CH_3N{=}C{=}O + H_2O \rightarrow CH_3NH_2 + CO_2$ $-\Delta H=38$kJ/mol
MIC + Water to Methyl Amine + Carbon Dioxide Rxn-3

Figure 2. The three major exothermic reactions involved in forming the residual in Tank 610
Note: Heat of reactions estimated via Aspen Ver. 8.6 with existing database for chemicals.

The nature of a runaway exothermic reaction is explained by the Arrhenius effect. The Arrhenius effect models the observation that as temperature rises, reaction rates increase. An activation barrier exists between reactants and products. This barrier lowers as temperature increases. A common rule of thumb is that the reaction rates doubles every 10°C. On an absolute temperature scale, that is a doubling of a reaction rate for every 2 to 3% rise in absolute temperature. Once things are initiated, if the heat is not removed, the mixture will accelerate into high temperature regions until the reactant is fully diminished. Secondary exothermic reactions can occur and be even more severe in consequence such as the T2 incident. [11,12].

Another chemical engineering challenge with MIC is its physical properties. One specific property is its volatility at ambient temperatures. At 20°C its vapor pressure is 348 mm Hg [9]. Its boiling point is 39.1°C, not far from the ambient temperature in central India in the summer. As the exothermic reaction generates heat, and if that heat is not removed, then the contents warm up, and the vapor pressures of all components increase. In a closed vessel, this means increased pressure.

In Bhopal, once the runaway reactions gained momentum, the pressure increased from a reported 2 psig at 22:20 on Dec 2, 1984 to 10 psig 40 minutes later [13]. One hour and 15 minutes later the pressure had exceeded 55 psig and the relief system was initiated. Temperature within the tank was reported above 100°C at the time.

Runaway reaction are well known to chemical engineers designing reaction and storage vessels. The effects of a MIC runaway was on the minds of the design engineers involved in the Bhopal pesticide plant. To prevent tank explosions, relief systems are used. The Bhopal storage tanks 610, 611, and 619 had such a relief system. In this instance a 3" diameter rupture disk followed by a spring relief valve with a telltale pressure in between existed on the storage tank. The telltale gauge is required practice now; however, in the early 80's it was optional. Why is a telltale good practice? Rupture disks can corrode over time and form small pin holes. Thus, the telltale detects this leak beforehand. This relief system was the only safety system that worked – the tank did not explode. When the plant was commissioned in February 1980 many safety systems were in place and functional.

2.2. What failed?

In one sentence: The local plant management neglected their safety systems. In detail, the plant was under extreme financial constraints. It was only able to sell about 1/3 of its design capacity to the Indian market. Management was looking for possibilities to make the plant profitable. This included making the cooling system non-operational on the MIC storage tank in June 1984 [8] . The purpose was to keep the contents within the tank below 0°C. The reaction rate of MIC with itself is low at cold temperatures, and thus, the release of heat of reaction could be managed with a cooling system. This also kept the vapor pressure of MIC below 200 mm Hg [**Error! Bookmark not defined.**]. Further, at lower temperatures, the Arrhenius effect works backwards. In other words, very low reaction rates and heat generation, keeping things under control. Finally, the tank was supposed to be water free. It was designed to hold only the intermediate MIC produced within the plant and then bled off to formulate the pesticide.

Further management decisions followed. Management did not react urgently to a necessary pipe repair needed on the flare system. The flare system was a last resort safety layer intended to burn off CO and related organic vapors before release to the surroundings. The flare was taken out of service several weeks earlier due to a corroded pipeline [14].

In between the flare and the relief system was a NaOH scrubber system. Although the capacity of this system was not intended for the neutralization of 40 tonnes of MIC, if operational, then a portion of the event would have been mitigated. The scrubber system was placed on standby with the reasoning being that it served as a cost savings. Another safety layer removed.

Cost savings reached down to the operating personnel and personnel cut backs. As personnel were let go, the replacements that moved into various open process positions were not trained for their job at the level done when the original crew hired in. This created a disgruntled work force. Disgruntled to the point that one or more may have committed a sabotage act (discussed below).

The local politicians share responsibility in this accident. Initially the plant was built on undeveloped agricultural fields outside the city limits on land leased from the state government of Madhya Pradesh [9]. The central population of Bhopal city was located 1 to 2 km southward. All of the surrounding lands were zoned for industrial use [9]. However, with the siting of a major chemical manufacturing plant at the city edge a revenue source appeared for many economic endeavors. Several shanty towns were built up around the plant near the fence line or across a dividing road. These shanty towns had over 30,000 inhabitants, and were the locations of the many of the deaths and the severest of injuries. Zoning laws were in place to prevent such shanty towns. However, local politicians, in order to reach favor with their newly found constituents, looked the other way regarding the enforcement of the zoning laws.

We see above several layers of protection compromised over time. Specifically, surrounding open grounds were inhabited, the relief system, in its entirety was compromise, and training non-existent. Terrible decision making.

2.3. How was the MIC contamined by water?

In one sentence: It does not matter [15]! For the benefit of a novice audience, two explanations evolved. Remember major safety systems were in place. They were compromised well before the accident.

2.3.1. The Water Washing Theory

In this scenario a worker on site was assigned the task of connecting a hose to a collection point for flushing. The normal procedure required disconnecting the RVVH line from the filter line beforehand and placing a blank near valve labeled 19 in Figure 3 [16]. This was not done. Further, complications occurred when one of the four nozzles did not show any water exiting during the wash cycle. This washing continued for several hours. It is contended that water backed up into the relief valve vent header RVVH line and passed through a jumper line that connected to the a process vent header and that was connected to a leaking valve attached to the MIC storage tank 610. Figure 4 shows the entire pathway. This theory was dismissed by Union Carbide Investigators based on hydrodynamic arguments and the lack of liquid water being found in any low points of the PVH.

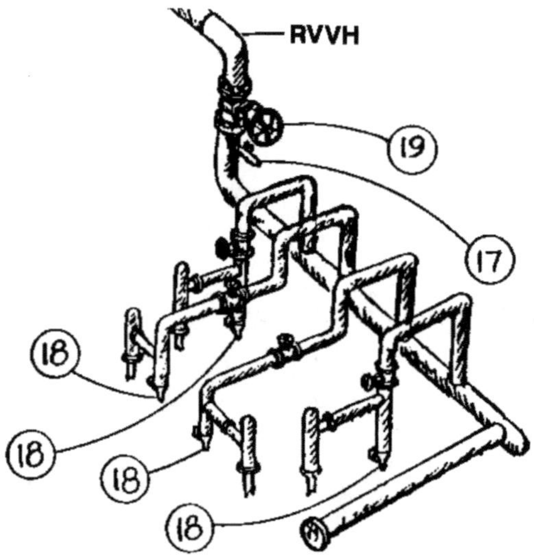

Figure 3. Filter system flushed. From [16]

2.3.2. Sabbatoge Theory

The Union Carbide investigation team concluded that water was intentionally added via a hose nearby the tank by a disgruntled worker. A missing gauge (labeled 10 in Figure 5) and a nearby water hose at tank 610 plus the testimony from an instrument supervisor lead to this conclusion. Kalelkar [16] presents a full discussion of this theory.

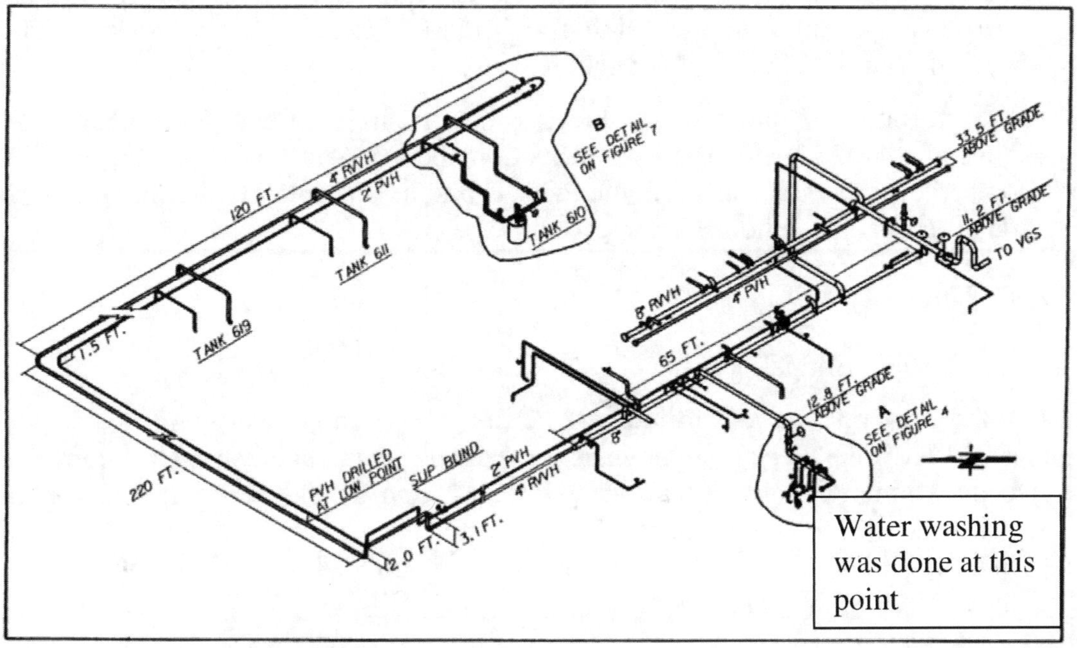

Figure 4. The layout of the PVH and RVVH. Significant distance and head exists from potential water source to storage tank. (adapted from 16)

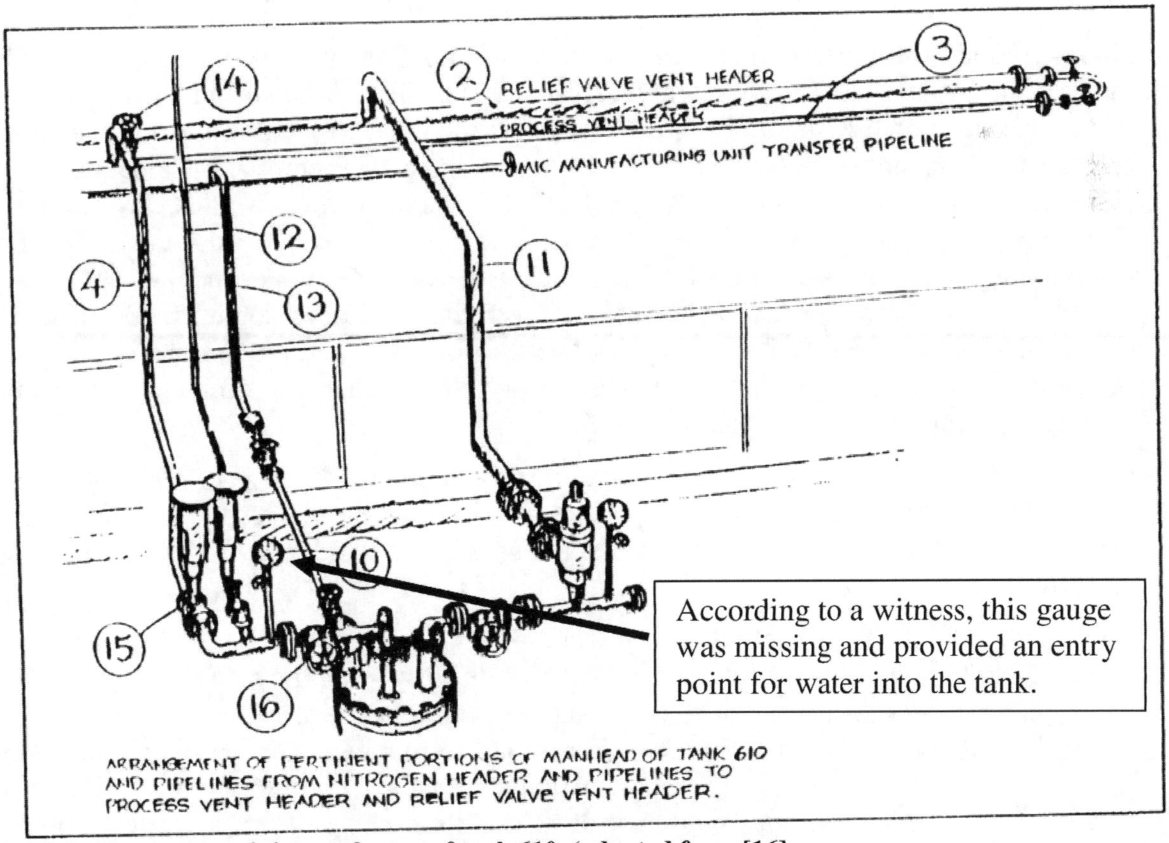

Figure 5. The process piping at the top of tank 610. (adapted from [16]

3. Impact of Bhopal on the Practice of Process Safety

It is interesting to note that the Bhopal accident sensitized the American community to chemical plant accidents. Previous accidents, such as Flixborough in 1974, Seveso, Italy in 1976 and a number of U.S. chemical plant accidents did not achieve this.

A large historical compendium could be written on what occurred after the accident. To complicate things, this history is filled with mis-information and frequent mis-statement of facts. The authors have done their best to substantiate the facts as best as possible. Only important and interesting items identified by the authors are shown below.

3.1. Historical Time Line

a) December 2 – 3, 1984, Bhopal, India accident occurs.

b) At 1 PM on December 3, Union Carbide held its first press conference on the accident at the Danbury Hilton Hotel. The company acknowledged that the accident had occurred at a plant owned by Union Carbide India Limited, with Union Carbide Corporation owning a 50.9 percent share. They announced plans to send medical assistance and technical experts to investigate the cause of the accident. They announced plans to halt production of MIC at the Institute, WV plant [17].

c) Several days after the incident, the then governor of Michigan, James Blanchard, announced that no MIC was used in Michigan. Several days later it was discovered that MIC was being shipped by tank truck from Institute, WV, to a chemical plant in western Michigan.

d) Union Carbide forms a technical team which assembles in India on December 6. The team spent 24 days at the site. However at the site, they were forbidden by the CBI (Central Bureau of Investigation, India) to inspect the tank, to interview plant personnel, and to examine plant records. Their initial focus was the safe removal of remaining MIC by using it in pesticide production. After about 17 days the team was allowed to examine Tank 610 and to take core samples from the remaining material in the tank [9]. The samples were taken back to the U.S. and an extensive experimental study was undertaken. Several years later the Indian government releases some 70,000 pages of documentation as a result of the legal discovery process [17]. The Union Carbide team believes the evidence supports the sabotage theory. The Indian Government believes the evidence supports the water washing theory.

e) On December 7, Warren M. Anderson, chairman and chief executive of Union Carbide, flew to India. He was immediately arrested by Indian authorities. After posting bail, he left the country never to return. Mr. Anderson died on September 29, 2014 at the age of 92 [18].

f) The accident marked the beginning of the end for Union Carbide. Prior to the accident, Union Carbide stock traded between $50 and $58. In early 1985 the stock traded at $32 to $40. Since the company was now worth more than its stock, it became the target of a hostile takeover by GAF Corporation. To respond to this takeover, Union Carbide sold its consumer products division, its most profitable division. In 1986 Union Carbide sold assets worth $3.3 billion and repurchased 38.8 million shares to protect the company from further takeovers. Union Carbide retained its commodity chemicals division, which other chemical companies were divesting (UN University archives, 2015). The downward spiral continued until the remaining Union Carbide assets were purchased by Dow Chemical in 1999.

g) In February, 1989, the Supreme Court of India mediates payment of $470 million from Union Carbide and Union Carbide India and dismisses the charges against Warren Anderson [19]. Union Carbide pays the settlement within 10 days of the order [17].

h) On the 20th anniversary of the accident, a person claiming to be a representative from Dow Chemical is interviewed by BBC World. The representative states that Dow Chemical has agreed to a multibillion compensation package for the Bhopal accident. Within 23 minutes, Dow shares in Frankfurt fell 4.24%, wiping $2 billion off of its market value. The BBC eventually discovers the representative to be a hoax, and the stock shares rebounded [20].

i) In September 2006, the Indian Welfare Commission reported that all claims had been adjudicated with no further cases pending. All settlement money had been distributed [[21]. However, lawsuits in both India and the U.S. continue to this day.

3.2. U.S. Industrial Response

Much of the industrial response in the area of process safety after the accident was directed towards actions to prevent another Bhopal accident.

a) In March, 1985, the American Institute of Chemical Engineers (AICHE), responding to industry concern about chemical plant safety and the Bhopal accident, forms the Center for Chemical Process Safety (CCPS) "dedicated to improving the ability of engineers to deal with process hazards." [22]. AICHE provides $1 million in seed money and establishes CCPS as an independent organization. CCPS focusses on "conducting research and providing objective, technical information on issues related to safety in the manufacture, handling and storage of toxic and reactive chemicals"[23]. Thomas Carmody, former product safety director at Union Carbide, is selected as its first director.

b) In 1985 CCPS forms the Undergraduate Education Committee to improve process safety education in undergraduate chemical engineering curricula. The committee decides to publish a book of problems that can be used throughout the curriculum. Reid Welker and Charles Springer of the University of Arkansas are identified as authors and contractors to write this book. The book is published in 1990 and free copies are distributed to all undergraduate chemical engineering students in the U.S. An instructor's book is also published providing solutions to all the problems and supplemental material- these materials are distributed free to all chemical engineering faculty in the U.S. at that time.

c) In 1988, the Chemical Manufacturer's Association (now the American Chemistry Council) establishes an industry practice called Responsible Care®. All companies who are members of the American Chemistry Council (ACC) are required to follow the principles of Responsible Care®. Today, this program includes process safety, product safety and security features. More information on this can be found at the ACC website (responsiblecare.americanchemistry.com).

d) In 1990, the American Petroleum Institute publishes API Recommended Practice 750, *Management of Process Hazards.* This recommended practice was the predecessor for the OSHA process safety management regulation to follow.

e) In 1991 the CCPS Undergraduate Education Committee forms Safety and Chemical Engineering Education (SACHE). SACHE provides many instructional modules in process safety. In the last ten years many of these materials are converted to web-based resources. Today, over 30,000 undergraduate students annually view these resources and receive certificates of completion.

3.3. U.S. Regulatory Response

a) In 1985 California passes the Hazardous Materials Management law.

b) In January of 1986, New Jersey passes the Toxic Catastrophe Prevention Act.

c) On October 17, 1986, Congress passes the Superfund Amendments and Reauthorization ACT (SARA) to amend the Comprehensive Environmental Response, Compensation, and Liability Act (CERCLA) of 1980. This included the Emergency Planning and Community Right to Know Act (EPCRA). The EPCRA act provides requirements for local and state emergency planning with respect to hazardous chemicals, public access to chemical hazards information in their communities, and reporting requirements for facilities handling hazardous chemicals. The act also established Local Emergency Planning Committees (LEPC) [24].

d) In September of 1989 Delaware passes a regulation for the *Management of Extremely Hazardous Substances.*

e) In 1990 Congress passes the Clean Air Act. This act contains Section 112(r) which eventually results in the EPA promulgating the final Risk Management Plan Rule on June 20, 1996. This regulation requires all facilities that use extremely hazardous substances to submit information on potential releases that may affect off-site plant areas. This includes submission of total extremely hazardous chemical inventory amounts, a detailed hazard assessment estimating the consequences of the release, a prevention program and emergency response.

f) In 1992 OSHA promulgates the regulation entitled 29 CFR 1910.119: *Process Safety Management of Highly Hazardous Substances.* This rule is based primarily on the API 750 consensus standard and has fourteen elements.

4. The plant site 30 years later

4.1.Report of suffering

The tragedy still lives on. The media continues to report of people affected by the Bhopal incident. About 100,000 people who were exposed to gas continue to suffer illnesses such as cancer, and have visual, respiratory, immune and neurological disorders [25]. Birth defects such as twisted limbs, brain damage, and musculoskeletal disorders was reported to Reuters from Bhopal Medical Appeal activist Sanitath Sarangi [25].

4.2.The plant still stands, rusting away

Figure 6 shows a photograph from Reuters news photographer Danish Siddiqui [26]. The view is from the MIC plant towards the flare tower. The pipes and pipe rack are still standing. Insulation surrounding the pipes is beginning to erode away. What is shown in this manuscript are evidence of chemicals left behind. Recent photographs shows bottles of chemicals sitting, collecting cobwebs, awaiting proper removal.

Figure 7 is a recent Goggle Earth image of the former Bhopal UCIL Plant. It still stands. What is amazing is the dense build up around the plant (a portion of the build up can be seen in the upper right hand corner) and that the city has expanded from 1 million to 1.8 million. We have often wonder why the plant remains. This statement was found from an on-line article [26]:

> Rather than seeing the factory demolished, [Rachna] Dhingra said the ICJB [International Campaign for Justice in Bhopal] would like to see the site preserved as a memorial. 'It has a symbolic importance. We want it to be restored and to set a precedent in terms of environmental rehabilitation and medical rehabilitation. To the people of Bhopal these are not just structures; it's memory – memory of pain, but also, one day, memory of justice.'

This may explain why the plant is being left behind.

Figure 6. A photograph of the plant, 2014, by Danish Siddiqui, Reuters [25]

Figure 7. Satellite Image from Goggle Maps. Accessed 4 Jan 2015 [27]

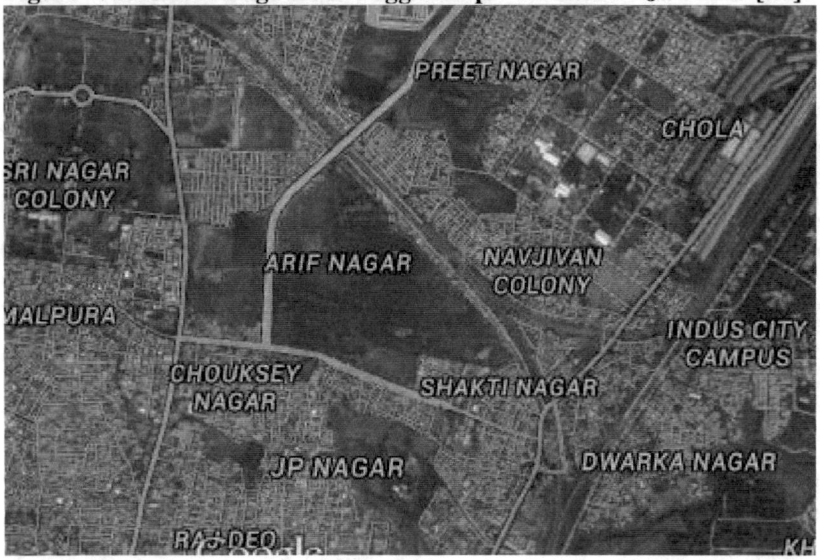

Figure 8. The surrounding communities now abutting the former UCIL Bhopal plant [27]

5. Conclusions

The authors conclude the following

1. After the Bhopal accident the United States was sensitized to chemical plant accidents.
2. Since a thorough investigation was not completed by a credible independent organization and no definitive report issued, speculation continues to run rampant on what actually occurred.

3. Union Carbide completed an extensive investigation of the incident. However, due to legal and liability issues most of this information was not published.
4. The accident was the beginning of the end for Union Carbide.
5. The accident resulted in considerable industry initiatives and government regulations related to process safety. This changed the practice of chemical engineering forever.

6. References

1. *Chem. Eng. News 62 (Dec 6, 1984) 6 and special C&E News Bhopal issue 63 (Feb 11, 1985) 14-65.*
2. http://www.aiche.org/ccps accessed 29 Dec 2014.
3. OSHA overview PowerPoint file https://www.osha.gov/pls/oshaweb/owadisp.show_document%3Fp_table%3DSTANDARDS%26p_id%3D9760 accessed 29 Dec 2014.
4. EPA, http://www.ecfr.gov/cgi-bin/text-idx?tpl=/ecfrbrowse/Title40/40cfr68_main_02.tpl accessed 29 December 2014.
5. http://www.csb.gov/ accessed 29 Dec 2014.
6. http://en.wikipedia.org/wiki/Methyl_isocyanate accessed 30 December 2014.
7. W. Worthy, "Methyl Isocyanate: The Chemistry of a Hazard," *Chem. Eng. News* 63 (6), (1985), 27-33, DOI: 10.1021/cen-v063n006.p027.
8. *Bhopal Methyl Isocyanate Incident Investigation Team Report*, Union Carbide Corporation, Danbury, Connecticut (March 1985).
9. T. D'Silva, *The Black Box of Bhopal, A Closer Look at the World's Deadliest Industrial Disaster*, Trafford Publishing, Canada, 2006.
10. S. Varadarajan, et. al *Report on Scientific Studies on the Factors Related to Bhopal Toxic Gas Leakage*, New Delhi, India (December 1985).
11. *T2 Laboratories Inc. Reactive Chemical Explosion* U.S. Chemical Safety Board,Available at http://www.csb.gov/investigations/completed-investigations/?F_InvestigationId=30. Accessed on 1 Jan 2015.
12. R. J. Willey, H. S. Fogler and M. B. Cutlip, "The integration of process safety into a chemical reaction engineering course: Kinetic modeling of the T2 incident," *Process Safety Progress* 30 (1), (2011), 39-44, DOI: 10.1002/prs.10431.
13. I. Eckerman, *The Bhopal Saga: Causes and Consequences of the World's Largest Industrial Disaster*, Universities Press (India) Private Limited, Hyderguda, Hyderabad, 2005.
14. S. Diamond,The Bhopal Disaster: How it Happened,The New York Times, A1, A6-A7, January 28, 1985
15. Personal quote from Dennis Hendershot.
16. A. S. Kalelkar, "Investigation of Large-Magnitude Incidents: Bhopal as a Case Study," The Institution of Chemical Engineers, I.Chem.E. Symposium Series No. 110, London, England, (1988), 561.
17. Browning, Jackson B., *Union Carbide: Disaster at Bhopal*, available at www.bhopal.com.
18. Martin, Douglas, "Warren Anderson, 92, Dies: Faced India Plant Disaster," *The New York Times*, October 20, 2014.

19. Endgame, *Chronology of the Union Carbide Corporation*, http://www.endgame.org/carbide-history.html, accessed 12 Jan 2015.
20 . CNN, "Bhopal hoax sends Dow stock down," 3 Dec 2004. http://www.cnn.com/2004/WORLD/europe/12/03/bhopal.hoax/index.html, accessed 12 Jan 2015.
21. Bhopal, *Union Carbide's Response Efforts to the Tragedy and the Settlement*, http://www.bhopal.com/UCC-Response-Efforts-to-Tragedy, accessed 14 Jan 2015.
22. Chemical Week, "AICHE points the way to chemical process safety," July 30, 1986.
23. CCPS Annual Report, September 1986.
24. EPA, http://www.epa.gov/superfund/contacts/infocenter/epcra.htm, accessed 12 Jan 2015.
25. http://www.buzzfeed.com/franciswhittaker/moving-images-showing-the-impact-of-the-1984-bhopal-disas#.iiXPQxxlz, Accessed 4 Jan 2015.
26. http://www.telegraph.co.uk/news/worldnews/asia/india/5978266/Bhopal-gas-disaster-the-legacy-lives-on.html, Accessed 4 Jan 2015.
27. www.google.com/maps Accessed 4 Jan 2015.
23. UN University Archives, http://archive.unu.edu/unupress/unupbooks/uu21le/uu21le0c.htm, Accessed January 12, 2015.

AUTHOR INDEX

AUTHOR INDEX